Winter Sports Resorts' Strategies to Adapt to Climate Change

Clémence Massart, Emmanuelle George,
Philippe Bourdeau, Anne-Marie Granet-Abisset and
Christophe Gauchon (eds.)

Winter Sports Resorts' Strategies to Adapt to Climate Change

General Trends and Local Responses

PETER LANG
Bruxelles · Berlin · Bern · New York · Oxford · Warszawa · Wien

Bibliographic Information published by the Deutsche Nationalbibliothek
The Deutsche Nationalbibliothek lists this publication in the Deutsche Nationalbibliografie; detailed bibliographic data is available online at http://dnb.d-nb.de.

Library of Congress Cataloging-in-Publication Data
A CIP catalog record for this book has been applied for at the Library of Congress.

Published with the support of LabEx Innovation & Transitions territoriales en Montagne

ISBN 978-2-8076-1311-9 (Print) · E-ISBN 978-2-8076-1312-6 (E-PDF)
E-ISBN 978-2-8076-1313-3 (EPUB) · E-ISBN 978-2-8076-1314-0 (MOBI)
DOI 10.3726/b16587
D/2020/5678/05

Peter Lang – Bruxelles · Berlin · Bern · New York · Oxford · Warszawa · Wien

This publication has been peer reviewed.

www.peterlang.com

Contents

II Strategies and Forms of Adaptation of Resorts to Global Changes

III Emerging Markets and Practices

List of Contributors

Coralie Achin
Université Grenoble Alpes, Labex ITTEM INRAE-LESSEM,

Michel Archambault
Université du Québec à Montréal, Department of Urban Studies and Tourism, Canada

Cenk Demiroglu
Umeå University, Suède, Department of Geography and Economic History

Yan Fang
Peking University, College of Urban and Environmental Sciences, Chine

Alison M. Gill
Simon Fraser University, Department of Geography, Canada

Tobias Luthe
École Polytechnique Fédérale de Lausanne, Suisse & University of Applied Sciences HTW Chur, Department Living Environment, Suisse

Andrea Macchiavelli
Université de Bergame, Italie

Sara Nordin
The European Tourism Research Institute (ETOUR), Sweden

Beltran Oriol
Universitat de Barcelona, Espagne

Ulrike Pröbstl
University of Natural Resources and Life Sciences, Institute of Landscape Development, Recreation and Conservation Planning, Vienne, Autriche

Daniel Scott
University of Waterloo, Geography and Environment Management, Canada & University of Innsbruck, Autriche

Chris Sheppard
Simon Fraser University, Department of Geography, Canada

Robert Steiger
University of Waterloo, Geography and Environment Management, Canada & University of Innsbruck, Autriche

Bo Svensson
Department of Humanities and Social Sciences, Mid-Sweden University, Sweden

Ismael Vaccaro
McGill University, Canada

Peter W. Williams
Simon Fraser University, Department of Geography, Canada

Doyeon Won
Texas A&M University Corpus Christi (United-States), Department of Kinesiology

Romano Wyss
École Polytechnique Fédérale de Lausanne, Suisse & University of Applied Sciences HTW Chur, Department Living Environment, Suisse

Michal Zemla
Pedagogical University in Cracow, Institute of Geography, Department of Tourism and Regional Research, Pologne

Introduction

For more than 30 years, the tourism sector has been changing as a result of climate change, an economic recession in industrialised countries, the emergence of new tourist destinations and the evolution of customer expectations. The purpose of the work presented here is to analyse the influence of these global trends on a tourism segment that has been hit particularly hard by the changes taking place around the world: winter sports resorts. Based on a systematic exploration of English-language literature over the past three decades, it examines how researchers identify the (sometimes different) ways in which resorts respond to these changes, particularly in terms of their position in a territory that has its own history, culture, constraints and resources. Therefore, the tension between globalisation (of pressures) and territorialisation (of the resorts' and their territories' response capacities) is the central thread of this work.

Winters sports market: background and current trends

Although Nordic skiing had already been practised for centuries in Scandinavia, it did not become popular in Europe, the United States and Canada until the end of the 19th century (Coleman, 2004). At first, it developed as a way to move more easily across snow-covered surfaces, and there were both practical (mail delivery, border defence, medical care etc.) and sports-related reasons (e.g. to reach the top of a glacier) to do so. During the interwar period, a new form of skiing – alpine or downhill skiing – developed in the Alps before spreading to the United States. World War II put the rise of this new sports practice on hold, and it was only in the 1960s that ski resorts really took off, when economic and demographic growth provided a favourable context for the development of mass tourism.

By bringing in significant revenue on a regular basis, winter tourism[1] was a godsend for mountain communities. Starting in the 1920s, and

---

1 Summer tourism had begun in certain valleys at the beginning of the 19th century;

especially after World War II, skiing allowed a double seasonality that would become an essential part of the economic and social structuring of the high alpine valleys. Because it supplements the revenue earned from agricultural activity, logging and mineral mining, winter tourism curtailed the migration that used to structure these territories (characterised by the salience of its climate) and link them to other (sometimes distant) areas. Since the 1960s, in the context of a widespread enrichment of the middle class that boosted mass tourism, winter sports emerged as the dominant economic activity and pigeonholed agro-pastoral activities and industrial jobs in the running of multiple activities. The huge influx of tourists and seasonal migrants also vastly increased the demography of these territories and turned the work cycles upside down in favour of the winter season and the regular rhythm of school holidays. These conditions required the development of varied infrastructures that have strengthened and made visible the (existing but inconspicuous) connection between these territories and global dynamics (Granet-Abisset, 1994).[2]

However, since the end of the 1980s, both in Europe and North America, the winter sports market has shown signs of slowing down (Williams and Fidgeon, 2000; Coleman, 2004), which resulted in the closure of many small resorts (Hamilton et al., 2003; Gauchon, 2010 Silberman and Rees, 2010). The interconnected factors of this slowdown can be divided into three parts. First of all, there is the economic factor: After a boom phase following the end of World War II, economic growth in the West started to taper off. According to some authors, this slump was accompanied by a reduction in spending on recreation (Barbier, 1993; Dawson and Scott, 2013). But this view is open to criticism since the tourism sector has been experiencing strong growth (for 2017, it is estimated at 7 % by the UNWTO).[3] It would be better to mention the rise in economic inequality that has made skiing an activity for the privileged.[4] Next, the environmental factor: climate change, which has

---

in its winter form, tourism essentially spread to all mountain regions during the 20th century.

2 Although the trends are presented here in broad outlines, they are more nuanced. For example, there are differences between the situations in Europe and the United States. See the historical studies on these territories for further details.

3 http://media.unwto.org/press-release/2018-01-15/2017-international-tourism-results-highest-seven-years (accessed 6 February 2018).

4 According to the French Observatory of Inequalities, 40 % of skiers are executives or in senior management positions, while only 9 % are labourers; more than 30 %

led to a decrease in the amount of snowfall, has become more and more apparent in the structuring of the winter sports system, whose weaknesses it has revealed (Gauchon, 2009) and whose sustainability it has called into question (Barbier, 1993). However, while the issue of climate change emerged at the end of the 1990s, the 2000s really made it clear how bad the situation was (Scott, Michael and Gössling, 2012). And finally, the cultural or societal factor: Tourists' preferences have always evolved and relate less and less to winter sports. In the northern hemisphere, after reaching its peak at the beginning of the 1980s (Coleman, 2004), skiing's appeal started to wane in the 1990s (Pechlaner and Tschurtschenthaler, 2010; Achin and George, 2016). There is no consensus on the reasons for this drop. Some authors say that ski resorts had to compete with the greater ease of access to warmer coastal regions, such as the Caribbean, especially thanks to low-cost flights (Gauchon, 2002; Pechlaner and Tschurtschenthaler, 2010). Others challenge this argument by pointing out that the ski resorts' losses did not lead to additional gains for the seaside resorts (Barbier, 1993). It is more certain that the demographic decline, an ageing population and the evolution of the relationship with the mountain have contributed to people shifting their focus from skiing to other leisure activities (Dawson and Scott, 2013; Hatt and Vlès, 2014; Kuscer, Mihalic and Pechlaner, 2016).

In order to deal with these global changes, ski resorts and the local populations that are largely dependent on their operations will have to adapt. However, responses vary according to not only the history of the territory that hosts the resort, as well as its resources, but also the cultural and political arrangement of the stakeholders. Indeed, as Higham and Hinch (2009) demonstrate, the globalisation of sport and tourism is reflected as much in the influence of cross-cutting factors as in the diversity of responses, which are themselves largely rooted in local cultures and issues. The academic literature dealing with these topics reflects this diversity, and in many respects it is difficult to generalise the results because they vary according to each resort's particular developmental trajectory (Achin and George, 2016). The adaptation strategies also depend on the particular type of resort: Large high-altitude resorts, which may benefit from the shrinking market, do not have to address the same challenges as smaller resorts at lower altitudes, which lack investment funds and

---

of skiers have an income of more than €3,100 per month. https://www.inegalites.fr/Les-sports-d-hiver-une-pratique-de-privilegies (accessed 26 February 2018).

are more directly affected by climate change (Bourdeau, 2007; Dawson and Scott, 2013; Falk and Hagsten, 2016). Depending on whether one's interest lies in supply or demand, adaptation is not considered in the same way either. Finally, the adaptations vary according to the type of change that is considered.

To date, very few studies on adaptation strategies have used concrete examples. They also tend to focus on the social structures, which are understood as a set of formal and informal norms and institutions that frame and guide the stakeholders' practices at the expense of individual initiatives; and yet, the latter play an essential role in innovation and adaptation (Wyss, 2013).[5] Lastly, they very often prioritise a type of pressure (that is, a type of adaptation) without grasping the overall dynamics of these changes and their respective influences and without taking the diversity of the territorial trajectories into account. In the rest of this introduction, we will lay out in detail the different pressures on the ski market and the responses that are given, as well as the debates and uncertainties that run through these fields of study and relate to the specificities of the territories under consideration.

Global changes and adaptation strategies

Environmental and technical adaptations

– modelling climate change

The most-studied kind of global change is climate change.[6] This phenomenon is tackled in two ways: through modelling, which is usually predictive and occasionally analogue (Rutty et al., 2017), and by using questionnaire-based surveys. Based on several temperature increase scenarios, the models estimate the number of days with snowfall and the snow depth at various altitudes. This approach can be enhanced by using questionnaires to identify how ski operators are responding to climate change. By contrast, the way in which visitors adapt to climate change has been studied to a lesser degree (Dawson and Scott, 2013),

---

5 For an example of individual innovation, see the article by Paget, Dimanche and Mounet, 2010.

6 A bibliographic search using the terms "ski resort" and "global change" mostly generates articles dealing with this aspect.

and studies focused on adaptation with respect to supply and demand are even fewer and farther between (Scott, Hall and Gössling, 2012; Jopp, DeLacy and Mair, 2010; Demiroglu, Dannevog and Aall, 2013; Rutty, Scott, Johnson, Jover, Pons and Steiger, 2015). The same applies to truly interdisciplinary studies, which combine natural sciences and humanities in equal parts (Demiroglu, Dannevog and Aall, 2013).[7] Finally, most of these studies focus on the Alps (51 %), along with other mountain regions in Europe, North America and Oceania (Demiroglu, Dannevog and Aall, 2013). In countries like China, where skiing is an emerging market (Yang and Wan, 2010), the assessment of climate change's effect on resorts is still in its infancy.

– identifying tourism operators' perception of climate change

Before exploring the adaptation strategies, studies have to focus on the way in which the actors in the ski industry perceive climate change. While these actors generally acknowledge that the phenomenon is real, they do not necessarily view it as the primary risk to their activity (Morrison and Pickering, 2013; Trawöger, 2014; Wyss, Abegg and Luthe, 2014). Changing demands and relationships with governments and other actors, like natural parks and economic partners, are perceived as just as critical as – and perhaps even more so than – global warming. Nonetheless, the ski industry has focused on adapting to climate change by producing artificial snow (Trawöger, 2014).

– producing artificial snow to guarantee operations at resorts

Artificial snow helps to guarantee resorts' period of operation and is widely used at most American, European, Australian and Japanese resorts. Increasingly, models are including the potential of producing snow in assessments of the impacts of climate change (Scott, McBoyle and Minogue, 2007; Falk, 2013; Demiroglu, Kučerová and Ozcelebi, 2015). Such snowmaking depends on temperature conditions and has both an economic and an environmental cost to it. For these reasons, artificial snow is better suited to large high-altitude resorts, which have lower temperatures and the necessary investment resources (Rutty et al., 2017). The phenomenon of concentrating and consolidating ski resorts,

---

7 Two exceptions are the works of Hamilton, Brown and Keim, 2007, and Spandre et al., 2019.

which started in the 1990s to the gain of high-altitude resorts, may be exacerbated by climate change and the technical response to address it (Gauchon, 2002; Scott, McBoyle and Minogue, 2007; Morrison and Pickering, 2013; Dawson and Scott, 2013; Beaudin and Huang, 2014). Producing artificial snow also depends on the availability of water and energy, which makes the technique susceptible to environmental criticism (Pickering and Buckley, 2010). In a context of rising temperatures and the depletion of natural resources, there is no guarantee that it can still be produced in sufficient quantities in the future.

Despite these limitations, the ski industry players are continuing to rely on this solution in the short term. They are hoping that the snowmaking techniques will be adapted to higher temperatures (Morrison and Pickering, 2013; Song-Yi, Chan, Jin-Han and Dong-Kun, 2015) and likely think that this solution is easier to implement than work on governance and stakeholder relationships. In France, the Snow Plan that Laurent Wauquiez, president of the Auvergne-Rhône-Alpes Regional Council, came up with in 2016, is indicative of this trend as it aims to equip the region's resorts with additional snow cannons costing a total of 200 million euros in investment over five years. The country's Provence-Alpes-Côte d'Azur region also has a Snow Plan (Smart Mountain PACA) aimed at equipping resorts with snow cannons by investing 7 million euros.

The stakeholders in the ski industry invest so much in snowmaking because they view natural and artificial snow as having the same value for skiers. This point, which has received little attention, is far from evident, however. According to Demiroglu, Turp, Ozturk and Kurnaz (2016), Turkish skiers have a moderate appreciation for artificial snow. Unbehaun, Pröbstl and Haider (2008) draw the same conclusions regarding Austrian skiers. Young skiers in South Korea prefer the quality of natural snow when choosing a resort (Won and Hwang, 2009). Skiers' origins also seem to play a role in how they view synthetic snow. In a study of 28 Austrian resorts, Falk (2013) shows that local skiers prefer natural snow. The author associates this preference with their level of competence, but we should note that local skiers can choose when (and, thus, on which surface) to ski, unlike tourists who booked their stay months in advance without knowing what the snow conditions would be. Richards (1996) also ties skiers' appreciation to their level of experience: The more experienced skiers are, the more they look for natural snow. According to other studies, however, technical adaptations like artificial snow are better accepted by skiers looking for a sports activity, in contrast to those who prefer a natural setting (Landauer, Pröbstl and Haider, 2012). This partial and sometimes contradictory information should be supplemented by further investigations.

– identifying skiers' response to climate change

More studies are seeking to identify skiers' response to the decline in natural snowfall. Three strategies seem to be available to customers: spatial substitution, temporal substitution[8] and product or activity substitution. Whichever strategy is chosen, it correlates with tourists' perception of climate change.[9] This issue is generally tackled by using questionnaires that ask skiers to imagine themselves in different snow scenarios. Beyond "perception", skiers' response to global changes is based on many criteria that are still poorly understood. This response seems to vary according to the regions considered (Scott, Dawson and Jones, 2008; Dawson and Scott, 2013). In Australia, for example, where natural snowfall is decreasing, tourists are likely to ski elsewhere or abandon it altogether (Pickering, Castley and Burtt, 2010). There are three major reasons why they mostly turn to New Zealand: the prices are better; the stay is exotic and other activities beyond skiing are on offer; and the mountains of New Zealand are more reminiscent of the Alps and better meet the expectations of experienced skiers than in Australia (Hopkins, Higham and Becken, 2013). By contrast, in the United States[10] and the Alps, tourists are less likely to make a spatial substitution. This is probably because these regions guarantee more snow, but the reason may also be tourists' travel habits and the distance that has to be travelled to find snow. In this regard, resorts that are situated close to urban centres have a significant competitive advantage (Dawson and Scott, 2013; Beaudin and Huang, 2014; Falk and Hagsten, 2016).

---

8 This strategy generally seems to get little consideration because it involves great professional and personal flexibility: Most trips always take place during the holidays.

9 The literature on customer adaptation to climate change is grouped under this theme of "perception".

10 See the study by Rutty, Scott, Johnson, Johnson, Jover, Pons and Steiger, 2015, on Ontario (Canada), which shows that skiers would increase their trips by a maximum of six hours for a long stay, always planned around their region. In Austria, this increase is only two hours (Unbehaun, Pröbstl and Haider, 2008).

– expanding the ski areas

To cope with decreasing snow resources, another line of adaptation involves moving the ski areas to higher altitudes. However, this solution runs into ecological, social and technical resistance (Morrison and Pickering, 2013) and even regulatory restrictions on expanding ski areas (Gauchon, 2002). Mountain regions above 3,000 metres are ecologically sensitive areas and would be damaged by human infrastructure. An altitude of 3,000 metres also implies a harsh climate that can cause many problems for skiing (storms, cold, wind etc.). By contrast, interconnecting ski areas has made it possible to expand the slopes' surface area and diversify tourist experiences while limiting new developments (Barbier, 1993; Tuppen, 2000; Marnezy and Gauchon, 2006). Nevertheless, as Martin Falk (2017) shows in a study examining Austrian resorts, the benefits of such connections are not homogeneous: They mainly benefit outlying villages and, in the absence of other innovations and new tracks, fade over time (especially because the cost for the tourist increases as the number of connections rises). In France, the increase in the number of large estates also gives rise to conflicts between developers and mountain protection associations (Gauchon, 2002).

– Winter sports resorts and environmental degradation

While ski resorts are impacted by environmental issues, they also play a role in causing them. In addition to the literature on adapting to climate change, there is less literature available on the mitigation measures that resorts can take and that generally involve transport and the energy efficiency of buildings (Clivaz, Gonseth and Matasci, 2015). Other environmental issues ascribed to resorts include waste, the loss of rare and fragile biodiversity and the destruction of natural and landscape resources (Pignatti, 1993; Glick, 2001; Caprio et al., 2011; Sato et al., 2014). All too often, these issues are ignored because of a long-term rationale based on making a quick profit (Arcuset, 2009; Innocenti, Merciu and Merciu, 2012).

Economic and commercial adaptation

– extending the resorts' period of operation: the issue of summer tourism

In order to adapt to global changes, resorts also rely on economic and commercial strategies. Seasonal diversification is a strategy being pushed particularly hard and, in some countries like France, is even promoted by public authorities (Achin and George, 2016). Indeed, by extending their period of operation, resorts can relieve themselves of their dependence on winter sports by branching out into other activities such as climbing, mountain biking, golf and grass skiing (Ghaderi, Khoshkam and Henderson, 2014). Diversification also encourages local actors to take control of their future by defining a multi-purpose territorial project (Bourdeau, 2008).

The studies that describe diversification strategies and assess their effects are still few and far between and do not enjoy consensus support. Some authors say that, thanks to global warming, summer tourism will increase in mountain regions but decrease in coastal and southern areas, which will become too hot. Therefore, consumer choices would be similar to those made in the 19th century, when mountain tourism was based on summer activities such as thermal treatments and mountaineering (Gauchon, 2002; Bourdeau, 2009). However, a study of German tourists rejects this optimistic hypothesis. Indeed, during some of the more recent heatwaves, only the inhabitants of nearby areas opted to go to the mountains (Pröbstl-Haider, Haider, Wirth and Beardmore, 2015). This topic still requires further investigation, especially because the results probably vary by country and, therefore, according to the intensity and frequency of the heatwaves.

There are also limits to resorts' ability to adapt to the summer season. Firstly, summer tourism does not generate as much income as winter tourism; this is true for both ski lifts and seasonal workers, who prefer to work at the coast during the summer (Bonnemains, 2014). In some areas like the Hautes-Pyrénées (France), summer tourism only concerns two months of the year, compared with five months for winter tourism (Gauchon, 2002). In places where it is actually increasing, as in Australia, it adds to the environmental pressures that already weigh heavily on mountain areas (Buckley, Pickering and Warnken, 2000). Finally, it is more suitable to medium-altitude than high-altitude mountain resorts, which, without snow cover, have unappealing landscapes because they have no vegetation. They have better natural and artificial snow cover and, therefore, are less likely to make seasonal adjustments anyway (Morrison and Pickering, 2013).

– identifying customer segments and trying to meet their expectations by diversifying the product range

During the wintertime, product diversification can also attract customers who do not ski at all or do very little of it. In recent years, interest in skiing has fallen in favour of other less sporting activities that focus more on well-being, culture and celebrations (Macchiavelli, 2009; Hatt and Vlès, 2014). However, this strategy is only possible for resorts with sufficient capital to make the necessary investments. Some authors also question this diversification strategy, arguing that tourists come to the resort primarily to practise a winter sport and enjoy a winter atmosphere (Unbehaun, Pröbstl and Haider, 2008). Additional activities might, therefore, be appreciated but will certainly not replace skiing, which remains these resorts' flagship product.

While they cannot attract all customer segments, resorts may seek to target (a) specific segment(s), develop a communication strategy for them,[11] assess their satisfaction and anticipate future consumer trends (Matzler et al., 2008; Ferrand and Vecchiatini, 2010). These adaptations, which are part of marketing, involve identifying different customer segments and their particular expectations. This approach is usually based on socio-economic criteria such as age, sex, marital status, income and level of education. Some studies show that, compared with clients who choose to visit the casino, go to the beach, on a cruise or into the wild, skiers are more likely to be male, young and single and have a high level of education and income and high demands regarding sports activities (Morrison, Braunlich, Cai and O'Leary, 1996; Morrison, Yang, O'Leary and Nadkarni, 1996).[12] In addition to these socio-demographic criteria, there may also be the frequency of stays, the surroundings or the place where people live and the countries' cultural particularities. In Finland, for example, cross-country skiing plays a bigger role than in the Alps, and the price is less important (Konu, Laukkanen and Komppula, 2011). While these studies have the merit of indicating overall trends, they are based on very broad scales that do not reflect the full range of situations. More contextualised studies, at the scale of a territory or a resort, would make it possible to add some colour to the tourist profiles.

---

11 For example, by partnering with tour operators, especially if it is a question of reaching foreign and emerging audiences.

12 Studies by the French Ministry of Youth, Sport and Voluntary Work indicate a similar consumer profile (Thiery, 2006).

It does not happen often that qualitative studies identify what it is that skiers pursue through their practice. For example, Richards (1996) shows that the demands of experienced skiers are different from those of more novice skiers, especially with respect to the quality of the slopes and the snow. By contrast, comfort and cost play a bigger role for inexperienced skiers. As a result, these two categories of skiers do not choose the same resorts, do not ski as often and are not willing to spend the same amount of money. However, further studies focusing on the drivers of practice still need to be done to help resorts build customer loyalty and assess the benefits of their investments.

Luthe and Schläpfer (2011) investigated the links between demand and sustainability and show that Swiss skiers are generally in favour of sustainability criteria, as long as they do not damage the skiing experience (artificial snow and grooming are the two main criteria in this regard). The authors also note that information received from NGOs can influence skiers' decision, but their sensitivity to sustainability remains relative to social categories. With regard to cross-country skiing and snowshoeing, other studies conducted in Switzerland and Austria have shown that skiers underestimate the environmental impact of their activities (Freuler and Hunziker, 2007; Sterl, Eder and Arnberger, 2010). In general, better information helps them to comply with the rules made by the nature managers, such as trails that are less disturbing to fauna and flora. However, the more skiers want to have a sporting experience and the more familiar they are with a trail, the less likely they are to adapt their habits to the managers' recommendations (Sterl, Eder and Arnberger, 2010). Therefore, beyond the social categories and the information received, their motivations appear to influence whether and how they adapt their behaviour to environmental issues. However, the analysis of the relationship between winter sports practices and sustainability should be developed and put into perspective in different contexts.

– encouraging non-skiers to ski

A few studies have focused on the reasons why some people do not ski (Williams and Fidgeon, 2000). They identify obstacles to the practice, as well as those people who, despite never having skied before, might be open to it: young families and "social adventurers", in other words, tourists who are curious about engaging in new practices. For these two categories, the biggest obstacles are cost, risk of accident and organisational cost, to which resorts can respond by developing products adapted to beginners (e.g. lessons) and breaking down their prices.[13] Other authors have focused on female customers, who ski less overall than men (Williams and Lattey, 1994). From a business point of view, the challenge is to develop tourism products that better meet women's expectations – for example, by placing greater emphasis on conviviality and the natural environment than on performance. This work should be linked to studies on how gender influences mountain sports. While the presence of women is greater than in other sports disciplines and is on the rise, gender parity is far from being reached. In both quantitative (two-thirds vs. one-third of practitioners) and qualitative terms, the roles of men and women continue to be different, and women's place in sports institutions is lower than that of men (Ottogalli-Mazzacavallo and Saint-Martin, 2009). As some American authors have argued, skiing remains firmly rooted in the male gender (Coleman, 2004; Stoddart, 2011).

– offering quality accommodation: the issue of land tenure

Another issue that resorts frequently have to deal with is the rental of private accommodation and their level of comfort, which affects the appeal of a destination. More broadly, this involves the issue of resorts' land tenure – a rather dull and little-studied issue that also varies by country: While Austria has favoured a hotel model, in France, most of the housing is second homes sold to finance the construction of infrastructure (Barbier, 1993). For reasons ranging from booking systems' lack of flexibility to the symbolic and emotional investments that second-home owners make in their homes (Palmer and Mathel, 2010), many people in France do not make their homes available to tourists. This phenomenon of "cold beds" obviously has a negative effect on the resorts as it limits the number of tourists that can be hosted and damages the resort's reputation. It is very different from American resorts, where all commercial activities (from hotels to restaurants and the sale of packages and equipment) are managed by the same company.

– financial measures: insure the resorts against unpredictable weather

Finally, financial measures can be taken to insure against the risks of climate change. Examples include snow-guarantee packages, as well as weather derivatives and insurance, which are financial instruments to

13 New skiers often struggle to calculate the total cost of a ski trip.

cover companies against unpredictable weather (Beyazıt and Koc, 2010; Demiroglu, Kučerová and Ozcelebi, 2015). Another example is the "Nivalliance" insurance system developed by France's National Union of Ski Lifts to compensate for the loss of income in terms of operating lifts during bad years. To respond to climate instability, some authors have also proposed geographical diversification and financial hedging (Tang and Jang, 2007).

Political and legislative adaptation

- collaboration issues between stakeholders: which governance models are best suited to ski resorts?

A third adaptation category involves public policies and regulations implemented to oversee the development of resorts, in particular by encouraging stakeholders to collaborate and coordinate their efforts by joining forces. Works mentioning issues of collaboration between the stakeholders can be classified under the theme of "governance". In the European context, this concept has been developed following the increase in the number of public/private partnerships. The latter have increased the complexity of the systems of stakeholders, and the arrangements for joining individual economic interests and a more collective interest that has played out at the resort and/or territory level (George, Achin, Fablet and François, 2016).

The governance models are largely contextualised in the sense that they depend on the history of the resorts, which in turn is linked to that of the territories and the communities. In France, for example, governance issues are all the more acute as public action has been renewed by a move towards a withdrawal by the state and the reassertion of the territories and, therefore, of more local management (George, Achin, Fablet and François, 2016). On a broader scale, governance issues are different in North America, where winter tourism has been developed by large groups, than in the Alps, where they result from small entrepreneurs whose organisational difficulties hamper the coherence and attractiveness of Alpine destinations (Pechlaner and Tschurtschenthaler, 2010). Eastern Europe demonstrates other forms of governance, which may vary from country to country but generally have unstable relations between private and public actors. In particular, resorts and other tourism stakeholders may have competitive rather than complementary relationships (Zemla,

2008),[14] which invalidates the definition of an attractive "tourist destination".[15] As for local populations, they seem to be consulted on rare occasions and benefit from tourism to a much lesser degree than the large private operators (Perović, 2008, for Serbia; Chrenka and Ira, 2011, for Slovakia).

In addition to these particular situations, however, there are some cross-cutting lessons to be learnt. First of all, before combining different tourism "products" to increase a region's attractiveness, private and public actors have to learn to collaborate (Escadafal and Boulin, 2016). Secondly, the relations between levels of political management must be properly articulated to meet the challenges of global change. The regional or national level seems to be better suited to dealing with long-term changes, while the municipal level reacts better to short-term changes (Luthe and Wyss, 2016). Local populations must also be involved, as they can help to mitigate the excesses of the resorts and promote the development of sustainable tourism (Gill and Williams, 2011). Finally, while they play a major role, these "internal" factors should not overshadow the significance of external pressures that the local actors cannot easily control (Gill and William, 2011).[16]

Each country's public policies and regulations also put specific pressures on the resorts. This is the case, for example, with civil insurance systems: While in Italy resorts are routinely the ones held responsible, which decreases their competitiveness because their fees – and, therefore, the package costs – increase (Izzo, 2014), liability is generally transferred to the skiers in Vermont, Colorado and California (Feldman and Stein, 2010). Environmental regulations also greatly affect the resorts and their competitiveness. Two types of instruments can be used to respond to different approaches: binding regulations and voluntary programmes

---

14 However, the Chinese resort of Wan Long, located 250 km north of Beijing, seems to have relied on its ability to complement other resorts in the region rather than on competition (Wan, 2008). Unfortunately, we do not have examples of the governance models and kinds of collaboration governing relationships between Chinese resorts.

15 Especially since, without resources, the resorts are incapable of diversifying (Nedzma Begic and Teoman Duman, 2013, on Bosnia and Herzegovina).

16 With regard to Whistler, these authors show that the evolution from a "pro-growth" model (focused on investors and political authorities) to a more corporate and sustainable model (that takes the populations into account) has been achieved thanks to both the Olympic Games and the local populations' demands.

based on standards and labels.[17] The particular effectiveness of these levers for action is still the subject of debate. While Briggs (2000) praises the voluntary involvement of resorts in Colorado,[18] Rivera, de Leon and Koerber (2006) consider this type of measure to be rather ineffective for the same area. According to them, firms only apply low-cost measures, such as saving resources, at the expense of more attractive measures like protecting wildlife.

Cultural and societal adaptation

The last category of change that we focus on involves cultural and societal issues and includes wage employment, the resorts' social sustainability and introducing younger generations to skiing. Although international literature has not examined them much, these aspects deserve more attention.

– What is social sustainability for resorts?

Several studies have focused on resorts' social sustainability: local populations' involvement in developing tourism and the benefits they get from it. These benefits can be weak as economic power often belongs to large groups such as Pierre & Vacances (Haimayer, 1989; Tuppen, 2000; Vaccaro and Beltran, 2009). The development of skiing has also competed for attention with other activities such as livestock farming (Laguna Marin-Yaseli and Lasanta Martinez, 2003; Ooi, Laing and Mair, 2015) or forestry (Price, 1987), which nevertheless contribute to the territories' sustainability by diversifying sources of income. Finally, the increase in the price of land, which has coincided with tourism development, has excluded former populations in favour of second-home owners who generally have little involvement in community management[19] (Loeffler and Steinicke, 2007; Nepal and Jamal, 2011).

---

17 According to Duglio and Beltramo, who focused on how they are applied in the Alps, none of the instruments used (standards, labels, audits) take all the aspects of (especially social) sustainability into account (2016).

18 The authors look at the case of resorts in the Western United States because, unlike the rest of the country, they are linked to the US Forest Service, which owns most of the land.

19 In some rare cases, second-home owners can be overly involved. See in particular the municipal elections of Villarembert-le Corbier (France) in 1983, in which the votes of second-home owners prevailed over those of the villagers (Collomb, 1989).

These phenomena are threatening the future of communities that revolve around the resorts. Firstly, they lack unity, and only close-knit communities can fight back against excess at the resorts (Ooi, Mair and Laing, 2016). Secondly, if they fail to renew their population, resorts might lack the resources to adapt. To date, these endogenous factors have received little attention in development policies, which have mostly emphasised exogenous factors (Mühlinghaus and Wälty, 2001). In the context of the difficulties characterising winter sports, however, the preservation of a diversified social fabric, the participation of local populations and tourism development planning are likely to become essential levers for innovation and adaptation (Gil and Williams, 1994).

– resort employees: guaranteeing quality and creativity

Besides being part of ensuring quality service, workers' skills and experiences also make the resorts more attractive (Shabanski, 2013). However, it is difficult to guarantee this appeal since most of the employees are seasonal workers for whom settling permanently at the resorts is a challenge. Their biggest obstacle is the often very high cost of housing (Marcelpoil and François, 2010; Granet-Abisset, 2014). Since development policies have rarely taken this issue into account (Gil and Williams, 1994), workers often have to live far from their workplace or even simply live in precarious circumstances in mobile homes. Social factors can also hinder seasonal workers in this regard. Residents tend to perceive them as rivals for access to resources (housing, jobs, love partners) and decry their festive lifestyle[20] (Ooi, Mair and Laing, 2016). In any case, keeping quality workers on staff is a major challenge for the resort. This is illustrated by the case of Delta (French Alps), a travel agency that specialises in business tourism, is highly successful and has internalised its professional skills (Paget, Dimanche and Mounet, 2010)

– teaching younger generations to ski

While it is sometimes cited as a major challenge for the sustainability of resorts, the issue of introducing the youth to skiing is never addressed.

---

20 Warne, Sinadinovic, Källmen and Vinberg (2017) attempt to correlate these risky practices with the professional and social precariousness of Swedish seasonal workers; however, they do not succeed in objectifying this relationship. In the end, they explain these risky practices by pointing to the resorts' particular sociability, which is very male-focused and favours community and festive life.

And yet, there has been a steady decline in the number of skiers in West European and North American countries. In some cases, this decline is due to the cut in public investment in snow classes, which is part of the loss of social tourism (Hudson, 2000). Other authors link this phenomenon to the closure of small resorts located between low and medium altitudes, which used to be true learning centres helping to spread the ski culture (Dawson and Scott, 2013). Although it may have initially benefitted the major resorts, the contraction of the ski market could have negative long-term effects.

Introducing people to skiing at a young age is all the more essential as it seems that only learning to ski as an adult hinders people's ability to ski (Williams and Fidgeon, 2000). As the skiers of tomorrow, the younger generation can also help to introduce their parents to the sport or encourage them to take it up again. China, where the ski market is rapidly growing, seems to be very attentive to participation rates among children (Wu and Qinghua, 2015); and in recent years, the French public authorities have taken measures to support such school trips (Marcelpoil and François, 2010). Lastly, even if they did not promote social tourism, the public authorities are now aware that they would still benefit from investing in initiatives aimed at promoting skiing among the youth, who are also the clientele of tomorrow.

Layout of the book

There are many complex and interconnected global changes, and the same is true of the adaptation strategies developed by winter sports resorts. Some adaptations have been studied in depth, albeit without a general vision, due to a lack of theoretical stabilisation or because the situations studied are rooted in particular cultures. Other issues have not been sufficiently addressed. This book cannot cover all the dimensions of global change and the adaptation strategies adopted to cope with it. In the three parts of the book, we wanted to highlight the areas that seemed essential to us to understand the challenges that winter sports resorts are facing: the economic and managerial models that characterise winter sports resorts; the adaptation strategies adopted by the ski industry and the forms they take; and finally, the emergence of new markets and new research areas. As they mobilise a concept that is conducive to a study of the ways in which resorts adapt to global change, the chapters that tackle these questions have a theoretical ambition that nevertheless forms part of the case studies embodying the subject, which makes it possible to illustrate the variability of the responses.

The first part of this book looks at the various forms that winter sports resorts can take, the viability of their economic models and the effectiveness of their governance models. In the first chapter ("Contemporary problems and challenges of Polish ski resorts – problems and opportunities of integrated and catalytic development theory implementation in a post-communist country"), Michal Zemla (Pedagogical University in Cracow, Poland) explores the specificity of Central and Eastern European resorts in relation to Western resorts. His thinking is grounded in four case studies in Poland: Bialka Tatrzańska and Wierchomla Trzy Doliny, two resorts constructed over the past 20 years in a capitalist economy, and Szczyrk and Wisła, which date to the country's communist past (specifically, the 1970s). In the second chapter ("Destination Governance and the Challenge of Balancing Economic Growth and Sustainable Development in Tourist Destinations – the Case of Åre"), Sara Nordin (European Tourism Research Institute, Sweden) and Bo Svensson (Department of Humanities and Social Sciences, Mid-Sweden University, Sweden) show how the notion of governance, which is being deployed more and more often in tourist development studies, can shed light on issues of sustainable development. In the case of Åre, which is the biggest resort in Sweden, the balance between economic development, social development and preservation of the environment requires, among others, that local authorities be closely involved in the decision-making processes, which is not self-evident when an economic actor is in a position of power. Next, Doyeon Won (Liverpool John Moores University, England) focuses on the development of skiing in South Korea. He shows that the involvement of economic actors has replaced that of public actors, especially because the activity is less profitable; he also examines the factors threatening the market at present ("Ski Resort Development in Korea: Past, Present, and Future"). In the third chapter, ("An Evolutionary Path of Resort Governance: The Case of British Columbia, Canada"), Alison Gill, William Peter and Chris Sheppard (Simon Fraser University, Canada) examine the effects of governance models on resorts' sustainability, resilience and competitiveness. Their study is based on an analysis of the development trajectories of resorts in British Columbia between 1975 and 2015 and highlights moments of inertia (conveyed through the notion of "path dependence") and creativity (associated with the concept of "path creativity"). Lastly, Beltran Oriol (University of Barcelona, Spain) and Ismael Vaccaro (McGill University, Canada) look at the trajectory of resorts built in the Spanish Pyrenees. They focus on understanding the factors pushing the government to support the winter sports market despite its very weak economic results in recent decades ("Skiing in the Catalan Pyrenees: between private ventures and public rescue").

The second part details the main adaptation strategies that the ski industry has developed. Andrea Macchiavelli (University of Bergamo, Italy) shows how the new trends in consumption, marked by an increase in short stays and recreational practices, can favour small resorts, and he examines their governance models ("Small mountain resorts between climate change and new expectations for demand"). In the second chapter ("How ski resorts in Quebec are transforming to deal with climate change"), Michel Archambault explores the changes affecting the resorts in Quebec and the diversification strategies that they have developed in response. In the third chapter ("From crisis to renewal: adaptative waves as an approach to understand systemic resilience in mountain tourism destinations"), Romano Wyss (École Polytechnique Fédérale de Lausanne, Switzerland) and Tobias Luthe (University of Applied Sciences HTW Chur, Switzerland) focus on the contrasting trajectories of two resorts (Ostana in Italy and San Bernardino in Switzerland) and the way in which they have (not) adapted to external pressures. To conceptualise the resilience of the trajectories, they propose the notion of an "adaptative waves model" and consider a tourism model based on complementarity between a region's economic sectors ("circular economy thinking"). In the last chapter ("Critical Reflections on Projections of Climate Change Risk for the Ski Industry"), Daniel Scott (University of Waterloo, Canada) and Robert Steiger (University of Innsbruck, Austria) provide a critical overview of the work focused on evaluating winter sports resorts' vulnerability to climate change.

The last part is dedicated to emerging markets and research areas that should be (further) developed in the coming years. In the first chapter ("Skiing, Climate Change, Regional Development and Terrorism: A GIS-based Suitability Analysis for Future Ski Tourism Development in Turkey"), Cenk Demigloru (Umeå University, Sweden) shows that in countries intending to develop new ski resorts, like Turkey and China, it is important to combine indicators that make it possible to identify the most appropriate sites. In Turkey, a country with significant potential, physical data (climate, topography and vegetation cover) are combined with socio-economic data (accessibility and security). In the second

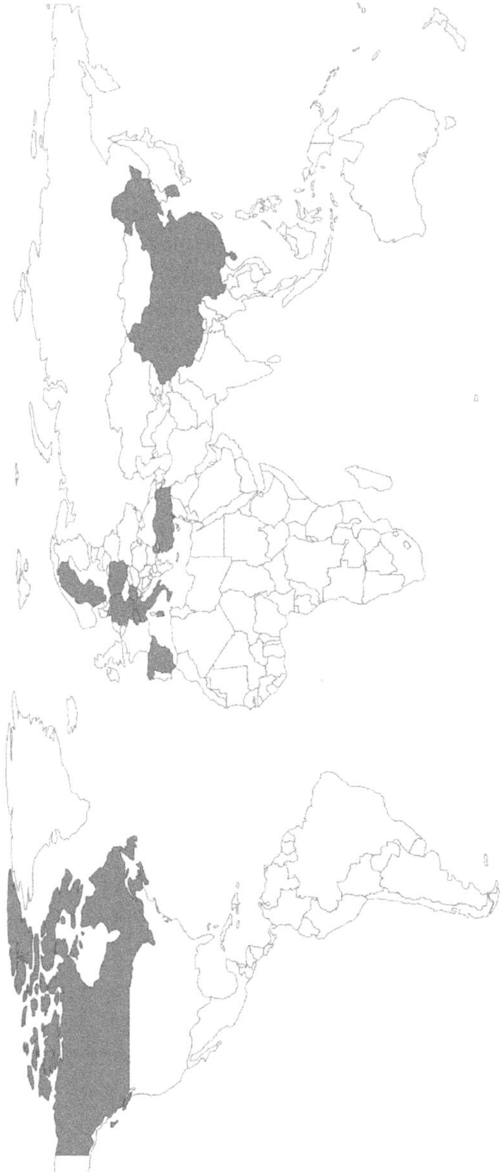

Fig. 1: Countries of cases studied in the book

chapter ("Ski tourism in China: historical development and future impact of climate change"), Yan Fang (Peking University, China), Daniel Scott (University of Waterloo, Canada) and Robert Steiger (University of Innsbruck, Austria) look at the development of skiing in China, as well as the factors that could threaten its remarkable growth to date. Using a literature review, Coralie Achin shows that tourists' adaptation to global changes is an issue that is still largely unknown, and yet, it is necessary so that resorts can align their strategies with tourists' responses and expectations ("The future of ski resorts in the light of climate change: a consideration from the tourist's point of view"). In the final chapter ("Environmental Management in Ski Resorts"), Ulrike Pröbstl (University of Natural Resources and Life Sciences, Austria) lays out Europe's EMAS certifications, which are based on a voluntary approach and external accreditation. Despite several (above all, economic) advantages, these environmental management approaches are still rare among winter sports resorts.

Bibliography

Achin, Coralie and George, Emmanuelle. 2016. "Acteurs publics et stations de moyenne montagne: regards croisés sur la mise en œuvre de politiques de diversification". In *Les instruments de l'action publique et les dispositifs territoriaux*, Némery, J.-C. and Thuriot, F. (eds.). Paris: L'Harmattan, 133–141.

Arcuset, Laurent. 2009. "Possible paths towards sustainable tourism development in a high-mountain resort". *Revue de Géographie Alpine* 97: 1–13.

Barbier, Bernard. 1993. "Problems of the French Winter Sport Resorts". *Tourism Recreation Research* 18: 5–11.

Beaudin, Laura and Huang, Ju-Chin. 2014. "Weather conditions and outdoor recreation: a study of New England ski areas". *Ecological Economics* 106: 56–68.

Beyazıt, Mehmet F. and Koc, Erdogan. 2010. "An analysis of snow options for ski resort establishments". *Tourism Management* 31: 676–683.

Bonnemains, Anouk. 2014. "Quelle capacité d'adaptation pour les stations de sports d'hiver de haute altitude des Alpes du Nord? Mise en regard de la vulnérabilité territoriale et du Plan énergie climat territorial Tarentaise Vanoise". *Sud-Ouest Européen* 37: 29–39.

Bourdeau Philippe (ed.). 2007. *Sports d'hiver en mutation*, éditions Hermès/ Lavoisier, Paris, 250.

Bourdeau Philippe. 2008. "The Alps in the age of new style tourism: between diversification and post-tourism?" In Borsdorf, A., Stötter, J., Veuillet, E. (eds.). *Managing Alpine Future*, Vienna: Austrian Academy of Science Press, 81–86.

Bourdeau Philippe. 2009. "Mountain tourism in a climate of change" In Jandl, R., Borsdorf, A., van Miegroet, H., Lackner, R., Psenner, R. (eds.), *Global Change and Sustainable Development in Mountain Regions, Alpine space – man & environment vol. 7*, Innsbruck: Innsbruck University Press, 39–52.

Briggs, James. 2000. "Ski resorts and the national forests: Rethinking forest service management practices for recreational use". *Boston College Environmental Affairs Law Review* 28: 79–118.

Buckley, Ralf, Pickering, Catherine M. and Warnken, Jan. 2000. "Environmental management for Alpine tourism and resorts in Australia". In *Tourism and Devlopment in Mountain Regions*, Godde, P. M., Price, M. F., Zimmermann, F. M. (eds.). 27–45.

Caprio, Enrico, Chamberlain, Dan E., Isaia, Marco and Rolando, Antonio. 2011. "Landscape changes caused by high altitude ski-pistes affect bird species richness and distribution in the Alps". *Biological Conservation* 144: 2958–2967.

Chrenka, Branislav and Ira, Vladimír. 2011. "Transformation of Tourist Landscapes in Mountain Areas: Case Studies from Slovakia". *Human Geographies – Journal of Studies & Research in Human Geography* 5: 1–13.

Clivaz, Christophe, Gonseth, Camille and Matasci, Cécilia. 2015. *Tourisme d'hiver: le défi climatique*. Lausanne: Presses polytechniques et universitaires romandes.

Coleman, Annie. 2004. *Ski Style. Sport and Culture in the Rockies*. Lawrence: University Press of Kansas.

Collomb, Gérard. 1989. *Du bon usage de la montagne. Touristes et Paysans*. Paris: L'Harmattan.

Dawson, Jackie and Scott, Daniel. 2013. "Managing for climate change in the alpine ski sector". *Tourism Management* 35: 244–254.

Demiroglu, O. Cenk. 2016. "An empirical study on the climate change perception and adaptation of winter sports tourists in Turkey". In

Proceedings of the Ulusal Alternatif Turizm Kongresi, Turkey: Erzincan University, 7–9 April 2016.

Demiroglu, O. Cenk, Dannevig, Halvor and Aall, Corlo. 2013. "The multidisciplinary literature of ski tourism and climate change". In *Tourism research: An Interdisciplinary Perspective*, Nazmi, K., Metin, K. (eds.). Cambridge: Cambridge Scholars Press, 223–237.

Demiroglu, O. Cenk, Kučerová, Jana and Ozcelebi, Oguzhan. 2015. "Snow reliability and climate elasticity". *Tourism Review* 70: 1–12.

Demiroglu, O. Cenk, Turp, Mustafa T., Ozturk, Tugba and Kurnaz, Mehmet L. 2016. "Impact of climate change on natural snow reliability, snowmaking capacities, and wind conditions of ski resorts in Northeast Turkey". *Atmosphere* 7: 1–12.

Duglio, Stefano and Beltramo, Riccardo. 2016. "Environmental Management and Sustainable Labels in the Ski Industry: A Critical Review". *Sustainability* 8: 1–13.

Escadafal, Alain and Boulin, Jean-Luc. 2016. "Numérique et tourisme: changement de paradigme ou simple évolution dans le management de destination?". In *Stations en tension*, Vlès, V., Bouneau, C. (eds.). Brussels: Peter Lang, 151–162.

Falk, Martin. 2013. "Impact of long-term weather on domestic and foreign winter tourism demand impact of long-term weather on domestic and foreign winter tourism demand". *International Journal of Tourism Research* 15: 1–17.

Falk, Martin. 2017. "Gains from horizontal collaboration among ski areas". *Tourism Management* 60: 92–104.

Falk, Martin and Hagsten, Eva. 2016. "Importance of early snowfall for Swedish ski resorts: Evidence based on monthly data". *Tourism Management* 53: 61–73.

Feldman, Eric and Stein, Alison. 2010. "Assuming the risk: tort law, policy, and politics on the slippery slopes". *Faculty Schalarship* 296: 258–303.

Ferrand, Alain and Vecchiatini, Daria. 2010. "The effect of service performance and ski resort image on skiers' satisfaction". *European Journal of Sport Science* 2: 1–17.

Freuler, Benjamin and Hunziker, Marcel. 2007. "Recreation activities in protected areas: bridging the gap between the attitudes and behaviour of snowshoe walkers". *For. Snow Landsc. Res* 81: 191–206.

Gauchon, Christophe. 2002. "Le tourisme dans les Alpes: pratiques, aménagement et protection". In *Question de géographie: les montagnes.* Pornic: Du Temps, 153–182.

Gauchon, Christophe. 2009. "Les hivers sans neige et l'économie des sports d'hiver. Un phénomène récurrent, une problématique toujours renouvelée". *Cahiers de Géographie* 8: 191–202.

Gauchon, Christophe. 2010. "Territoires "dé-touristifiés" des montagnes françaises: quels enseignements?", In *Tourismes, Patrimoines, Identités, Territoire*, Bataillou, C. (ed.). Perpignan: PUP, 473–484.

George, Emmanuelle, Achin, Coralie, Fablet, Gabriel and François, Hugues. 2016. "Entre permanences et bifurcations: une lecture territoriale des destinations touristiques de montagne". *Mondes du Tourisme* special edition "La Troisième Révolution touristique": 15 pages.

Ghaderi, Zahed, Khoshkam, Mana and Henderson, Joan C. 2014. "From snow skiing to grass skiing". *Anatolia* 25: 96–107.

Gill, Alison and Williams, Peter M. 1994. "Managing growth in mountain tourism communities", *Tourism Management* 15: 212–220.

Gill, Alison and Williams, Peter M. 2011. "Rethinking resort growth: understanding evolving governance strategies in Whistler, British Columbia". *Journal of Sustainable Tourism* 19: 629–648.

Glick, Daniel. 2001. *Powder Burn. Arson, Money, and Mystery on Vail Mountain.* New York: Public Affairs.

Granet-Abisset, Anne-Marie. 1994. *La route réinventée. Les migrations des Queyrassins aux 19ème et 20ème siècle.* Grenoble: PUG

Granet-Abisset, Anne-Marie. 2014. "Tourisme et pluriactivité: Les "nouveaux" saisonniers des stations alpines depuis les années 1960". In *Les transformations économiques, techniques et sociales des zones touristiques: une approche comparative (19e-20e siècles)*, Laurent, T., Cédric, H., Marc, G. (eds.). Lausanne: Alphil, 257–274.

Haimayer, Peter. 1989. "Glacier-skiing areas in Austria: a socio-political perspective". *Mountain Research and Development* 9: 51–58.

Hamilton, Lawrence, Brown, Benjamin C. and Keim, Barry. 2007. "Ski areas, weather and climate". *International Journal of Climatology* 27: 2113–2124.

Hamilton, Lawrence, Rohall, David E., Brown, Benjamin C., Hayward, Gregg F. and Keim, Barry D. 2003. "Warming winters and New

Hampshire's lost ski areas: an integrated case study". *International Journal of Sociology & Social Policy* 23: 52–73.

Hatt, Emeline and Vlès, Vincent. 2014. "Mutations socio-environnementales et perspectives d'adaptation des stations de montagne pyrénéenne". *Sud-Ouest européen* 37: 1–27.

Higham, James and Hinch, Tom. 2009. Sport and Tourism: Globalization, Mobility and Identity, Elsevier, Amsterdam.

Hopkins, Debbie, Higham, James E. S. and Becken, Susanne. 2013. "Climate change in a regional context: relative vulnerability in the Australasian skier market". *Regional Environmental Change* 13: 449–458.

Hudson, Simon. 2000. *Snow Business. A study of the international ski industry*. Andover: Cengage Learning EMEA.

Innocenti, Piero, Merciu, Cristina and Merciu, George-Laurentju. 2012. "Mountain tourism a challenge for sustainable development. Case study". *Economic Insights – Trends & Challenges* 64: 62–72.

Izzo, Umberto. 2014. "Safety and liability in the practice of skiing: how law and cognitive sciences shape the driving factor of winter tourism". *Entertainment and Sports Law* 12: 1–13.

Jopp, Ryan, DeLacy, Terry and Mair, Judith. 2010. "Developing a framework for regional destination adaptation to climate change". *Current Issues in Tourism* 13:6, 591–605.

Kim, Song-Yi, Park, Chan, Park, Jin-Han and Lee, Dong-Kun. 2015. "Estimating effects of climate change on ski industry – The case of ski resorts in South Korea". *Journal of Environmental Impact Assessment* 24: 432–443.

Kuscer, Kir, Mihalič, Tanja et Pechlaner, Harald. 2016. "Innovation, sustainable tourism and environments in mountain destination development: a comparative analysis of Austria, Slovenia and Switzerland", *Journal of Sustainable Tourism*, 25:4, 489–504

Kuscer, Kir, Mihalic, Tanja and Pechlaner, Harald. 2017. "Innovation, sustainable tourism and environments in mountain destination development: a comparative analysis of Austria, Slovenia and Switzerland". *Journal of Sustainable Tourism* 25: 489–504.

Konu, Henna, Laukkanen, Tommi and Komppula, Raija. 2011. "Using ski destination choice criteria to segment Finnish ski resort customers". *Tourism Management* 32: 1096–1105.

Laguna, Marín-Yaseli M. and Teodoro, Lasanta M. 2003. "Competing for meadows: a case study on tourism and livestock farming in the Spanish Pyrenees". *Mountain Research and Development* 23: 169–176.

Landauer, Mia, Pröbstl, Ulrike and Haider, Wolfgang. 2012. "Managing cross-country skiing destinations under the conditions of climate change – Scenarios for destinations in Austria and Finland". *Tourism Management* 33: 741–751.

Loeffler, Roland and Steinicke, Ernst. 2007. "Amenity migration in the U.S. Sierra Nevada". *Geographical Review* 97: 67–99.

Luthe, Tobias and Schläpfer, Félix. 2011. "Effects of third-party information on the demand for more sustainable consumption: a choice experiment on the transition of winter tourism". *Environmental Innovation and Societal Transitions* 1: 234–254.

Luthe, Tobias and Wyss, Romano. 2016. "Resilience to climate change in a cross-scale tourism governance context: a combined quantitative-qualitative network analysis". *Ecology and Society* 21: 234–254.

Macchiavelli, Andrea. 2009. "Le tourisme alpin. Conditions et ambivalences face à l'innovation". *Journal of Alpine Research | Revue de géographie alpine* 97: 84–95.

Marcelpoil, Emmanuelle and François, Hugues. 2010. "Territoires de montagne et développement durable" (chapitre 20). In *Développement durable et Territoires*, Zuindeau, B. (ed.). Lille: Éditions Septentrion, 253–264.

Marnezy, Alain et Gauchon, Christophe. 2006. "Les domaines skiables reliés dans les Alpes françaises". *Cahiers de Géographie* 4: 115–124.

Matzler, Kurt, Füller, Johann, Renzl, Birgit, Herting, Stephan and Späth, Sebastian. 2008. "Customer satisfaction with Alpine Ski areas: the moderating effects of personal, situational, and product factors". *Journal of Travel Research* 46: 403–493.

Mijatov, Maja, Ivkov-Džigurski, Anđelija, Pivac, Tatjana and Košić, Kristina. 2016. "The leisure time aspects in a ski centre – Kopaonik Mountain case study (Serbia)". *Journal of the Geographical Institute 'Jovan Cvijic' SASA* 66: 291–306.

Morrison, Alastair M., Braunlich, Carl G., Cai, Liping A. and O'Leary, Joseph T. 1996. "A profile of the Casino Resort Vacationer". *Journal of Travel Research* 35: 55–61.

Morrison, Alastair M., Yang, Chung-Hui, O'Leary, Joseph T. and Nadkarni, Nandini. 1996. "Comparative profiles of travellers on cruises and land based resort vacations". *Journal of Tourism Studies* 7: 99–111.

Morrison, Clare and Pickering, Catherine M. 2013. "Perceptions of climate change impacts, adaptation and limits to adaption in the Australian Alps: the ski-tourism industry and key stakeholders". *Journal of Sustainable Tourism* 21: 173–191.

Mühlinghaus, Sabine and Wälty, Samuel. 2001. "Endogenous development in Swiss mountain communities: local initiatives in Urnäsch and Schamserberg". *Mountain Research and Development* 3: 236–242.

Nepal, Sanjay K. and Jamal, Tazim B. 2011. "Resort-induced changes in small mountain communities in British Columbia, Canada". *Mountain Research & Development* 31: 89–101.

Ooi, Natalie, Laing, Jennifer and Mair, Judith. 2015. "Sociocultural change facing ranchers in the Rocky Mountain West as a result of mountain resort tourism and amenity migration". *Journal of Rural Studies* 41: 59–71.

Ooi, Natalie, Mair, Judith and Laing, Jennifer. 2016. "The transition from seasonal worker to permanent resident: social barriers faced within a mountain resort community". *Journal of Travel Research* 55: 246–260.

Ottogalli-Mazzacavallo, Cécile and Saint-Martin, Jean (eds.). 2009. *Femmes et hommes dans les sports de montagne. Au-delà des différences*. Grenoble: C.N.R.S./Maison des Sciences de l'Homme/alpes, C.R.I.S. – LARHRA.

Paget, Élodie, Dimanche, Frédéric and Mounet, Jean-Pierre. 2010. "À tourism innovation case. An actor-network approach". *Annals of Tourism Research* 37: 828–847.

Palmer, Adrian and Mathel, Virginie. 2010. "Case Study: causes and consequences of underutilised capacity in a tourist resort development". *Tourism Management* 31: 925–935.

Pechlaner, Harald and Tschurtschenthaler, Paul. 2010. "Tourism policy, tourism organizations and change management in Alpine regions and destinations – A european perspective". *Current Issues in Tourism* 6: 508–539.

Perović, Maša. 2008. "Green development and participation-implications for countries in transition: Case study of Stara Planina, Republic of Serbia". *Spatium* 17–18: 38–44

Pickering, Catherine M. and Buckley, Ralf C. 2010. "Climate response by the ski industry: the shortcomings of snowmaking for Australian resorts". *Ambio* 39: 430–438.

Pickering, Catherine M., Castley J. Guy and Burtt, Michelle. 2010. "Skiing less often in a warmer world: attitudes of tourists to climate change in an Australian Ski Resort", *Geographical Research* 48: 137–147.

Pignatti, Sandro. 1993. "Impact of tourism on the mountain landscape of central Italy". *Landscape and Urban Planning* 24: 49–53.

Price, Martin. 1987. "Tourism and forestry in the Swiss Alps: parasitism or symbiosis?", *Mountain Research and Development* 7: 1–12.

Pröbstl-Haider, Ulrike, Haider, Wolfgang, Wirth, Veronika and Beardmore, Ben. 2015. "Will climate change increase the attractiveness of summer destinations in the European Alps? A survey of German tourists". *Journal of Outdoor Recreation and Tourism* 11: 44–57.

Richards, Greg. 1996. "Skilled consumption and UK ski holidays". *Tourism Management* 17: 25–34.

Rivera, Jorge, de Leon, Peter and Koerber, Charles. 2006. "Is greener whiter yet? The sustainable slopes program after five years". *Policy Studies Journal* 34: 195–221

Rutty, Michelle, Scott, Daniel, Johnson, Peter, Jover, Eric, Pons, Marc and Steiger, Robert. 2015. "The geography of skier adaptation to adverse conditions in the Ontario ski market". *The Canadian Geographer / Le Géographe canadien* 59: 391–403.

Rutty, Michelle, Scott, Daniel, Johnson, Peter, Pons, Marc, Steiger, Robert and Viella, Marc. 2017. "Using ski industry response to climatic variability to assess climate change risk: An analogue study in Eastern Canada". Tourism Management 58: 196–204.

Sato, Chloe F., Schroder, Meliesa, Green, Ken, Michael, Damian R., Osborne, William S. and Lindenmayer, David B. 2014. "Managing ski resorts to improve biodiversity conservation: Australian reptiles as a case study". *Ecological Management & Restoration* 15: 147–154

Scott, Daniel, Dawson, Jackie and Jones, Brenda. 2008. "Climate change vulnerability of the US Northeast winter recreation – tourism sector". *Mitigation & Adaptation Strategies for Global Change* 13: 577–596

Scott, Daniel, Hall, Michael C. and Gössling, Stefan. 2012. *Tourism and climate change. Impacts, adaptation and mitigation.* London and New York: Routledge.

Scott, Daniel, McBoyle, Geoff and Minogue, Alanna. 2007. "Climate change and Quebec's ski industry". *Global Environmental Change* 17: 181–190.

Shabanski, Philip. 2013. "Can a small winter resort be managed and marketed like a large one: a case of Bulgarian Ski Resort Bansko". *Problems of Management in the 21st Century* 8: 94–101

Silberman, Jordan A. and Rees, Peter W. 2010. "Reinventing mountain settlements: A GIS model for identifying possible ski towns in the U.S. Rocky Mountains". *Applied Geography*, 30: 36–49.

Spandre, P., Francois, H., Verfaillie, D., Lafaysse, M., Deque, M., Eckert, N., George, E., Morin, S., 2019. « Climate controls on snow reliability in French Alps ski resorts », *Scientific Reports*, doi 10.1038/s41598-019-44068-8

Sterl, Petra, Eder, Renate and Arnberger, Arne. 2010. "Exploring factors influencing the attitude of ski tourers towards the ski touring management measures of the Gesäuse National Park". *Eco. Mont* 2: 31–38.

Stoddart, Mark C. J. 2011. "Constructing masculinized sportscapes: Skiing, gender and nature in British Columbia, Canada". *International Review for the Sociology of Sport* 46: 108–124.

Tang, Chun-Hung and Jang, SooCheong. 2007. "Weather risk management in ski resorts: Financial hedging and geographical diversification". *International Journal of Hospitality Management* 30: 301–311.

Thiery, Patrick. 2006. "Les sports d'hiver: pratiques et pratiquants". *Bulletin de statistiques et d'études du ministère de la Jeunesse, des sports et de la Vie Associative* 06–06: 1–5.

Thulemark, Maria and Hauge, Alte. 2014. "Creativity in the recreational industry. Re-conceptualization of the Creative Class theory in a tourism-dominated rural area". *Scandinavian Journal of Public Administration* 18: 87–105.

Trawöger, Lisa. 2014. "Convinced, ambivalent or annoyed: Tyrolean ski tourism stakeholders and their perceptions of climate change". *Tourism Management* 40: 338–351.

Tuppen, John. 2000. "The restructuring of winter sports resorts in the French Alps: Problems, Processes and Policies". *International Journal of Tourism Research* 2: 327–344.

Unbehaun, Wiebke, Pröbstl, Ulrike and Haider, Wolfgang. 2008. "Trends in winter sport tourism: challenges for the future". *Tourism Review* 63: 36–47.

Vaccaro, Ismaël and Beltran, Oriol. 2009. "The mountainous space as a commodity: the Pyrenees at the age of globalization". *Revue de Géographie alpine* 97: 1–13.

Vlès, Vincent and Bouneau, Christophe. 2016. *Stations en tension*. Brussels: Peter Lang.

Wan, Yan-ru. 2008. "Innovation in market: a case study of Wan Long Ski Resort". *Management Science and Engineering* 2: 40–46.

Warne, Maria, Sinadinovic, Kristina, Källmen, Hakan and Vinberg, Stig. 2017. "Risky consumption of alcohol and drugs among employees at ski resorts". *Nordic Studies on Alcohol & Drugs / Nordisk Alkohol- & Narkotikatidskrift* 34: 201–216.

Williams, Peter W. and Fidgeon, Paul. 2000. "Addressing participation constraint: A case study of potential skiers". *Tourism Management* 21: 379–393.

Williams, Peter W. and Lattey, Christine. 1994. Skiing constraints for women. *Journal of Travel Research* 33(2): 21–25.

Won, Doyeon and Hwang, Sunhwan. 2009. "Factors influencing the college skiers and snowboarders' choice of a ski destination in Korea: a conjoint study". *Managing Leisure* 14: 17–27

Wu, Benny and Qinghua, Wei. 2015. *China Ski Industry White Book*, 46 pages.

Wyss, Romano. 2013. "Coopération pour l'adaptation climatique dans le tourisme". *Journal of Alpine Research | Revue de géographie alpine* 101: 1–10.

Wyss, Romano, Abegg, Bruno and Luthe, Tobias. 2014. "Perceptions of climate change in a tourism governance context". *Tourism Management Perspectives* 11: 69–76.

Yang, Jianming and Wan, Chunyan. 2010. "Progress in research on the impacts of global climate change on winter ski tourism". *Advances in Climate Change Research* 1: 55–62

Zemla, Michal. 2008. "Failures in building partnership for success in the competitive market: the case of Polish Ski Resorts". *Managing Global Transitions* 4: 421–444.

I

The Economic and Managerial Models of Ski Resorts

1.1

Contemporary Problems and Challenges of Polish Ski Resorts – Problems and Opportunities of Integrated and Catalytic Development Theory Implementation in Post-Communist Country

Michal Zemla

Introduction

Ski resorts around the world vary in size, organization, age and development processes, due to different local environmental, social, economic, or legal conditions. However, some classifications might be applicable to most of the resorts located in different countries. Pearce's division into catalytic and integrated resorts is one of the most popular and functional. However, it was developed based on the analysis of resorts set in the context of Western, mainly European economies, thus it is difficult to describe the development of ski resorts in post-communist countries using the same models. Ski resorts in countries like Poland, Slovakia and the Czech Republic were established in the communist times and had to face challenges arising from the systemic changes. Also, many new and modern ski resorts were established in those countries after the fall of communism. The aim of the chapter is to present how the theory of integrated and catalytic models can be used to describe the contemporary ski resorts development in Poland and better understand the challenges and problems which those resorts had to and still have to confront. In order to present the contemporary problems and development directions in Polish ski resorts, four different resorts were selected. Białka Tatrzańska and Wierchomla Dwie Doliny (Two Valleys) are resorts built from scratch in the last twenty years, but Białka Tatrzańska matches the catalytic model, while Wierchomla represents integrated resorts. Szczyrk and Wisła are resorts built and developed in the 1970s.

Catalytic and integrated ski resorts

A "ski resort" is a term that has not been clearly defined in the literature on the subject. It is most commonly used to refer to a town or village with infrastructure for skiers (Gill, Williams, 1994). Some authors, however, use this term to define a complex of lifts and slopes under common management, where a combined ski ticket can be used (Tuppen, 2000). Flagestad and Hope (2001b) introduced a much clearer term – a "winter sports destination". In this paper, it is understood as a tourist destination of a special kind, identified by unique tourism infrastructure and the way in which visitors spend their time there. Assuming that a tourist destination represents physical, historical, and ethnographic features distinguishing it from other places, so that it could attract non-residents and develop one or more different types of tourism (Goncalves, Águas, 1997), a winter sports destination may be defined as a geographical, economic and social unit consisting of all those firms, organizations, activities, areas and installations which are intended to serve the specific needs of winter sports tourists (Flagestad, Hope, 2001b). This approach emphasizes the complex character of winter sport destinations, with a clear indication of the role of supplementary services (hotel accommodation, gastronomy, entertainment) and environment, as well as allows us to view the quality of the ski resort's product as a concept which is broader than the quality of ski services alone, according to the concept of the tourism value chain (Weiermair, 1997). Here, the term "a ski resort" will be used as a synonym of "a winter sports destination", and the term "a ski area" will be used in the second possible meaning of the "ski resort" (Williams, 1996). In that sense, a ski resort is a type of a tourist destination. It is an area of consumption, configured for a targeted service type, a ski tourism product.[1] Winter sports resorts are often specifically designed as ski products; the resort activities essentially revolve around skiing and the whole product development and packaging are focused around this activity (Paget, Dimanche and Mounetm, 2010).

A ski resort is a complex system involving a wide variety of actors (Paget, Dimanche, Mounet, 2010) that can establish a variety of links among them. There are different ways of presenting the process of evolution and the origins of contemporary winter sports resorts in Europe and around the world. A common approach presents several (usually four) generations of ski resorts in Europe (mainly in France), ranging from the oldest and located in inhabited areas (first generation resorts) to the fully integrated ones, built in the wilderness (Hudson, 2000). However, the difference between resorts categorized as representing younger and older generations is not only connected with the evolution of thinking about ski resorts development. Pearce (Pearce 1978; WTO 1995), who presented his division into catalytic and integrated resorts[2], argued that integrated resorts are typical of the French Alps while catalytic resorts are common in the Austrian Alps, and the reasons for this situation are natural, social and legal.[3] Integrated resorts are those which have been carefully planned as individual entities. They have been created by a single operator who coordinates all resort activities (Paget, Dimanche, Mounet, 2010). Most of them are carefully planned to function efficiently and provide an interesting environment for tourists (Pearce 1996, p. 18). Catalytic resorts occur in areas with residential and economic communities and are developed with the participation of many

---

1 The reason of some misunderstandings between the notion of terms: « a ski resort » and « a ski area » is relation between them very different in Alpine countries and in Central European countries. In Alps, where huge ski areas dominate, one ski area often connect few destinations (ski resorts), while in Poland few small, separated ski areas can be located in one ski resort.

2 Original Pearce's idea regarded winter sports resorts, but later, other authors as well as Pearce himself also used that classification to categorize golf resorts, seaside resorts etc. See for example (Pearce, 1996; Inkson, Minnaert, 2018; Sharma; WTO 1995; Inskeep, Kallenberger 1992).

3 Even though integrated resorts are typical of France and catalytic resorts of Austria, there are examples of integrated resorts in Austria and catalytic ones in France. La Plagne is often presented as a typical example of an integrated resort in France and sometimes that kind of resorts are labelled as La Plagne type resorts (Inkson, Minnaert, 2018), but some older, first generation French resorts, like Chamonix, are typical examples of the catalytic model of resort development.

local actors (Inkson, Minnaert, 2018). Their localization further to the north and at lower altitudes is responsible for the fact that building a ski resort with reasonable snow conditions in Austria often does not require expensive investments in wild high-mountain areas, as it is usually the case in France. An even more important difference between typical Austrian and French resorts regards the local inhabitants' participation in the resort development. Typically, catalytic resorts are built step by step (Pearce identified three typical phases of a catalytic resort development – Pearce 1989), with a strong support of local labor, capital and authorities. As a result, local people also benefit economically from the ski resort development. In contrast, an integrated resort is often developed as a one-stage investment by one, huge, external company, managing ski infrastructure at several destinations. The localization of a resort at high altitude, far away from human settlements, almost completely excludes local people from drawing economic benefits from the resort development. However, integrated resorts are perceived more favorably as regards their potential to match customer needs, as all the elements of the product are under the same management. With the development of catalytic resorts, coordination is always required but sometimes it is not efficient (Inkson, Minnaert, 2018). The next difference between catalytic and integrated resorts concerns the size of the ski infrastructure and other facilities. Small hotels and boarding houses established by the local capital are typical of catalytic resorts, while huge hotels and apartment houses built by one investor are commonly found at integrated resorts.

The organizational and social differences between those two types of winter sports resorts are emphasized even stronger by Flagestad (2002) and Flagestad and Hope (2001a; 2001b), who examined the differences between the corporate and the community model of winter sports destinations, where the former resembles integrated resorts and the latter – catalytic resorts. Looking from the perspective of social costs and benefits, Kurek criticizes integrated development of a ski resort as very harmful for the environment, and not giving any substantial benefits to the local community (Kurek 1999). In his interpretation of that concept, a private external investor is treated as a profit-oriented company, ignoring the social and environmental costs of their actions. However, if a company can recognize the connection between their long-term profit and the quality of the socio-ecological environment, this model of resort development might be the proper way to achieve sustainability. Inskeep and Kalleberger (1992) state that an integrated approach to resort development helps prevent serious environmental, socio-economic and marketing problems, often associated with unplanned tourism destinations. Similarly, Kaspar points to the examples of New Zealand's ski resorts with the strongly dominant position of one ski operator as a potential model for European resorts aiming at sustainable development, due to relatively easy management (Kaspar 1992). Flagestad and Hope note that this easiness of management should also ensure better business performance (Flagestad and Hope 2001b), and WTO experts emphasize better potential response to skiers' needs at integrated resorts (WTO, 1995). However, Paget, Dimanche and Mounet (2010) point out that the current French law (so called, the Mountain Law of 1985) imposes the responsibility to manage and develop tourism in each resort on the municipality in which the resort is situated. It then has the opportunity to take on management duties (direct management) or delegate them to a private operator. In the latter case, the delegation may involve only the operation of lifts (partial delegation) or both, the operation of lifts and management of ski slopes (full delegation). This regulation also helps integrated French resorts to function as a network of economic, independent tourism enterprises, where actors revolve around the organizing authority and private operator (Paget, Dimanche and Mounet, 2010).

The development of the winter sports industry in Poland

The models and evolutionary schemes presented above might be perceived as universal tools of analysis of development of ski resorts, but they quite well depict primarily the situation in the Western world of the 20th century. However, the situation might be very different when we consider the countries from the former communist block. Winter sports were also popular in most European communist countries and ski infrastructure could be found in the Soviet Union, Czechoslovakia, Poland, Bulgaria, Romania, etc. Still, the rationales for those investments, the ways of conducting investments and the localization criteria were completely different. The economic crisis and the transformation process at the end of the century also played an important role in shaping the present state of the winter sports industry in those countries. Poland is an example of a country displaying specific features. For many people, the phenomenon of skiing in Poland can be quite surprising. Poland is usually perceived as a typical lowland country and this is generally correct. Only about 3 % of the country lies over 500 m above the sea level. These mountains are also relatively low, with only one range (the Tatras) exceeding 2000 m. Otherwise, most Polish summits reach the altitude of about 1000 m. Still, this 3 % covers almost 10,000 km^2 – nearly half of the whole surface of Slovenia. It is also worth mentioning that the areas situated in the mountain regions include about 20 % of the accommodation capacity in Poland (Żemła, 2008). This shows how important mountains are for tourism in Poland. Being a typical lowland country, Poland generates a relatively large demand (according to public polls, 20 % of Poles can ski – TNS OBOP, 2004) for skiing services – an estimated 0.5 % of the global demand (Hudson, 2000)[4]. The relatively low level of ski infrastructure development in Polish mountain areas has resulted in a surplus of demand over supply and, consequently, long queues to lifts, which have been among the most frequent customers' complaints for many years (Żemła, 2004). Skiing is popular with Polish people, but they practice it rather occasionally and they are not very proficient skiers. This is due to the fact that Polish ski areas have always been very small, even in comparison to those in the neighboring countries. This limited supply causes difficulties as regards providing high quality services and maintaining West European standards of ski area development (Żemła, 2004). The relatively large number of suppliers has resulted in an urgent need for cooperation, which is often very difficult to establish (Żemła, 2008). There are many ski destinations consisting of more than one ski areas. However, those areas function unconnectedly, due to separate slope, ski lifts and ticketing systems.

There are more than 100 localities in Poland with winter sports infrastructure (Krzesiwo, 2014). Polish ski resorts are spatially split between two large mountain ranges: the Sudety and the Carpathian Mountains. The latter are generally higher than the former and make up 75 % of Polish mountain areas (Żemła 2004). Apart from four major ski localities in the Sudety (Sienna–Czarna Góra, Zieleniec, Karpacz and Szklarska Poręba), the most important Polish winter sports destinations

---

4 Unfortunately, there are no any statistical data presenting the size of Polish winter sports market, eg. the number of people visiting Polish ski resorts or the number of Polish people participating in winter sports. In some press releases some estimations of those numbers are offered, but they are divergent, not reliable and often prepared for particular purposes.

are located in the Carpathian Mountains. Szczyrk, Białka Tatrzańska, Wisła and Zakopane seem to be the most popular destinations in the Polish Carpathian Mountains, but only Zakopane and Wisła have become recognizable names among international tourists, due to the fact that they are associated with ski-jumping competitions. Generally, the distribution of ski areas in the Carpathian Mountains is not even. Three massifs: the Śląski Beskid, the Sądecki Beskid and the Tatras contain almost all most important Carpathian winter sports destinations in Poland.

The general ski resorts development scheme in Poland in the 20th century was similar in most localities. Big investments were conducted mainly in the 1970s, by state-owned industrial companies, without a proper consideration of the local social, economic and ecological conditions. Those were typically relatively big systems of slopes and t-bar lifts, in a few cases including also single chairlifts. At the same time, several small resorts were established by local people at numerous destinations. Those were often individual small t-bar lifts with single slopes for beginners. The 1980s were the time of political and economic instability and almost no new investments were undertaken. The early 1990s brought a political and economic transformation in Poland, which radically changed the conditions of running a business and caused slight economic stagnation. Those difficult economic conditions found Polish ski resorts using obsolete and insufficient infrastructure, resulting in enormous queues to lifts (Krzesiwo, 2014). At the same time, at the end of the 20th century, the domestic demand started to grow quickly and the queues became even longer, which resulted in a sharp increase in the number of outbound trips (Żemła, 2008). The new century brought another unfavorable phenomenon to Polish ski resorts operators, which was the global warming effect, causing problems with natural snow cover in many winter seasons (Krzesiwo, 2016). Another predicament was the strong protests of Polish ecological organizations and the local conflicts between land owners and lifts owners. Nevertheless, the last ten years have witnessed substantial investments, both in the existing and new resorts (Baran, 2010). Those investments have significantly increased the number of lifts and slopes, as well as their standard. Currently, almost all ski resorts in Poland are equipped with modern high-speed chairlifts, the percentage of slopes equipped with artificial snow making infrastructure is close to 100 % and the percentage of slopes illuminated in the evening is only slightly lower (Krzesiwo, 2014).

Case studies

In order to present various ways of the development of Polish ski resorts, four examples were selected: Szczyrk and Wisła in the Śląski Beskid, Białka Tatrzańska in the Tatras and Wierchomla Two Valleys in the Sądecki Beskid (fig. 2). Each of them represents a very different model of development. The last two are new resorts established after the fall of communism; the development of Białka Tatrzańska is rather similar to the catalytic model, while Wierchomla might be classified as an integrated resort. Wisła and Szczyrk have been important winter sports destinations since the 1970s. However, Szczyrk was greatly invested in by Silesian coal mines, which made it the biggest Polish winter sports destination, still holding its position today. At the same time, Wisła was a place with many very small, locally owned ski areas. This difference determined the whole development process at those destinations.

Szczyrk is a small town (slightly over 5000 inhabitants), located in the Śląski Beskid, in Silesian Voivodeship, Bielsko-Biała County (Polish: powiat), in the narrow Żylica Valley (fig. 3). Even though it has municipality rights, Szczyrk resembles a tourism village with buildings clustered around the only road, parallel to the river. It is surrounded with the tallest mountains of the Śląski Beskid. Skrzyczne – the highest point of the whole range, a little south to the town center, creates favorable conditions for the development of winter sports, and the northern face of Skrzyczne is the highest elevation (over 700 meters) in the Polish Beskid mountains. Those favorable conditions were appreciated in as early as the 1950s, when the Polish Ministry of Sport established a modern ski area, with a two-section single chairlift. In the 1960s, two t-bar lifts were added. However, the time of the biggest investment was the beginning of the 1970s, when Silesian coal mines established the Czyrna-Solisko ski resort on the northern face of Małe Skrzyczne (Żemła, 2010). This was a unique system of thirteen t-bar lifts, serving over twenty kilometers of slopes. The resort without any further investment had remained the biggest ski area in Poland till 2017, when it was modernized and maintained its status. At the same time, a few smaller ski areas were created in Szczyrk, mainly due to local people's investments. They included Beskidek, Beskid, Kaimówka and Biały Krzyż. This development of winter sports infrastructure was followed by numerous hotels, restaurants and other elements of tourism infrastructure. Again, most investments in big hotels were made by external, state-owned industrial companies. Towards the

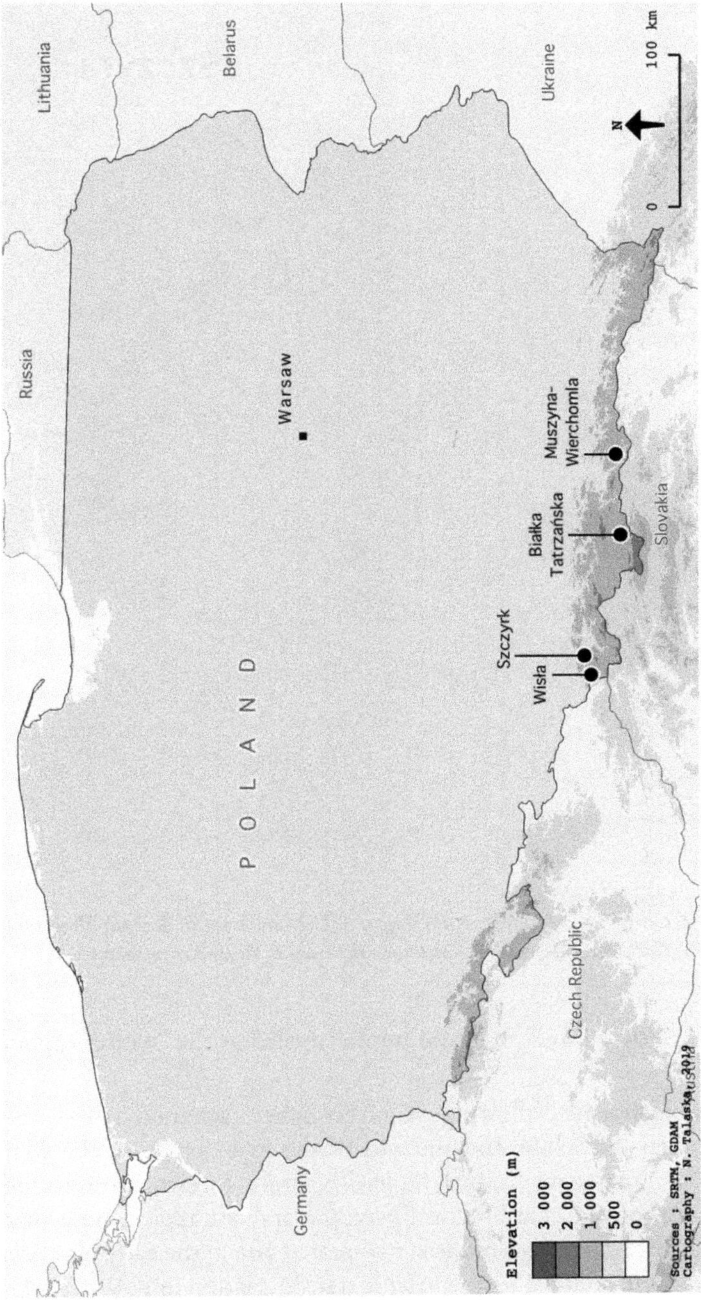

Fig. 2: Selected winter sports destinations of Southern Poland
Szczyrk – losing and re-establishing the status of a leading ski resort in Poland

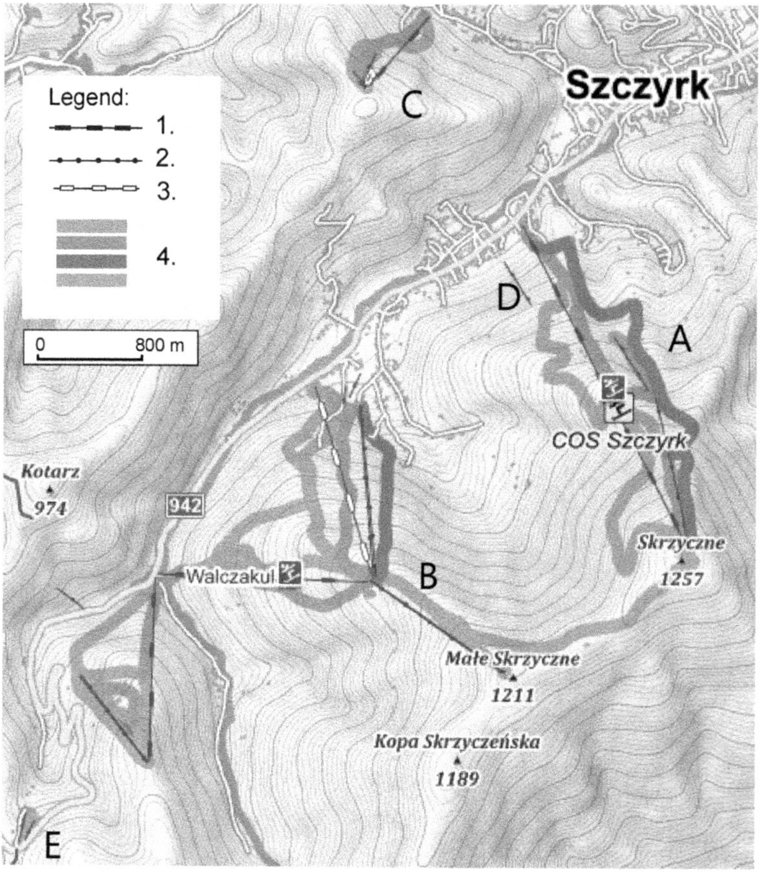

Fig. 3: Ski areas in Szczyrk
1. chair lifts; 2. t-bar lifts; 3 gondola; 4. ski slopes; A. COS ski area; B. Szczyrk Mountain Resort ski area; C. Beskid ski area; D. Kaniówka ski area; E. Biały Krzyż ski area

end of the 1970s, Szczyrk was commonly labelled as the "winter capital of Poland".

Unfortunately, the following years brought stagnation. Over the period between the 1980s and the end of the first decade of the 21st century, only one new lift was built. The only investment was the modernization of old chairlifts on Skrzyczne and the replacement of a single-seated lift by a not-so-modern two-seated one in the early nineties. At the time of the political and economic transformation in Poland in the

1990s, Szczyrk remained the biggest winter sports destination in Poland in terms of size, but it offered only old lifts and slopes which were not equipped with snow guns or lights. It was a time when Szczyrk started to lose its prestige and was labelled in the press as a "ski museum" and not as the "capital" any more. The stagnation caused many problems resulting from the communist style of investing. The lack of proper spatial planning brought terrible traffic jams and problems with parking places. Some of the hotels were privatized, but others were closed down and fell into ruin. However, the most serious problem was connected with the fact that in the 1970s, the rights of the local land owners were not properly protected. That concerned the land on which lifts and the slopes of Czyrna-Solisko ski area were built. The mistrust between the GAT venture running the ski area (created by a state-owned coal mine) and the owners of the land first stopped all investment talks and then led to an open conflict. As a result, nine out of thirteen lifts were closed down for most of the winter seasons between 2000 and 2010. This eventually ruined the position of Szczyrk as the leading Polish ski resort. Consequently, the incomes of the owners of hotels and other hospitality and gastronomy facilities plummeted and many of them were desperately trying to sell their businesses at a low price. Also the Beskidek ski area declined. The conflict evoked huge emotions, as both parties – GAT and the association of land owners, were not transparent and not playing fair.

New perspectives opened at the beginning of the next decade. The Czyrna-Solisko ski area was finally privatized and the buyer was a company established by the local land owners. This was a breakthrough, but at first only in terms of image and organization. The new owner did not have sufficient means to make urgent investments. In 2014, the ski area was sold once more and this time the buyer was the Tatry Mountains Resorts – a Slovak giant running several huge resorts in Slovakia and the Czech Republic. The TMR developed a plan of large investments, which were to restore the status of the best Polish ski resort to the area. During the winter season 2017/18, a completely new ski area (now known as the Szczyrk Mountain Resort) was re-opened. A new gondola and two high-capacity chairlifts appeared along with new slopes and modernized artificial snowing and lightening systems. At the same time, Centralny Ośrodek Sportu (COS – Central Sports Resort), running the ski area on the Skrzyczne mountain, modernized its chairlifts again by introducing modern, fast, four-seat chairlifts. Another important year was 2015, when the old Beskid ski area was acquired by a private company from

Katowice, specializing in heavy constructions. The new owner created a modern small area with a fast four-seat chairlift, three slopes and a revolutionary snow making system, efficient also in warm temperatures.[5] Another breakthrough connected with the 2017/2018 winter season was the introduction of a common ticketing system for all there ski areas in Szczyrk, as well as free ski buses which should help to solve the problem of traffic jams. Since 2010, many new good hotels have been opened in Szczyrk, and the gastronomic offer has been broadened, but Szczyrk still lacks a satisfactory après-ski offer and pedestrian areas for evening strolls.

Wisła – the development and modernization of the winter sports infrastructure

Wisła is a close neighbor of Szczyrk, located in the Śląski Beskid. It is a town in Cieszyn County, Silesian Voivodeship, in southern Poland, with a population of about 12,000, near the border with the Czech Republic (fig. 4). Wisła is the Polish name for the Vistula River, which is the longest river in Poland and has its source in the mountains near the town. It has a complex spatial structure, embracing numerous settlements in many valleys of the streams flowing to the main Vistula River. Wisła received municipality rights earlier than Szczyrk and nowadays it features typical urban buildings in the town center surrounding the market square. Already in the late 19th century, Wisła became a popular tourist destination, especially as regards spa tourism (Pawlak, 2014). Nowadays, it has extended its offer over the summer season, which is yet another difference between Wisła and Szczyrk. In the communist times, Wisła was developed as an all-year-round destination, while Szczyrk was rather a typical winter destination. In the 1970s and 80s, Silesian coal mines and other state-owned industrial companies established several ski areas there, including Soszów and Cieńków. Moreover, local people, mostly land owners, built a few ski lifts, which usually functioned as small ski areas with just one short t-bar lift and an easy slope. At the same time, huge investments in hotels were made by industrial companies and brand new hotel quarters were established, making Wisła one of the locations in Poland with the highest rate of the tourism function (e.g. Deffert or Schneider). At the end of the communist era, Wisła was a

---

5 This system provided perfect snow conditions for the opening of the ski jumping competition season in Wisła, in October 2017.

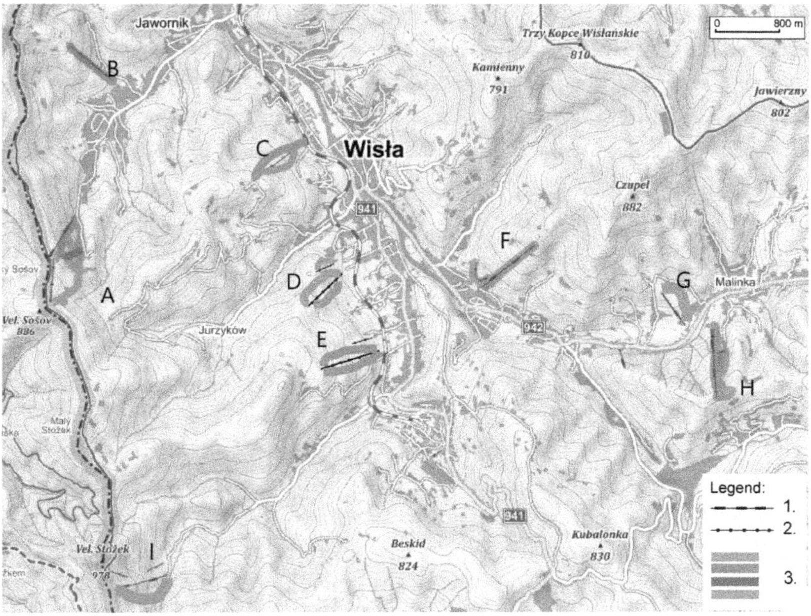

Fig. 4: The most important ski areas in Wisła
1. chair lifts; 2. t-bar lifts; 3. ski slopes; A. Soszów ski area; B. Kiczora ski area; C. Skolnity ski area; D. Siglany ski area; E. Pasieki ski area; F. Nowa Osada ski area; G. Klepki ski area; H. Cieńków ski area; I. Stożek ski area

typical mass tourism destination with summer tourism dominating and underdeveloped winter sports offer. However, because of its localization close to big cities and a large number of tourists, the ski areas located there were still very popular.

Nowadays, it can be said that Wisła went through the transformation time successfully as, apart from some temporary problems, none of the hotels or ski areas were closed down or declined. Nevertheless, establishing Wisła as a leading winter sports destination took a long time and a lot of effort of many people. The hotels and ski areas established by industrial companies were privatized step by step, though the Cieńków ski area is still run successfully by a venture owned by a coal mine holding. The new ski areas owners were both internal and external, small and medium-sized companies; one of the problems they faced was finding financial resources for necessary investments. The first chairlift in Wisła was built in 2004, in the Stożek ski area, but this was an old and slow lift, used earlier in Austria. Three years later, in the Soszów ski area, a new and modern chairlift was constructed, and next year – a similar one appeared in the Cieńków ski area. This situation, where there were three major ski areas in Wisła, lasted for a few years, with a big investment boom between 2012 and 2014. New, fast and high-capacity chairlifts were constructed, e.g. in Nowa Osada, Skolnity, Stok or Klepki, replacing the very old t-bars, often already closed down. All the ski areas are equipped with efficient snowing and illumination systems. All together, at present, there are 17 ski areas in Wisła, seven with chairlifts. This sharp increase in the number and quality of the ski infrastructure in Wisła was a challenge for new organizational solutions. Each of the ski areas has a different owner and there is a need for cooperation among them rather than fierce competition. In 2014, a new program was launched – the Wisła Skipass, which is a combined ticketing system, including 11 ski areas (eight from Wisła and three from neighboring Ustroń). The tickets are valid for at least two days, as individual ski areas are located far away from each other and skiers are unlikely to use two or more of them during one day. This offer is strongly supported by the local authorities and hospitality establishments, as it helps to attract tourists for winter holidays, while scattered, small ski areas were attractive mainly for one-day skiers from the nearby Katowice and Cracow agglomerations. The next move, which was already made in Szczyrk, is the introduction of public transport for skiers, often travelling 10 kilometers from their hotels in the town center to remote ski areas. Also, the large number of small ski areas with slopes of moderate difficulty makes Wisła more attractive to families with children and intermediate skiers, while the offer for advanced ones is not satisfactory.

Białka Tatrzańska – local effort to establish the best ski resort in Poland

Białka Tatrzanska is a village in the community (Polish: gmina) of Bukowina Tatrzańska, within Tatra County, Małopolska Voivodeship, in southern Poland, close to the border with Slovakia (fig. 5). It lies approximately 6 kilometers north of Bukowina Tatrzańska, 16 km north-east of Zakopane, and 75 km south of the regional capital Cracow. The village has a population of 2,200 inhabitants. In contrast to Wisła and Szczyrk, it is a brand new destination on the tourist map of Poland. Until the end of the 20th century, it had been a peaceful village with no winter

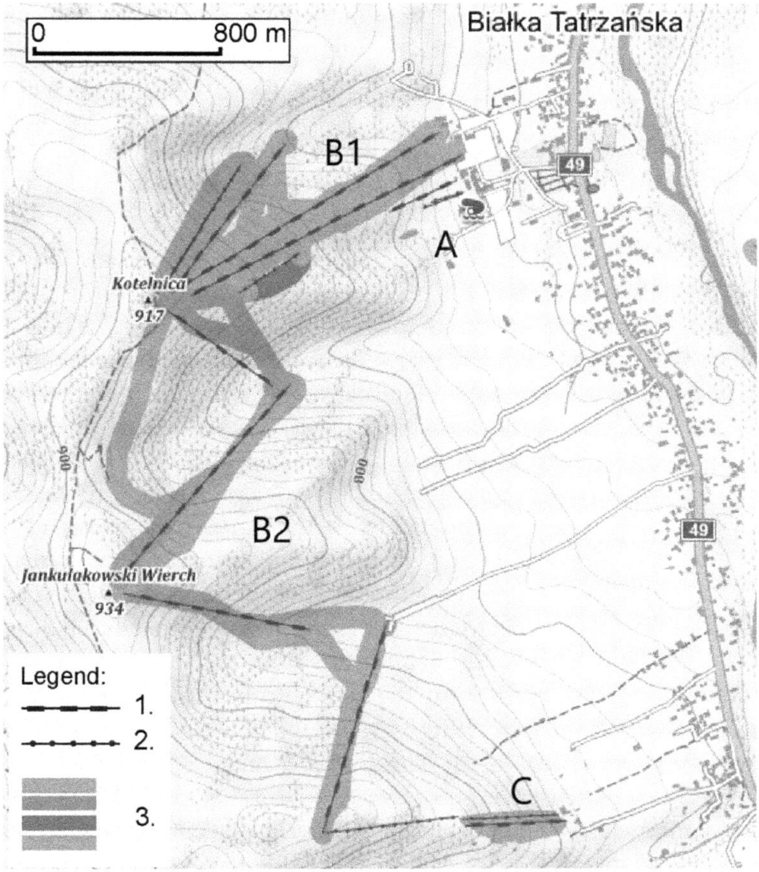

Fig. 5: Ski areas in Białka Tatrzańska
1. chair lifts; 2. t-bar lifts; 3. ski slopes; A. Bania ski area; B1. Kotelnica ski area the first part; B2. Kotelnica ski area the new part; C. Kaniówka ski area

sports offer and almost no tourism. It is located relatively close to the Tatras – Polish most beautiful and most visited mountain range. However, so remote, with hills rather than mountains, it could not attract many tourists and was losing competition not only with popular Zakopane, but also with Poronin, Biały Dunajec and Bukowina Tatrzańska. However, the growing popularity of Zakopane, and later also of the neighboring villages, made them overcrowded, the prices increased considerably and the tourists started to search for cheaper and more peaceful options in villages like Białka. The rapid development of tourism in the village started in the late 20th century. The number of private rooms and B&B establishments offered by the local inhabitants started to grow and the first two small ski areas – Kaniówka and Bania were built. Because of slight denivelation, Białka was perceived as attractive only for beginners, and such small, not very crowded ski areas seemed suitable for those natural conditions.

The first step towards changing this perception was taken in 2000, when the local people established a new company, which was intended to operate a new ski area – Kotelnica. This time, the area offered a four-seat chairlift (opened in 2003) and two wide slopes, using the whole available denivelation. It is easily noticeable that this investment preceded those in Wisła and Szczyrk, and at that time Kotelnica offered one of the few and most modern chairlifts in Poland. However, Polish skiers and the media were even more impressed by two other facts. The first was the extraordinary care for the quality of the slopes, equipped with the most modern snowing, grooming and illuminating systems available. It was enough for the press to announce Kotelnica as the best ski area in Poland. However, an absolutely unique fact was that built directly next to the Bania ski area, Kotelnica was from the very beginning not a competitor but a partner for the two already existing ski areas. All the areas unified their ticketing systems and the owners of Bania and Kaniówka participated in the creation of Kotelnica. The media were delighted to report that it was possible to develop the ski area due to the common effort of many families living in Białka and owning small pieces of land on the Kotelnica Mountain. This was even more shocking, because so far Tatra highlanders had been commonly perceived as greedy, selfish and never sacrificing their own business for the sake of common good. With each year, the position of Białka Tatrzańska as the best ski destination in Poland was becoming stronger, which was confirmed by several prizes awarded to the Kotelnica village and ski area. The Kotelnica ski area owners' plan was to add one more modern chairlift every year, until Kotelnica and Kaniówka were joined by a system of lifts and slopes, which was finally achieved in 2017. All together, the three ski areas offer nine modern chairlifts, twelve t-bar or carpet lifts and over fourteen kilometers of slopes, all of which are illuminated and artificially snowed. The resort is also famous for the high quality of slope maintenance. Offering a large number of lifts and slopes joined in the "ski swing" way, Kotelnica became an interesting option for advanced skiers as well, even though most of the slopes are not difficult. The problems caused by the huge popularity of the resort include the congestion on the only road crossing the village, the lack of parking places and long queues to lifts in the top season. The success of the ski resort also changed the role of tourism in the village. Year after year, new, bigger and more luxurious hotels have been opened, almost all of them owned by local inhabitants. Another symbol of the successful transformation of Białka Tatrzańska into a tourism resort is the Bania Thermal Baths complex, which was one of the first thermal baths in Polish mountains. It was opened next to the Bania ski area and the Bania Hotel, and is perceived as one of the best places of its kind in Poland. Over the last few years, many other attractions have been added to the tourism offer of Białka to enhance its attractiveness in winter and during the summer season. They include well maintained and illuminated cross country skiing slopes and cycling routes. Białka also developed cooperation with the ski resorts in the neighboring villages, issuing the first combined ski ticket, valid in several ski areas. The successful actions of Białka's entrepreneurs were also an inspiring example for actions taken in the neighboring villages, especially in Bukowina Tatrzańska, where several land owners joined together to open another complex of thermal baths.

Wierchomla Mała – a remote destination established by an external investor

Wierchomla Mała is a remote village located in an isolated valley in the Sądecki Beskid, in the Piwniczna community, Nowy Sącz County, Małopolska Voivodeship (fig. 6). Even though this mountain range is among the most popular, crowded and developed mountain tourism areas in Poland, the most popular spa resorts (Szczawnica, Krynica and Muszyna) are located on other sides of the mountain, without the possibility to cross the range in a car or train. The only important tourism destination located close to Wierchomla is a small spa resort – Piwniczna, but reaching this place requires a little longer journey from large Polish cities.

Until the end of the 20th century, Wierchomla had not been regarded as a winter sports or tourism destination at all. It was selected by a businessman from Warsaw, who started buying out land for a new ski area in 1995. In 1997, the first three t-bar lifts were installed and a modern ski area lifted off in 2000, when a high-speed chairlift was opened as

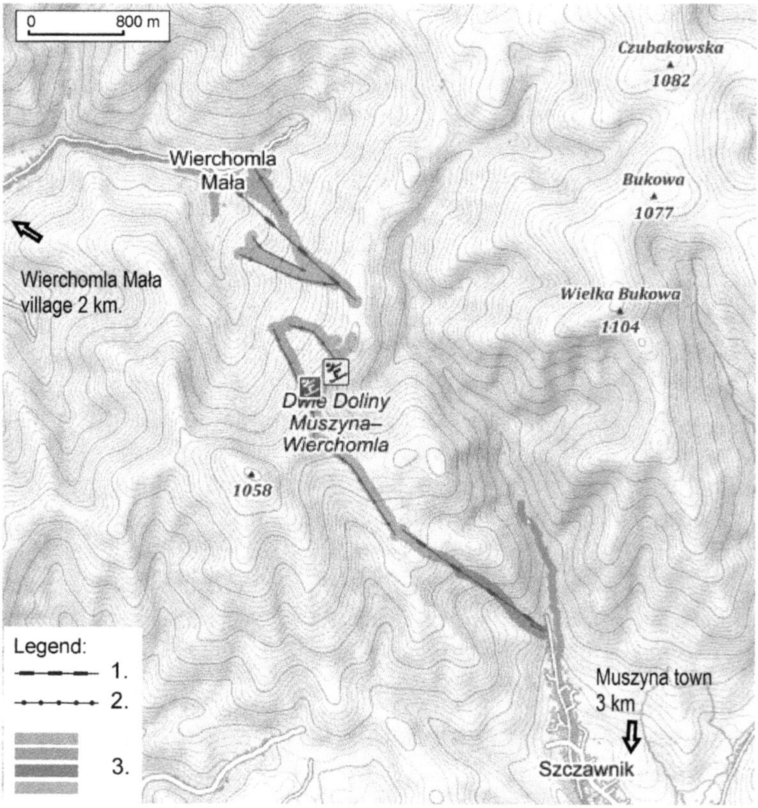

Fig. 6: Muszyna-Wierchomla Two Valleys ski area
1. chair lifts; 2. t-bar lifts; 3. ski slopes

one of the first lifts of that kind in Poland. It is highly symptomatic that the ski area was not built in the village but in the upper part of the valley, a few kilometers above the settlements. Establishing the ski area accelerated building several accommodation amenities in the village. Still, the distance between the village and the ski area prevented Wierchomla from turning into a typical tourist resort. Because of its remote location, the ski area was not suitable for one-day visits and the accommodation offer was limited. The solution was the opening of a new, spacious, three-star, fully equipped hotel at the lower station of the chairlift, in 2006. At the same time, the company operating the ski area received European Union funds for further development. In 2007, a new part of the ski area was opened. New lifts and slopes joined Wierchomla with the Muszyna Valley, providing good access to the ski area for the many guests visiting Muszyna and the neighboring locations. New lifts did not affect the nature of the ski area. The bottom station in the Muszyna Valley is also located a few kilometers above the settlements. Nowadays, the ski area is labelled as Muszyna-Wierchomla Two Valleys (Dwie Doliny) and perceived as one of the most popular and modern ski resorts in Poland. Further investment after 2007 involved expanding the hospitality and gastronomic base, as well as improving the quality of slope preparation. In 2011, the area was awarded the title of the best ski area in Małopolska voivodeship, surpassing even Białka Tatrzańska.

Conclusions

It is possible to find numerous similarities between the developments of ski resorts in Poland and in Alpine countries in the 21st century. Depending on the local conditions, the catalytic or the integrated approach might be applied as a successful model. However, this is mainly true as regards the resorts developed from scratch after the fall of communism. Much more complicated is the situation of the resorts developed during the communist era, which are adjusting their offers to the free market and open borders conditions.

The examples of Białka Tatrzańska and Wierchomla presented above confirm that both models – catalytic and integrated – might be an efficient way to success in Poland. Between the two of the most successful ski resorts in Poland, one possesses very strong features of the catalytic model, while the other one – of the integrated model. However, it does not mean that when establishing a new ski resort in Poland, it does not matter whether it will follow the catalytic or the integrated path. Local natural and social conditions determine the choice of a particular model and that was the case of both Wierchomla and Białka. Even though Białka was a peaceful village before becoming a ski resort, it was located very close to one of the most important tourist destinations in Polish mountains and it was densely inhabited. Also, the small denivelation of the hills surrounding the village made it even technically impossible to build any ski infrastructure without integrating it into the existing settlement. Those two facts made the catalytic model much more suitable for this place. However, local entrepreneurship and sound leadership allowed that model to be implemented in a very effective way.

A different situation can be observed in Wierchomla. The remoteness of this small village surrounded with relatively high mountains made it possible to locate ski infrastructure far away from inhabited areas. Also, the low level of the local inhabitants' entrepreneurship (at least initially), combined with a very low level of tourism traffic before establishing the resort, as well as the fact that the village was rather poor, clearly showed the role of and the need for an external investor to create the resort. Low prices of the land could have additionally encouraged the investor to choose this particular place. Local authorities based in Piwniczna, several kilometers from Wierchomla, were not very active as regards the development process of the ski resort, and remained focused on spa activity – probably the key economic activity for the community of Piwniczna as a whole.

Wisła and Szczyrk were popular ski destinations in Poland still before the 1990s, which brought several additional profits but also posed some challenges when developing those resorts later. Those profits and challenges varied and played different roles in the development of Wisła and Szczyrk. Szczyrk, proclaimed as the Polish winter capital in the 1980s, was widely perceived as the best and the biggest ski resort in Poland at that time. It has continued to be perceived in this way ever since, especially that up till now there have been no other resorts in Poland outnumbering Szczyrk in terms of the number and length of slopes and lifts. This is at least partially due to the particularly favorable skiing conditions in Szczyrk. This huge number of lifts and slopes was also an important heritage of Szczyrk in the capitalist system. However, the strong and positive image was quickly ruined due to local disagreements, temporary blockage of the largest ski area and complete lack of new investments. Old and dilapidated lifts appeared to be one of the most important factors preventing the development of the resort. Another important problem was connected with the fact that the lifts and slopes were built in the 1970s and earlier, without proper planning or with planning that was proper for that moment and not for the 21st century. The lack of respect to formal aspects, which was typical of investments during the communist era, was the main cause of local conflicts between the owners of lifts built in places which were not their property and the local land owners. Accessibility by private cars was not viewed as a major factor of a ski resort development in the 1950s, when the first chairlifts were constructed, and not even in the 70s, when most lifts were built. As a result, the present-day modern center with chairlifts on Skrzyczne, replacing the old lift and based in town, does not offer any parking space, because the closest vicinity of the lower station is densely built-up. There are some parking places available on the private grounds around, but they are very expensive and not sufficient regarding their size. Also, the single, narrow road along the town is vulnerable to huge traffic jams. Those problems appeared to be too difficult to solve only locally, and finding an external investor to buy and modernize the largest ski area might be seen as a real breakthrough. The development of Szczyrk as a ski resort followed the integrated model, emphasizing the role of a big, external company. However, some features of catalytic development can be observed as well, such as: integrating the ski infrastructure into the existing settlements, the local self-government's engagement, or the local inhabitants' participation in the ski business, (accommodation facilities owners' in particular). Other features are not typical either of the catalytic or the integrated models. They include the cooperation of up to three external big companies managing big ski areas within one resort, the existence of a large number of big accommodation facilities run by external companies, but not the same that own the ski areas.

The road which Wisła travelled to regain its position on the Polish winter sports market in the 21st century was very different. In contrast to Szczyrk, it has never been a home of huge ski areas. Wisła was never really dependent on winter sports, either, attracting tourists with a variety of activities. Towards the end of the 20th century, Wisła was offering a number of small ski areas in remote sections of the town, located far away from one another. Actually, the present shape of Wisła is much the same. The rapid development in the last ten years has brought an increase in the number and quality of slopes and lifts rather than a major reorganization of the resort. Only two slight changes are worth mentioning here: new ski areas have been established in the town center, and almost all ski areas have been linked with a common multi-day ticketing system, making Wisła much more attractive for longer skiing holidays. Due to the small size of the ski areas and the majority of accommodation facilities, at first glance it may look as if the contemporary development of Wisła as a ski resort has followed the catalytic model, but remembering that some ski areas are run by external companies, this statement is not so obvious any more.

The examples presented above show four different ski resorts found in one country, even within one mountain range. We can also observe four completely different paths of development. The resorts established in the

21st century usually closely follow a catalytic or an integrated model. It is much more complicated to describe the development of resorts established earlier, in communist times, which had to revise their offer and re-establish their market position in this century. The development of Szczyrk, also in the communist times, was mostly following the integrated model, while Wisła has always been closer to the catalytic one, though the implementation of that model has never been accurate – neither decades ago nor in present times. Additionally, those resorts had to struggle with different problems inherited together with winter sports infrastructure after the communist system had ended.

One thing shared by all four resorts is the fact that they were not neglected by ecologists. A very strict ecological law, long legal procedures and even the activity of many influential ecological organizations, over interpreting nature protection rules were pointed by almost all operators of Polish ski areas as the main problem holding back the development of winter sports in Poland. In many localities, the conflicts with ecologists were very severe (Krzesiwo, 2014). Actually, the conflict between ski developers and ecologists became ideological and both parties did not play fair and used emotional and populists arguments. The only example of a severe problem concerning the four ski resorts in question is that the plan to link the existing ski areas in Wierchomla and Krynica with lifts and slopes must have been abandoned, as this might result in potential ecological damages.

References

Baran, Marek. 2010. *Koleje linowe i wyciągi narciarskie. Historia i dzień dzisiejszy*. Łódź: Dom wydawniczy Księży Młyn.

Flagestad, Arvid. 2002. *Strategic success and organisational structure in winter sport destinations: a multiple stakeholder approach to measuring organisational performance in Scandinavian and Swiss case studies*. Öestersund: ETOUR.

Flagestad, Arvid and Hope, Christine A. 2001a. ""Scandinavian Winter"; Antecedents, concepts and empirical observations underlying a destination umbrella branding model". Tourism Review 56(1/2): 5–12.

Flagestad, Arvid and Hope, Christine A. 2001b. "Strategic success in winter sports destinations: a sustainable value creation perspective". *Tourism management* 22(5): 445–461.

Gill, Alison and Williams, Peter. 1994. "Managing growth in mountain tourism communities". *Tourism Management* 15(3): 212–220.

Gonçalves, da Conceição, Fernando, Vitor, Águas, Roque and Manuel, Paulo. 1997. "The concept of life cycle: an application to the tourist product". *Journal of Travel Research* 36(2): 12–22.

Hudson, Simon. 2000. *Snow business. A study of the international ski industry.* London: Cassel.

Inkson, Clare and Minnaert, Lynn. 2018. *Tourism management: an introduction.* London: Sage, Thousand Oaks.

Inskeep, Edward and Kallenberger, Mark. 1992. *An integrated approach to resort development.* Madrid: WTO.

Kaspar, Robert. 1992. "Ski tourism in New Zealand. Toward sustainable management". In *Strategies for Reducing the Environmental Impact of Tourism.* Pillmann, W. and Predl, S. (eds.) Vienna: International Society for Environmental Protection. 368–382.

Krzesiwo, Kinga. 2014. *Rozwój i funkcjonowanie stacji narciarskich w polskich Karpatach.* Kraków: Uniwersytet Jagielloński.

Krzesiwo, Kinga. 2016. "Społeczne i gospodarcze aspekty rozwoju turystyki narciarskiej". *Przedsiębiorczość-Edukacja* 12: 233–244.

Kurek, Włodzimierz. 1999. "Rozwój turystyki i jej wpływ na środowisko przyrodnicze Alp". *Turyzm* 9(1): 27–36.

Pawlak, Anna. 2014. "The interwar spa resort Wisła – the degree of cultural environment preservation". *Technical Transactions. Architecture 4-A*: 61–71.

Pearce, Douglas. 1978. "Tourist development: two processes". *Travel Research Journal 1*: 43–51.

Pearce, Douglas. 1996. *Tourism development.* London: Harlow.

Sharma, Kumar K. 2004. *Tourism and socio-cultural development.* New Delhi: Sarup & Sons.

TNS OBOP. 2004. *Białe szaleństwo po polsku.* Paper Online. Cited 06.01.2004 on: http://www.tns-global.pl/centrum/zal/news/TNSOBOP_narty.doc.

Tuppen, John. 2000. "The restructuring of winter sports resorts in the French Alps: Problems, processes and policies". *The International Journal of Tourism Research* 2(5): 327–344.

Weiermair, Klaus. 1997. "On the concept and definition of quality in tourism". *AIEST Editions* 39: 33–58.

Williams, Peter. 1996. "Sustainable Alpine tourism development: towards a self improvement approach". In: *Alpine tourism. sustainability reconsidered and redesigned.* Weiermair, K. (eds.). Innsbruck: University of Innsbruck. 126–138.

WTO. 1995. *National and regional tourism planning.* Madrid: WTO.

Żemła, Michał. 2004. "Product quality policy of Polish ski resorts". *Hotellink* 4, 275–284.

Żemła, Michał. 2008. "Failures in building partnership for success in the competitive market: The case study of Polish ski resorts". *Managing Global Transitions* 6(4): 421–444.

Żemła, Michał. 2010. "Konkurencyjność polskich miejscowości górskich na rynku turystyki sportów zimowych – przypadek Szczyrku". In: *Konkurencyjność miast i regionów na globalnym rynku turystycznym.* Sala, Józef (eds.). Warszaw.

1.2

Destination Governance and the Challenge of Balancing Economic Growth and Sustainable Development in Tourist Destinations – The Case of Åre, Sweden

Sara Nordin and Bo Svensson

Introduction

The notion of destination governance has received notable attention among scholars in recent years (Zhang and Zhu, 2014) and received an increasingly central role in destination development since it was first introduced in the early 2000s (Svensson, Nordin and Flagestad, 2005; Beritelli Bieger and Laesser, 2007). It has even to some extent replaced, and definitely complemented related terms such as destination management and destination planning (Volgger, Pechlander and Pichler, 2017).

Destination governance is a complex matter, including various aspects such as resource dependencies, multi-actor-complexity and the public-private dimension of destinations (e.g. Svensson, Nordin and Flagestad, 2005). The perspective here is that destination governance is not only complex, it is also a central issue in many, if not all, places where tourism development is on the agenda.

Even though academic research into destination governance has increased and developed in later years, there is reason to look more closely into how well it can handle other challenges than economic growth at destinations. In our case, questions of sustainability – including both the social and environmental dimensions, along with the economic one – is a growing concern as new visons of destination development emerge. The challenge of balancing economic growth and sustainable development at a destination will be the focus of this chapter along with an exploration of the need for an understanding of destination governance that is broader in scope than what generally has been the case so far.

Destination governance

Earlier research (Nordin and Svensson, 2007) has suggested that a focus on the governance perspective is important in progressing our understanding of destination development. A few simple observations about the functioning of tourist destinations motivate this statement. First, there is the multi-actor complexity of destinations, i.e. destinations are usually run by a constellation of actors, and not by a single organization of any kind. In this network constellation, the various actors may have diverse interests and different perceptions of reality. Understanding their interactions, potential agreements and joint problem solving may therefore be important. A need to negotiate shared purposes and exchange resources such as information, trust and expertise may be the rationale for this interaction. Secondly, the resource dependencies between actors are important characteristics of the destination that need to be understood in any attempt at revealing what goes on.

Thirdly, the public-private dimension of this interaction, i.e. the formal and informal relationships between the local government and the industry are often highly important to map in order to get a clear picture of development processes taking place in the destination in question. Our suggestion has been, and still is, that destination governance is a relevant and important perspective in approaching destination dynamics, since it draws attention to the network complexity of destinations. Not least the public-private linkages of this interaction is a crucial component for understanding the policy processes at hand (Rhodes, 1996).

Governance has over the last decades become one of the most frequently used and theorized concepts in the social sciences. There seems to be a model of governance in almost any situation. Our focus here is on how actors work together to provide governance at a tourist destination, sometimes conceptualized as "interactive governance" (Peters, 2017). In this chapter governance will be applied to investigate how well actors in a destination adapts to new needs and situations, in this case the emerging need to balance economic, social and environmental dimensions of sustainability.

Sustainability

The sustainability concept has gained increasing recognition in the tourism sector, as in many other, not least since the Brundtland Report of 1987 which advocated sustainable development. In a tourism context, Weaver (2006, 2012) acknowledges that all tourism entails cost and associates sustainability with "strategic management that strives to minimize the direct or indirect costs of a given activity whilst concurrently maximizing the attendant benefits, both locally and globally" (2016:1030). With the increased focus on costs, the coining of terms such as "overtourism" and "tourismphobia" have emerged in more recent time (Goodwin, 2017; Dredge, 2017; Milano, 2017). However, previous research has also acknowledged the need for alternative and more environment and host-friendly practices also in the matters of development, planning, and policies with rapidly growing tourism economies (Saarinen, 2006). Saarinen (2014), moreover, points to a need for stronger governmental and inter-governmental policies and regulations in order to have tools for setting the limits to growth in tourism. We have, for example, seen destinations such as Whistler working with methods such as The Natural Step to try to balance extensive tourism development with the needs of local citizens (Gill and Williams, 2011). In many destinations, however, there is still a prevalent focus on economic growth. This was for a long time also the case in the mountain resort of Åre, Sweden and according to some, still is.

The case study

The empirical contribution in this chapter is based on a single case-study of the Swedish resort of Åre (fig. 7). Data were drawn from authors' empirical research in the destination, spanning over a15 years period.

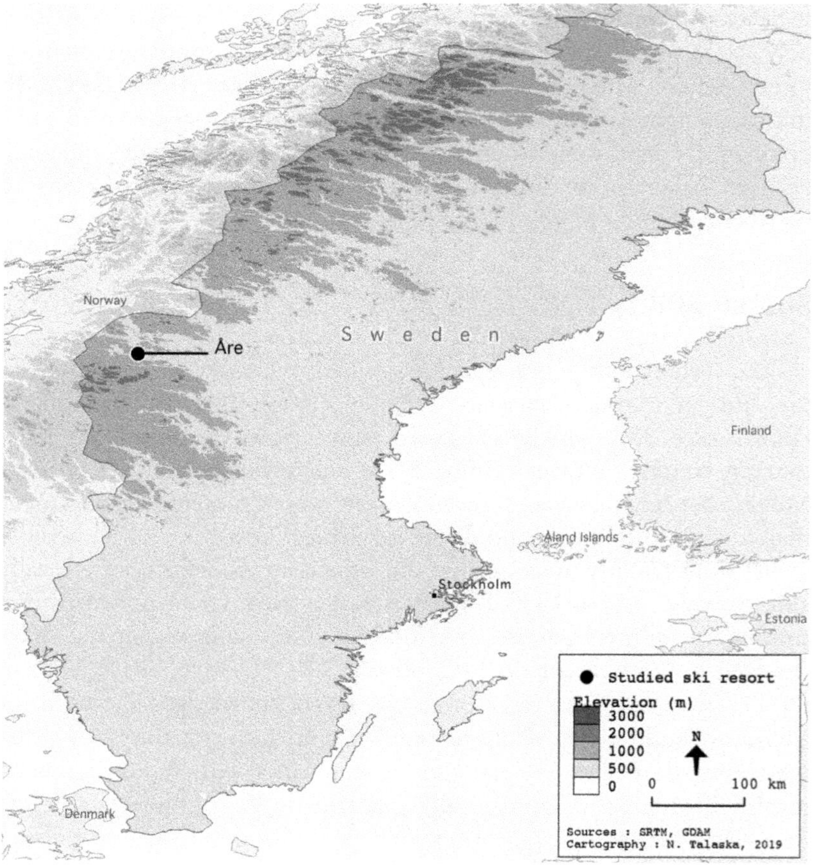

Fig. 7: Studied ski resort in Sweden

Interviews were complemented by archival documents and newspaper articles.

Åre is the premier ski destination in Sweden, perhaps even in northern Scandinavia. Over the last ten, fifteen years Åre has changed its strategy from a strong winter focus to transforming into an all year-round based tourist destination, with far more to offer the visitors than just skiing experiences (Nordin and Svensson, 2007). However, the expansion of tourist activities to include a more differentiated offer and the extension of seasons, also means a harder pressure and exploitation of nature. More lifts, facilities and hiking and biking trails have been developed, along with increased numbers of visitors. Åre has also arranged large events, such as the Alpine World Ski Championships of 2007 and 2019, adding to its strong development as a tourist destination.

Governance in the early 2000s

The more comprehensive research on governance structures in Åre goes back to the early 2000s, when Åre was facing a period of extensive growth and arranging the Alpine World Ski Championships of 2007 (Nordin and Svensson, 2007). The millennium marked a strong rejuvenation of Åre. This was possible with the entrance of new actors, both foreign and domestic. During this phase Skistar, Scandinavia's leading operator in Alpine skiing, positioned itself as the dominating destination actor and Åre began reaching out to new customers and markets.

At the time, Åre developed a well-recognized model of public-private cooperation called the Vision 2011 group. It was a platform where politicians and senior officers informally met with industry to discuss strategic matters regarding Åre's future. It was described as one of the main drivers taking the destination into a new and perhaps stronger growth phase than ever before. We, of course, need to acknowledge the complexity of this and that there are several various factors working together to create growth at a destination as well as the fact that there are limitations to what can be said about the causality between governance form and growth. It is also interesting to recognize that it very well could be the growth phase that created a need for this kind of public-private network and new form of governance and not the other way around. In any case this solution was something new in the destination.

At the time, given the economic prosperity of the destination and its ability to compete globally, the destination governance of Åre and the public-private interplay was often presented as a "success story" to inspire other destinations and rural communities also striving for economic growth. The conclusion was that Åre successfully had adjusted its strategies and forms of governance and fostered a tourist destination characterized by growth and development.

There were, however, a few perceived issues with the new governance form. It was a closed, self-appointed body with no representative mandate, largely lacking democratic and transparent features, such as open protocols. It was rather based on close personal relationships and mutual trust and it was sometimes described as effective and result-oriented, but at the cost of the involvement of the wider destination and local community.

A broader involvement

The criticism eventually called for a new governance structure. When the 2020 vision was launched, the process involved a larger number of actors based in the destination with several public meetings. Approximately 60+ companies and about 150 other destination actors met continuously for about a year to develop a vision that was more broadly anchored in the destination.

Although growth was the main objective, the 2020 vision involved objectives with concern for the environment, including the following goals: 50 percent will use eco-certified electricity of which 30 percent is wind generated, 100 percent of the companies shall aim to reduce the consumption of energy, 100 percent of the new company cars shall be eco-cars, of which 50 percent shall be electric /hybrid cars by 2014, 55 percent of the winter guests shall travel by train and 50 percent of the employees shall be able to commute using public transport.

Though more democratic and including, this process was, however, described as a less efficient way of working and, moreover, resulted in fewer outcomes and not all objectives of the vision fulfilled by 2020. This was also the case with the sustainability work in the destination.

In 2012 the destination of Åre was, however, granted funding from a government agency under the Ministry of Enterprise and Innovation, to engage in a project to develop a more sustainable destination development. This led to the start of a more serious approach to sustainable development, although the involvement of private actors on the whole was described as unsatisfactory. The main reason for this was described as an underestimation of the complexity of communicating to destination actors what sustainability was all about and to engage them in a process to develop a joint platform. It took a lot longer than perceived to start a process of this kind in the destination and was hard to develop a work plan and priorities that a majority engaged in and agreed upon.

Sustainable development

It was not until the Vision of 2035, launched in 2015, that sustainability was given a more prominent role. "Åre – the most attractive European alpine all-year destination" from the 2020 vision was replaced by "Åre – Europe's most attractive and sustainable alpine year-round destination", emphasizing the focus on a sustainable development. Companies were encouraged to certify their work as environmentally sustainable and the objective of the vision is that 80 percent of the companies should be certified by 2035. The destination also started a process with the Global Sustainable Tourism Council[1] and the second environmental objective is to continuously increase the destination's results at the Global Sustainable Tourism Review. Nevertheless, the 2035 vision involves objectives relating to further economic growth as well, such as increased number of visitors, more job opportunities and to further develop the tourism sector. There is, however, a strong progress towards a balance between an economic and a sustainable development, also including social dimensions.

The local government has, for example, been recognized for their successful work in integrating immigrants and asylum seekers in Åre, which has created a win-win situation for the municipality with a faster integration at the same time as the tourism sector has been able to benefit and fill the need for labor.

The 2035 vision is again a more closed body, involving fewer actors. Around the same time as it was launched, a local destination marketing organization (Åre Destination AB), was formed. This is a new platform for collaboration and development with more than 200 companies, but with Skistar as majority owner and hence to some extent in an even more dominating position.

Åre has grown a tourist destination since the early 2000s, both in terms of size (including more companies) and in number of visitors. Even though informal networks will always play a significant role, people at central and powerful positions have changed, not least at the political arena, but also with the entrance of new actors from other regions. This has affected the trust between destination actors more generally, but also

---

1 The Global Sustainable Tourism Council (GSTC) is managing the GSTC Criteria, the global baseline standards for sustainable travel and tourism; as well as acting as the international accreditation body for sustainable tourism certification.

more specifically between members of the vision 2035 group. During the same period, social media has given way to a growing use of new information channels, which has created an increased visibility and transparency, but also made information travel fast and hence harder to control.

An increasing number of visitors also put a greater pressure on natural resources and nature, along with a growing usage of water and a larger disposal of garbage. "People are worried now. They question if Åre can continue to grow. Back then that was never an issue", a company leader reflected when he explained why sustainability received a larger focus in the new vision compared to the early 2000s.

The vision 2035 group to greater extent consists of public actors, whereas the private company representatives have diminished in numbers. Smaller companies (along with larger) are represented by the destination company, whereas Skistar has a given role in the group and is as a company not as criticized and questioned as when they first entered the destination as an outsider. Events such as arranging the Alpine World Ski Championships are, furthermore, described as a uniting factor in the destination.

Conflicting goals

An interesting issue is how residents that are not particularly interested in tourism development feel about the strong growth. Throughout history, it has been widely debated in Åre how to use tax money and what to give priority to in terms of community development. Should schools and elderly care give way to strong tourism development? What happens when the locals cannot keep up with the increasing housing prices? There are a number of issues of this kind in Åre that become particularly evident when the tourism destination and the village to such a large extent correlate.

From the very beginning, there have been conflicts of interest in Åre with farmers who were forced to sell land at the mountain. Still today, various tourism activities and facilities take up and affect the use of land and public space, they can disturb the natural environment and transform a rather peaceful village into a tourism resort with plenty of visitors, from time to time disturbing occurrences and an overall different character. Increased transportation, the exploitation of resources causing for instance drainage and impacts on sedimentations, higher noise levels and disturbances on wildlife are just a few examples of possible effects. This also relates to sustainability issues, where increasing the number of visitors also has negative consequences.

On the other hand, Åre is characterized by a development that is rare in the larger region that it is situated in – in the northern inland of Sweden. This positive progress naturally also affects the citizens. Downhill skiing and other tourist activities generate job opportunities and attract large volumes of visitors to the destination, which contributes to economic growth. Some of the local residents could have been forced to move without this growth. Taking the cluster perspective (Porter, 1990), Åre has been successful in attracting more tourism-related businesses and industries in other fields, population growth, increased tax incomes and it has successfully been able to transform from a winter sports destination into a more or less year-round based resort. This can partly be explained by the fact that the community and the destination to such a large extent correlates, which means that the high correlation has both positive and negative effects. The conflicting objectives, however, calls for a collaboration between actors to enhance better decision-making, planning and policy processes.

Central actors

As mentioned before, strong network relations and trust have led to unconventional public-private coalitions and joint approaches in the destination of Åre. This is also in line with the general development from government to governance (Rhodes, 1997; Dredge, 2006), emphasizing network solutions. Nevertheless, there are changes over time in terms of influence. What we can see here is that for mountain resorts with the need of heavy investments in ski infrastructure, public support is essential in developing from a small village into an international ski resort. This is evident in Åre, who goes through a path of development with over time, a transfer of influence from public towards strong private actors. Skistar is described to have brought new levels of professionalism to the destination, but they also listen just as much, if not more to the customers and shareholders as to the village residents and businesses.

In Åre, respondents describe how Skistar (then named Sälenstjärnan) first came to the destination and basically did things their way. You could either accept their terms or be on your own – the latter perhaps not being the best option, considering Skistar's strong position and possession of the main attraction during the winter. This approach is, however, soon after described as slowly changing into a more open attitude on Skistar's behalf, clearly realizing that even a major actor in a dominant position depends on the destination as a whole. This can be looked upon in the light of cluster theory, which emphasizes the interdependencies between the actors and yet highlights the importance of a cluster animator (sometimes also referred to as a "network broker") as a driver of development. The interdependencies can be considered even stronger in a situation where the driver of development cannot control the product they are selling since they are partly delivered by subcontractors, as production and consumption often take place simultaneously. Again, with the ski product and the destination overlapping with the village of Åre, conflicts of interest are bound to occur.

While there has been a good dialogue between Skistar and the other major actors of influence in the Vision 2011 group and the following groups, other actors that were not part of the constellation largely express that they cannot influence a powerful actor such as Skistar and that the lacking balance of power clearly affects the residents and local businesses. Others express that the "old" Åre is long lost and that this solution is necessary and good for the destination. Skistar is also, as mentioned previously, one of the dominant owners of the new destination company. One conclusion that can be drawn here is that it may even be necessary to entrust private actors to a larger extent when it comes to the management of international ski resorts. Yet it is also evident from our research that the major private companies need to be locally anchored and a part of the destination.

It is also clear that over time, external influences have increased in Åre. This is not only an effect of Skistar's entrance but started long before. The destination has on the whole been successful in attracting external actors and investors, which has led to a transfer of power and influence increasingly from locals to external stakeholders. This is a common occurrence in a destination developing from a small place into an international resort. However, there are certain issues to keep in mind during the process. There is, again, the balance between protecting the old identity and perhaps small-scale charm and a possible mass-tourism. When decisions are made, to an increasingly larger extent, further away from the destination, informal networks based in the destination with, as some describe it, "their hearts in the village", and informal destination governance solutions such as the Vision 2011 group seem to play a very important role in keeping influence locally. With the strong interdependencies in mind, it is also important for the dominant actor to be part of formal and informal networks at the destination level to access local knowledge and information as well as to be able to achieve the local cooperation needed in order to act as a united destination with high-quality-services and products for the guests.

What about the local government then? The old model of destination governance very much resembles "governance without government" (Rhodes, 1996), even if the local government all along was part of the inner circle described previously. Regulations handled by the local government were sometimes put aside to the benefit of grasping possibilities to reach a higher economic growth. Somehow, the role of the local government seems to have been that of "the follower" or supporter of major business stakeholders, and very much at their service. Issues of economic growth in the destination were the major concern of the first more formalized vision group, and of the local government. As described under the section of conflicting goals, an increasing number of questions remained unattended. Today this is clear to both political and public officials representing the municipality. The local government now seeks to alter its role in the destination and its development broadly defined. A growing population and a business development not entirely related to tourism calls for this. Questions of social and environmental concern need an upgrading and the local government is starting to take a leading role in this.

Reflections and conclusions

When it comes to sustainability issues and to be able to create the form of destination management Weaver (2016) referred to with a tourism development resulting in more advantages than costs, and to be able to balance economic growth with a sustainable destination development, our case study implies a need for a new form of governance. This governance solution is one where the local government needs to take a stronger lead in the sustainability work and rules and regulations cannot time and again be overruled in the name of economic growth.

Earlier processes serve as a reference point, and even if no formal evaluation has taken place, at least the local government sees a different future. Although its shape and process are unclear at this point, some steps illustrate this. One such is that there is now a representative of the local government working closely with the staff of the destination management organization (DMO). The local government is also reorganizing its within office activities to become less dependent on a single person and they have developed a structure with more of a team responsible for the work towards the destination, and other parts of its territory, in a more coherent way. Unsolved questions relating to traffic and the environment more broadly understood calls for attention. There is also a broader spectrum of business actors in the destination, including IT-related services, tourism, restaurant constellations and permanent housing developers, calling for attention. From the local government perspective, the situation is different, but challenging, and one that many rural municipalities and tourism dominated areas envy.

What the future will bring in terms of a new governance model remains to be seen, but it will remain an interesting topic. Some have called more detailed consideration of agency in governance research (Peters, 2014), which certainly seems appropriate in our destination, and not only concerning the municipality, but also business interests and local citizens. This would require a more detailed consideration of the acting bodies as sources of action into the analysis. This means to go beyond relations among actors and the outcome of this interaction, to look at the involved as sources of action to better understand the actual compatibility of the different actors and their interests. This seems important for a deepening understanding of the needs for change, both on behalf of individual actors and of the destination governance arrangement as such.

We also see comparative studies on destination governance as a way of strengthening the theoretical and empirical foundations of destination governance as an analytical tool. At last, it seems inevitable that governance approaches to sustainability at destinations, is an urgent matter.

References

Atkinson, Michael M. and Coleman, William D. 1992. "Policy networks, policy communities and the problems of governance". *Governance* 5(2): 154–180.

Beritelli, Pietro, Bieger, Thomas and Laesser, Christian. 2007. "Destination governance: using corporate governance theories as a foundation for effective destination management". *Journal of Travel Research* 46(1): 96–107.

Dowding, Keith. 1995. "Model or metaphor? A critical review of the policy network approach". *Political studies* 43(1): 136–158.

Dredge, Dianne. 2006. Policy networks and the local organisation of tourism". *Tourism management, 27*(2), 269–280.

Dredge, Dianne. 2017. "Institutional and policy support for tourism social entrepreneurship". In *Social Entrepreneurship and Tourism. Philosophy and Practice*, Sheldon, Pauline J. and Daniele, Roberto (eds.). Springer, Cham, 35–55.

Gill, Alison. M. and Williams, Peter. W. 2011. "Rethinking resort growth: Understanding evolving governance strategies in Whistler, British Columbia". *Journal of Sustainable Tourism* 19(4–5): 629–648.

Goodwin, Harold. 2017. "The challenge of overtourism". *Responsible Tourism Partnership Working paper 4.*

Milano, Claudio. 2017. "Overtourism and tourismphobia: global trends and local contexts". *Barcelona, Ostelea School of Tourism and Hospitality.*

Nordin, Sara and Svensson, Bo. 2007. "Innovative destination governance: the Swedish ski resort of Åre". *The International Journal of Entrepreneurship and Innovation* 8(1): 53–66.

Peters, B. Guy. 2017. "Is governance for everybody?". *Politics and Society* 33(4): 301–306.

Porter, Michael E. 1990. "The competitive advantage of nations". *Harvard business review March-April: 73–93.*

Rhodes, R. A. W. 1996. "The New Governance: Governing without Government", *Political studies*, 44(4), 652–667.

Saarinen, Jarkko. 2006. "Traditions of sustainability in tourism studies". *Annals of tourism research, 33*(4), 1121–1140.

Saarinen, Jarkko. 2014. "Critical sustainability: Setting the limits to growth and responsibility in tourism". *Sustainability* 6(1): 1–17.

Scharpf, Fritz W. 1997. "Introduction: the problem-solving capacity of multi-level governance". *Journal of European public policy* 4(4): 520–538.

Svensson, Bo, Nordin, Sara and Flagestad, Arvid. 2005. "A governance perspective on destination development-exploring partnerships, clusters and innovation systems". *Tourism review* 60(2): 32–37.

Volgger, Michael, Pechlaner, Harald and Pichler, Sabine. 2017. "The practice of destination governance: A comparative analysis of key dimensions and underlying concepts". *Journal of Tourism, Heritage & Services Marketing* 3(1): 18–24.

Weaver, David B. 2006. *Sustainable tourism: Theory and practice*. London, Routledge.

Weaver, David B. 2012. "Organic, incremental and induced paths to sustainable mass tourism convergence". *Tourism Management* 33(5): 1030–1037.

Zhang, Hanyu and Zhu, Mingfang. 2014. "Tourism destination governance: A review and research agenda". *International Journal of e-Education, e-Business, e-Management and e-Learning* 4(2): 125–128.

1.3

Ski Resorts Development in Korea: Past, Present, and Future

Doyeon Won

Introduction

South Korea (hereafter, Korea) has distinct four seasons but, historically, winter sports have not been popular for several reasons. While Korea is now considered one of the developed countries and sport powerhouses in the world, Korea had not been an affluent country until the late 1990s (e.g., Korea became an OECD member country in 1998) and its athletic prowess at the international competitions were mainly based on the highly controlled 'elite' sport system as opposed to the 'sport for all' system (Chung and Won 2011; Won and Chiu 2018). Consequently, the development of Korean leisure sport industries, including ski resorts, has been slower than anticipated. Also, the delayed development concerning social overhead infrastructures in the Northeastern region and mountain areas has been a significant constraint for both developers and winter sport participants. In terms of natural environments, most of the potential and current ski slope areas are below 600 meters in altitude, and the length of each slope is shorter than those in other international ski destinations. Due to the high cost incurred to consume skiing and snowboarding activities, these sports have been considered as overly extravagant sports because consumers must pay both sport-related expenses (e.g., lift tickets, equipment purchase or rentals, and ski suits) and travel-related expenses (e.g., transportation, lodging, and amenities). Still now, all the ski equipment is imported from foreign countries as there is no ski equipment manufacturer in Korea.

Traditionally, it has been primarily Japan (and China) in the East and Turkey in the West when it comes to Asian winter destinations for skiers and snowboarders due to multiple options in the respective countries and longer skiing seasons. For example, a skiing season in Japan runs from late-December to late-March while a season in Korea runs from mid-December to late-February. Japan also hosted the 1972 Sapporo and 1998 Nagano Winter Olympics events, and thus Japan has had opportunities to build the country as a winter destination in East Asia. In response, Korea has tried to utilize existing and upcoming ski resorts as tourist attractions and maximize the hype around the 2018 Winter Olympics to build its brand as a solid winter sport destination country. Fortunately, Korea is now considered as an up-and-coming winter sport destination, especially for Asians due to the accessibility and recent investment around the PyeongChang Olympics (Carter, 2017). For example, the Alpensia, YongPyong, and Jisan Forest resorts were selected as some of the top 10 ski destinations in Asia.

As such, Korea has seen a substantial record of growth on ski resorts. This chapter discusses how ski resorts in Korea have been developed over the last four decades and what future challenges lie ahead to be 'more' attractive winter destinations. In understanding the developmental process of Korean ski resorts, one of the major aspects to take into is the actors (or sectors) who led the development across different time periods because there have been different groups of forces that played as a catalyst of the development. Along with the actors, some of the major supply and demand side factors include: projectability (e.g., profitability, project costs, ease of getting governmental approvals), real estate values (e.g., pre-development affordability and post-development earning potential), consumer demands (e.g., domestic popularity of ski and snowboard, sport tourism from other Asian countries, accessibility from the major metropolitan areas), domestic geopolitics, Winter Olympics (and nationalism sentiments), and environmental concerns.

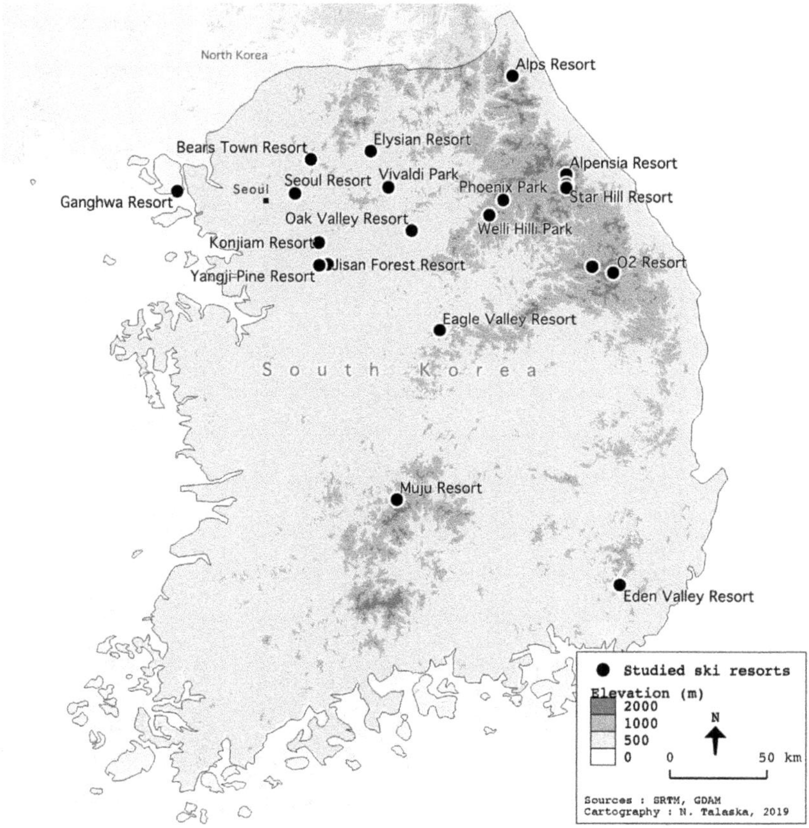

Fig. 8: Ski resort location map in Korea

Ski resorts in Korea: overview

In Korea, the first ski resort, the YongPyong Resort, was opened to the public in 1975. Since the first ski resort, there have been 19 ski resorts ever built in Korea with one additional ski resort being built (see Fig. 8 and Tab. 1). The new ski resort in Ganghwa is currently under construction to be completed by 2020 and opened to the public for the 2020/21 ski season. With one defunct or two temporarily closed resorts (the Seoul, Eagle Valley and Alps resorts), there are 16 active ski resorts in Korea as of the 2018/19 ski season. Also, there have been some discussions to push forward for developing a ski resort in the Southwestern region in

Tab. 1: Ski Resorts in Korea (as of 2019)

Province	Resort	Year opened	Size (Km^2)	Elevation in meters (Difference)	Slopes (Km)	Trails (#)	# Chair Lifts (Gondola)	Visitors in 2007/08	Visitors in 2014/15
	Star Hill Resort	1982	1,02	250–450 (200)	3,4	4	5	85 260	52 296
	Yangji Pine Resort	1982	1,79	80–230 (150)	4,8	8	6	238 299	187 433
	Bears Town Resort	1985	0,25	300–630 (330)	15,0	7	8	342 130	256 765 (in 2013)
Gyeonggi	Seoul Resort (*Defunct* since 2008)	1993	2,78	250–400 (150)	2,0	4	3	Defunct	Defunct
	Jisan Forest Resort	1996	0,03	100–304 (204)	8,0	7	5	683 017	414 459
	Konjiam Resort	2008	3,68	177–497 (320)	7,2	9	5	383 403 (in 2009)	372 502
	YongPyong Resort	1975	11,03	700–1 438 (738)	24,2	28	14 (1)	604 911	439 175
	Alps Resort (*closed* since 2007)	1984	0,47	Max 1 500	9,4	6	5	94 180 (in 2011)	113 111
	Vivaldi Park	1993	4,36	270–580 (310)	8,0	10	10 (1)	871 912	728 477
	Phoenix Park	1995	7,30	680–1 080 (400)	15,3	21	9 (1)	778 646	467 791

Gangwon	Welli Hilli Park	1995	5,26	573-896 (323)	12,3	18	9 (1)	522 018	362 773
	Elysian Resort	2002	0,20	300-473 (173)	6,0	10	6	290 729	347 530
	High1 Resort	2006	8,81	690-1 367 (677)	21,0	18	9 (3)	741 220	620 953
	Oak Valley Resort	2006	2,81	173-435 (262)	6,3	7	3	423 034	426 307
	O_2 Resort (*closed* b/w 2013-2018)	2008	6,12	845-1 420 (575)	11,0	19	6 (1)	86 372 (in 2009)	63 742 (in 2013)
	Alpensia Resort	2009	1,78	775-970 (195)	4,2	6	3	94 180 (in 2011)	113 111
N. Chungcheong	Eagle Valley Resort (*closed* since 2016)	1990	2,01	300-380 (80)	3,7	9	4	60 189	20 592 (in 2013)
N. Jeolla	Muju Resort	1990	12,91	800-1 614 (814)	24,0	34	14 (1)	754 081	376 658
S. Gyeongsang	Eden Valley Resort	2006	0,23	200-400 (200)	4,0	7	3	246 064	206 974
Incheon	Ganghwa Resort	2020	0,02	Max. 340	unk.	4	unk.	n/a	n/a

Note: Data were adopted from the Korea Ski Resort Business Association (SBAK).

Korea, temporarily named as the Bonghwa–Munsoosan Ski Resort, by the local government. The initial plan, developed in 2009, was cancelled due to the waterworks-related environmental issue and now the new plan aims to build a ski resort by 2021 at the same site. For the 2018 Olympics, the Jungsun Alpine Ski Slopes (at Mountain Gariwang) was specifically developed for the 2018 Winter Olympics but not for the public use.

As shown in Tab. 1, there are five active and one defunct ski resorts in Gyeonggi Province, nine active and one closed ski resorts in Gangwon Province, one closed ski resort in North-Chungcheong Province, one ski resort in North-Jeolla Province and one ski resort in South-Gyeongsang Province. Of the 19 ski resorts, the Muju Resort is the largest one with 1,291,015 m^2 of the slope areas, followed by the YongPyong (1,103,449 m^2), High1 (880,838 m^2), and Phoenix Park (729,549 m^2). On the other hand, the Star Hill resort is the smallest with 102,026 m^2 of the slope areas. The number of ski slopes ranges from four to 34 and the number of ski lifts ranges from three to fourteen. Only seven ski resorts are equipped with gondola lifts while other 12 resorts only have chairlifts.

The developmental process of ski resorts in Korea can be divided into four time periods or generations: The first period, from the 1970s to 1980s, was the dawn and the introduction stage of Korean ski resorts, when it began the construction of the first wave of ski resorts, including the YongPyong resort (open in 1975) and other four resorts built in the early to mid-1980s. The second period, during the 1990s, was the growth stage of the ski resort development, when the private sector was still the primary driving force. The second wave of ski resorts included the Muju resort (1990) and six other resorts built in the early to mid-1990s. The third period, during the 2000s, was the pre-Olympics stage of the ski resort development, when Korea tried to host a Winter Olympics event and participated in the bidding process for three consecutive times. This period can be called the 'post-IMF era' when Korea had a significant financial crisis and received rescue funds from the International Monetary Fund (IMF). The third wave included the development of the High1 resort (2006) and six others, built during the 2000s. The fourth and last period, after the 2018 Winter Olympics. The fourth wave includes one on-going construction in Ganghwa, planned to open in 2020, and another one to be built by 2021.

The first period (the 1970s and 1980s)

The first organized ski competition was in 1947 at the Jiri-mountain and the first Olympics event in which a Korean skier participated was the 1960 Winter Olympics. However, the first ski resort that was ever built in the South Korean soil is the YongPyong resort that was built in 1975. Until 1975, ski enthusiasts had to use some undeveloped slopes in either Jiri or Taebaek mountain regions. Thus, the year 1975 is one of the critical milestones in the history of Korean ski resort development as both ski enthusiasts and the general public could have access to a ski resort for developed slopes, lifts, condominiums, and other amenities for the first time. Initially, skiing was considered as a luxury sport due to the high cost for user fees, skiing equipment, accommodations, and transportation. For example, the number of registered personal vehicles in 1975, 1980, 1985 and 1990 were only 80,000, 25,000, 560,000, and 2,070,000, respectively (Statistics Korea 2019). As these numbers indicate, only a limited number of the general public were able to enjoy skiing by the time when the YongPyong was first introduced in 1975. Over time, skiing became an aspirant sport and several private investors became interested in building new ski resorts as they witnessed the success at the YongPyong.

In addition to the physical geography, one of the unique aspects regarding ski resort development in Korea is human geography. About 25 % of Korean population live in the Seoul Metropolitan Area (SMA), and 50 % of Koreans reside in the Greater Seoul-Incheon-Gyeonggi area (aka., Seoul Capital Area; SCA). Thus, the development of ski resorts has two-folds. The first group of ski resorts has been developed where the natural settings for ski resorts, such as snowfalls, are advantageous, mainly around the Taebaek and Sobaek mountain ranges in the Northeastern region. While the access to the ski resorts in these areas has been dramatically improved over the past few decades, these ski resorts, such as the YongPyong and Alpes resorts, were located distant from the SCA and mainly attractive to (traditional) skiers and ski-tourists, as opposed ski-excursionists and snowboarders. The second group of ski resorts, such as the Bears Town resort, was developed near the SCA, mainly within one or one and a half hour driving distance from Seoul, to attract casual skiers and excursionists from the SCA. In addition to the fact that the number of populations residing in Seoul is large, people in Seoul were the early adopters to ski when ski was introduced in the 1970s and became more popular in the 1980s. Typically, the ski resorts near Seoul/SCA were smaller in size with fewer trails. Also, due to the technical limitations in snowmaking in earlier days, the quality of snow on slopes was not attractive to more serious skiers.

During the first period, most of the ski resorts were developed by the private sector, including individual investors and syndicates. Since these developments were initiated under the developmental government, the process of getting governmental or municipal approvals was relatively easier for the developers. The environmental issues and collateral damages due to the disrupted developments (in some cases) were not the most important consideration during this developmental age. Instead, the main issue concerning ski resort development was the lack of interest because ski was in the introduction stage of the product life cycle.

As previously noted, ski resorts developed in better natural settings came with greater sizes and slopes, but the access to those ski resorts and accommodations were not sufficient enough to satisfy skiers. For example, the Alpes resort boasted the highest snow quality because of its north most location and altitude (600 meters above sea level) among the Korean ski resorts and, thus, had an average snowfall greater than 1 meter during ski seasons until the early 2000s. However, the Alpes resort was not the favorite option for skiers mainly due to the limited options for accommodation and amenities. Thus, many of the earlier ski resorts realized the need to build membership hotels and condominiums to attract a greater number of casual and serious skiers and tourists. For example, the Yangji Pine resort was opened in 1982 but its first facility for accommodation, Youth Hostel, was opened in late 1984 and additional condominiums for accommodations were built in 1996. Until 1996, the Yangji Pine resort was not able to accommodate the needs of family skiers and tourists.

Of the five ski resorts developed during the first generation, most of them gradually expanded their resorts by adding additional trails, developing additional golf courses and building accommodations (for both users and developers) and also experienced changes in ownership. During this time, the YongPyong was merged by one of the large corporations, SsangYong Corporation in 1982 because larger corporations expected a high potential for developing or purchasing ski resorts due to the possible financial gains through real estate developments (e.g., luxury condo projects, membership hotels, country club memberships). Consequently, the first-generation resorts later faced a greater level of competitions as the second wave of ski resorts were developed in the 1990s.

The second period (the 1990s)

During the late 1980s and early 1990s, the ski was picked up especially by college students and young adults (i.e., Generation X). Also, the earlier success of some ski resorts, such as the YongPyong and Bears Town sent a very positive signal to potential developers and investors. Until then, investors and construction companies in the leisure and tourism sectors were more interested in developing safer investment options, such as seaside resorts and golf courses. In addition, the developmental government was supportive of building new ski resorts and willing to make some compromises, such as environmental impacts. One of the major reasons that the developments in the Northeastern region (mainly, Gangwon province) was well received by the central and municipal governments and the regional public was related to the demographic changes and domestic geopolitics. The population in Gangwon had been in decline from 1.87 million during its peak year in 1975 to 1.58 million in 1990 and 1.47 million in 1995. Also, the triad area, in which Hwengsung, Pyongchang and Hongchon counties are located, had suffered from aging populations and lacked infrastructure and investments. Thus, the developers could work with politicians in the Northeast region and made developmental activities as a political agenda. Consequently, they could easily receive governmental approvals and financial and administrative support.

During the second period, the private sector led the developmental projects with the support from the centralized government with a development-at-all-cost mentality. For example, the plan to develop the Muju resort was approved in 1984 even though the ski resort, including family hotels, golf courses, and a zoo, was to be built within the national park areas for the benefits of a private company, the SsangBangWool corporation. The Ministry of Works (now the Ministry of Construction and Transportation) revised the national parks schemes and made a change of zoning for this development project. Of the seven ski resorts opened during the second generation, three bigger ones built in more naturally advantageous locations were developed by large corporations such as Hyundai, Bokwang (formerly part of Samsung Group) and SsangBangWool corporations while other four ski resorts closer to the

SCA were developed either by mid-sized construction companies or so-called resort groups. Also, one of the four ski resorts developed by a construction company, the Eagle Valley resort, was later purchased by a larger corporation, called the Sajo Corporation.

The conglomerates and large corporations in Korea tend to expand their business areas and diversify their business portfolio by penetrating new sectors because there were no governmental regulations and restrictions. The ski resort business was one of the targeted business areas for several conglomerates due to the boom in the national economy in the late 1980s and 1990s and dramatic bursts in real estate markets. Skiing at that time was considered as a hot cultural item and conspicuous consumption while the increased gross national income (GNI) facilitated the general public's increased interest in skiing. Consequently, the investors were able to recover their investments and make profits from the sales of associated condominiums and club memberships.

All seven ski resorts opened during this time period were initially designed exclusively for skiers because snowboarding was no fully introduced in the 1980s. When snowboarders first tried to use ski resorts, many of ski resorts in Korea either prohibited snowboarding or allowed them to use only one or two trails. As the popularity of snowboarding became bigger, many of them later modified their ski slopes and added snowboard-only trails and pipes. With the increased interest in snowboarding, the total number of people who visited ski resorts during the 1990s was gradually increased and the annual accumulated visits in the 1998/99 season were 2.44 million with an increase of 19 % from the previous season. As noted, the second-generation ski resorts were exclusively developed by the private sector. All of them were developed, invested, owned and operated by the private sector developers. However, from the later 1990s, Korea suffered a national financial crisis, so-called the International Monetary Foundation (IMF) crisis, which was resulted from the 1997 Asian Financial Crisis. The SsangBangWool Corporation went bankrupt and sold the Muju resort and the Phoenix Park resort barely survived with the help from their related companies such as the Samsung Life Insurance Corporation and Samsung Fire and Marine Insurance Corporation. Consequently, investments by the private sector were dramatically decreased and, thus, there was no ski resort developed and opened during 1995 and 2002.

The third period (the 2000s)

During the IMF era (from 1997 to 2001), the investment and developmental activities concerning ski resorts by the private sector were temporarily shrunk. However, Korea hoped to host the 2010 Winter Olympics and, thus, the nation had to build a winter sport portfolio by promoting winter sports and building winter sport facilities, including ski resorts. As briefly mentioned, the Northwest region has suffered the shrinking population and a lack of industrialization. The social and economic fallout of the demographic changes has been detrimental to this region and the working-age population had been gradually decreased due to the lack of employment opportunities. Thus, the political geography worked positively on this region when to get support from the central government (and the general public) to be a host city of a Winter Olympics. The geopolitics worked favorably when competing against Muju county in North-Jeolla province to become a candidate city and continuously getting support to participate in the Olympic bids three times for the 2010, 2014 and 2018 Winter Olympics. The basic argument supporting Gangwon-province's bid was that Gangwon should have an opportunity to have a major catalyst for developing its region and receive national attention and resources for social overhead capitals.

Consequently, two of the four ski resorts built in Gangwon were planned and developed by the municipal government and other two ski resorts were also developed in close collaboration with the municipal governments. One additional factor in this private-public partnership is the re-introduction of the local self-government system and decentralization (i.e., municipality) in 1995. The local governments and non-governmental organizations became more involved in the developmental activities concerning ski resorts. Of the seven ski resorts developed, four resorts were opened in Gangwon, two in Gyeonggi, and one in South-Gyeongsang province. The Eden Valley R\resort was the first and the only ski resort developed in the Southwest region of the country.

The High1 resort, opened to the public in 2006, was developed by the Gangwon Land, which owns two hotels, two condominiums, a golf country club, a water park and a casino. The Gangwon Land is a government enterprise established by the Mine Reclamation Corporation (established and controlled under the Korea Ministry of Trade, Industry and Energy) and its flagship business is the Gangwon Land Casino. Currently, 51 %

of the Gangwon Land's stocks are owned by the Mine Reclamation Corporation, Gangwon-province Development Corporation, and four counties with the abandoned mine areas, namely Jungsun, Taebaek, Youngwal and Samchuk counties in Gangwon-province. The project was supported by the 'Special Act on Development of the Abandoned Mine Land' (hereafter, the Special Act), which was passed in 2000 to rejuvenate formerly populated areas with the coal mining business with peak years in the 1970s and 1980s. Consequently, the Gangwon Land project was financially supported by the special funds for the abandoned mine area districts and it became a land-grant resort in the middle of the national forest reserve. Thus, Gangwon Land was able to build the 3rd largest ski resort with the greatest number of ski gondolas in Korea.

The Alpensia resort was developed by a similar mechanism and was open to the public in 2009. This resort is owned by the Gangwon-province Development Corporation and, thus, was granted land in the national forest reserve and financially supported by the public funding sources. Previously mentioned, the Gangwon-province Development Corporation also owns the share of the High1 resort. Similarly, the O_2 resort was developed and built by the collaborative effort between the Taebaek Tourism Development (a state-owned enterprise; SOE) and a consortium led by the Kolon Industries Corporation to rejuvenate Jungsun county. For the O_2 resort, the Taebaek Tourism Development was able to secure a site within the national forest reserve and received the national and provincial development funds.

As mentioned, the three ski resorts, the High1, O_2 and Alpensia, were built to primarily develop the local and regional economy and to strengthen Korea's bid to host a Winter Olympics event (Won and Chiu 2018). Most of the ski resorts developed during the 3rd period were planned during the early 2000s when Korea initially aimed to host the 2010 or 2014 Olympics (to be prepared for the IOC sessions in 2003 and 2007). Korea planned to strengthen its bidding portfolio by developing multiple snow-sports facilities and associated infrastructures. Other four ski resorts developed during the 3rd period also were well-supported by the general public due to the nation's wish to host a Winter Olympics and by their respective municipal or regional governments due to the changes in the local government system (from the centralized governance to the local self-governance system). Evidently, these developmental activities contributed to the successful bid for the 2018 PyeongChang Olympics. However, the increased number of ski resorts, especially in

Gangwon-province, has created stiff market competitions among ski resorts. For example, the O_2 resort was built only 20 kilometers apart from the High1 resort and, consequently, the smaller and less attractive O_2 resort has struggled in drawing skiers and snowboarders to the property and making profits. Also, the popularity of skiing and snowboarding, in general, has been in decline after its peak year in the 2007–08 season. The Alpes resort went bankrupt in 2007 due to the heightened market competition and its inability to attract skiers and snowboarders and to find condominium buyers. Relative to other ski resorts in Gangwon, the Alpes was equipped with outdated facilities and poor infrastructure. The decreasing participation in ski and snowboard is still the major issue that the majority of ski resorts face as of today.

The 4th period (the 2010s) and post-PyeongChang Olympics

The latest addition to the ski resort industry was the Alpensia resort opened in 2009. Ever since, there has been only some discussions or plans of developing a few new ski resorts across the nation or to re-develop defunct ski resorts such as the Alpes and Eagle Valley resorts. As mentioned, the popularity of skiing and snowboarding has been in decline since its peak season in the 2007/08 with about 6.6 million visitors (SBAK, 2016). With an average annual decrease of 8 %, the total visitors in the 2014/15 season were decreased to 5.1 million. Since the 2014/15 season, the SBAK does not publicize the data regarding annual visitors to protect their business assets (e.g., market value and reputation). As mentioned, the O_2 resort has been closed from the 2013/14 season until its reopening in the 2018/19 season. The Alpes resort has been closed since the 2006/07 season with the two failed redevelopment plans. The YangJi Pine resort made an application to commence corporate rehabilitation proceedings in 2015 and was sold to an equity company in 2017.

Also, the global climate change has impacted on the operation of ski resorts due to the snowfall shortage and consequent shorter ski seasons. The shortfall of snow results in an increased operating cost for snowmaking and a decreased operable ski terrains and slopes. For example, the Eagle Valley resort has been closed since the 2016/17 season because of low snowfall after its first operation in 1990. Especially, ski resorts nearing the SCA are having a harder time due to the geo-climate location.

The only development project in progress is the Ganghwa resort, which will be the only ski resort located in the west of Seoul. The resort is a part of the designated 'Tourist Complex' and is planned to be completed by 2020 for the 2020/21 season. There are 39 designated tourist complexes in Korea as of 2016, which are selected by the Ministry of Culture, Sports and Tourism (MCST) to promote the nature-based tourism in their respected regions. Of the ski resorts, the Alpensia, Oak Valley, Phoenix Park, YongPyong, Welli Hilli, and Vivaldi Park are located in the tourist complex zones. These ski resorts have been benefited from the 'Tourism Promotion and Development Fund Act' and "Tourism Promotion Act of 2009".

Another talk on the development plan is in Bonghwa of North-Gyeongsang province to target skiers and snowboarders from the southeastern region. However, the plan can only be proceeded if a dam is constructed in this area for municipal water supplies. After resolving the waterworks issues, the Bonghwa county plans to tap private capitals to build the planned ski resort and tourism complex. Since the Lee Myung-Bak government (2008–2013), attracting and utilizing private-sector capital has become a preferred way of building infrastructure (e.g., highways) and public facilities (e.g., tourism complex) to minimize financial risks. Consequently, both the Ganghwa and Bonghwa resorts are extensively relying on private sector funding.

Factors affecting the choice of ski resorts in Korea

Selecting a ski resort is a multi-attribute task for many skiers. Thus, skiers in Korea also consider various factors such as (the number and variability of) ski trails, amenities, snow quality, overall physical environment, accessibility, lodging and crowding because users want to pursue such values as entertainment, safety, accomplishment and socialization (Bae 2009; Won, Bang and Shonk 2008; Won and Hwang 2009). The factors that are directly related to the core product include snow condition, hill and trails, crowding, lift ticket specials, the type of ski slopes and other considerable factors include an accessibility to the ski site, the price of food, facilities and accommodation (lodging), nightlife and entertainment options (Siomkos, Vasiliadis and Lathiras 2016; Won et al. 2008).

While some of the ski resorts have inherited competitive advantages over other competing ski sites (e.g., Canada's Whistler Blackcomb), most of the Korean ski resorts have tried to address the needs and wants of skiers and, more recently, snowboarders. Previously mentioned, the ski resorts built in more naturally advantages areas emphasize the snow quality and various hill and trails while try to build additional quality lodging facilities and provide other entertainment options for tourists in order for them to maximize consumer utilities. For example, the Phoenix Park added spar and water park while continuously expanded the ski slopes and trails. Similarly, the YongPyong has expanded its operations including an indoor waterpark, multiple golf courses, cable car tours, extreme leisure sport activities, and resort condominiums. On the other hand, ski resorts within the closer proximity to the SCA or other metropolitan areas have attracted skiers and snowboarders by mainly emphasizing the closeness from home or work, lift ticket specials, and extended operating hours. For instance, the Konjiam resort operates its lifts until 4:00 am to accommodate the needs of excursionists and younger users.

For the ski industry, one of the major target consumers has been college students and young adults. For the sustainable success of the ski resorts, building a long and loyal relationship with this population is even more important for the survival of the ski industry. Also, there has been a shift in user profiles in that snowboarding (and other newer snow sports) has become a preferred winter sport among the younger generations. Thus, it has become imperative to understand what influence the choice of younger generations. Overall, college-aged skiers and snowboarders consider snow quality, crowding, and accessibility to ski sites to be important when choosing a ski destination (Hwang and Won 2009). More specifically, skiers mainly concern about the snow quality, followed by crowding and the range of slopes, while snowboarders seriously consider both the snow quality and "travel distance" to ski destinations, indicating subtle differences between two groups (Hwang and Won 2009). Gradually allowed from the 1990's, now most of the ski resorts allow both skiers and snowboarders to use all of the slopes and trails in their respective resort. Also, some ski resorts have built snowboard-only trails and pipes in reflecting the changes in consumer preferences. Especially, ski resorts with lesser accessibility try to offer large-scale snowboard-only slopes, and trails to compete against the ski resorts near the SCA. For example, the Phoenix Park resort has snowboard-only parks and the Welli Hilli resort has a superpipe for snowboarders. The Vivaldi Park uses free shuttle buses to attract skiers and snowboarders without personal transportations.

Also, the importance of understanding the subculture of snowboarders should be well recognized by the ski resort managers in Korea as younger generations prefer snowboarding over skiing. For example, the expression of lifestyle identity and the pursuit of creativeness are some of the key values that snowboarders seek from their snowboarding activities (Shin et al. 2013). While it has been lessened, there are conflicts between skiers and snowboarders due to the compatibility of both activities and the differences in lifestyle between traditional skiers and snowboarders (Makens 2001; Won et al. 2008). In addition to developing snowboard-only slopes and adding a superpipe, some of the ski resorts near the SCA have actively promoted social activities and accommodate the meetings and gatherings of social membership groups for snowboarders to satisfy snowboarders' needs. For example, the Jisan Forest resort is well-known for social club activities for snowboarders and younger users because the resort offers various events and promotional activities for these social clubs.

One of the issues that ski resorts should consider is the climate change issue and related practices. A more prevalent trend in Korea is the consumers' increased awareness of sustainability. As previously noted, a few ski resorts in Korea had to temporarily close their operations due to the shortage of snowfalls and consequent expenses of snowmaking. A higher number of skiers and snowboarders will select ski resorts that implement eco-friendly business practices and green marketing.

A decline in ski and snowboard participation

Skiing (including snowboarding) is still one of the popular sports and, for many, one of the desirable sport activities that people want to try. The primary participants of these two sports are males, adolescents and young adults, and residents in the SCA (Korea Gallup, 2013). When people were asked to choose a sport activity to participate if they were given financial affordance, the ski was ranked as the 8th aspirant sport activity for Koreans, preceded by golf, swimming, yoga, horse riding, dance sports, bodybuilding and cycling (Korea Institute of Sport Science, 2018).

However, contrary to the wishful thinking of the proponents for the Winter Olympics and state-level sport administrators, the popularity of skiing and snowboarding has been declined over the past 10 years or so. As mentioned, the number of annual ski and snowboard participants has been gradually decreased from the 2007/08 season (6.6 million) to the 2014/15 season (5.1 million)[1]. The Ministry of Culture, Sport and Tourism (MCST) reported similar figures about the number of skiers and snowboarders who took the slopes: 6.9 million in the 2011/12 season to 4.9 million in the 2015/16 season. More recently, it was estimated that only around 4.8 million Koreans visited ski resorts in the 2016/17 winter ski season (Hanrahan, 2018). The estimated occupancy rates of the ski resorts were 12.8 % in 2013 and 10.7 % in 2015 (Korea Institute of Sport Science, 2018). While the occupancy rates of the ski slopes were also influenced by the number of newly developed ski slopes and resorts, the decline in the occupancy rate indicates an alarming trend for the financial sustainability of Korean ski industry.

There are several (possible) reasons for the declining participation of skiing and snowboarding in Korea. Firstly, the birth rate in Korea has been dramatically decreased over the last few decades. Thus, the population of ages 0–14 has been gradually decreased since the 1970s, from 13.7 million in 1970 to 10.9 million in 1990 to 6.8 million in 2010 (Statistics Korea 2016). Also, the estimated population of this age category is projected to be 6.1 million in 2030 and only 4.7 million in 2050 (Statistics Korea 2016). Previous statistics and studies in Korea indicate that skiing and snowboarding are more popular among younger populations (Gallup 2013). Thus, this change in demographics could be considered one of the general reasons that many ski resorts in Korea are having problems attracting skiers and snowboarders to their slopes.

Secondly, the changes in the environmental climate and the general public's perception concerning environmental issues have negatively contributed to a decline in ski participation. Global warming has caused the shortened duration of ski seasons and forced a few ski resorts to be either temporarily or permanently closed due to reduced snowfalls. Consequently, skiers and snowboarders are given a smaller window of opportunities to enjoy their preferred winter sports. Also, greater public

1 After seven straight-year decline in participation, the Ski-Resort Business Association of Korea (SBAK) stopped publicizing any information about its participation figures.

awareness of sustainability has resulted in some negative sentiments on skiing and snowboarding activities due to the environmental consequences associated with the ski resort development and operation. Ironically, the PyeongChang Olympics was the critical event that people realized the environmental issues caused by winter sport facilities. Some of the facilities built for the Olympics in the "Protected Area for Forest Genetic Resource Conservation" were used only for this particular event. For example, the Alpine downhill courses at Mount Gariwang were built by destructing the virgin forest in this area. There has been a public uproar over the construction of ski resorts and unsustainable operation of ski resorts as well (e.g., carbon footprint and waste).

Thirdly, more of younger populations in Korea are less physically active and prefer playing technology-based activities such as video gaming and argumenta reality (AR) or virtual reality (VR) experiences. Thus, skiing has become a less preferred option as a leisure-time activity for them. This advance in information and communication technology (ICT) comes as a double-edged sword for the ski industry. As well known, Korea is often considered the top-ranked country in the world when it comes to ICT development (International Telecommunication Union 2017). While the ICT can offer much to the ski industry, the Korean ski industry has not embraced the benefits of ICT to their service provision and operations. If the ski industry does not make timely and greater use of ICT, the gap between what the industry can offer and what consumers want will be widened.

Lastly, the perceived and actual costs associated with skiing have been and will be one of the major constraints for potential skiers. As mentioned earlier, skiing has been considered one of the high-end sports and also considered as a sport only for younger populations or risk-takers due to the perceived physical or financial risks (e.g., personal injury, traffic accidents). To this end, some ski resorts offer deep discounts to attract skiers using the promotional deals in collaboration with credit card companies. Also, the market competition has been intensified due to the greater number of ski resorts (i.e., the supply) and, thus, consumers have a wider range of options to choose from. However, the ski industry has failed to communicate better with their potential consumers and, consequently, the industry has been unsuccessful to reverse the negative participation trend.

The other issue with the consumer cost is related to ski clothing and equipment. In general, many Koreans are sensitive to fashion trends, pursue conspicuous and hedonic consumption in comparison to the Western counterparts, and prefer foreign (luxury) brands (Park, Rabolt and Jeon 2008). Indeed, the cost of luxury goods is on average 1.14 times the global average price (Exane 2017). Similarly, the price tags for the preferred foreign ski apparel and equipment brands are higher than most people can afford. Even worse, there has been no equipment manufacturer (i.e., ski brands such as the Salomon) and all equipment has been imported with higher price tags. As such, both the perceived and actual cost for skiing has played a negative role in attracting new skiers and stimulating demand.

Conclusion and suggestions

Starting from the 1970s, Korea has made substantial progress in developing ski resorts and positioning the nation as an international ski destination. Over the last four decades, there have been several factors and forces that influenced the development of ski resorts, such as the development actors, real estate, consumer demand, environmental concerns and nationalism (e.g., Olympics). Also, the unique characteristics of human and physical geography resulted in the development of two groups of ski resorts.

The initial development (the first period from the 1970s to the 1980s) of ski resorts in Korea was to embrace recreational skiers' emerging interest in this new sport. The five ski resorts built during this time period were developed by the private sector with an aim to generate wealth from the associated property development. During the second period (the 1990s), mostly conglomerates or specialty private companies were the main actors with the synchronized support from the centralized government with the developmental doctrine. Rather rapid development of ski resorts was temporarily interrupted as Korea faced a major financial crisis, so-called the IMF crisis from 1997 to 2001. During the third period (the 2000s), a heightened nationalistic sentiment, hoping to host a Winter Olympic Games, and geopolitics intensified by the self-government system played a key role in developing ski resorts and related infrastructure. Consequently, the public sector (with some private-public partnerships) was the driving force during this time period. Since the opening of the Alpensia resort in 2009, the fourth period during the 2010s has not seen any addition, except one resort being developed in the west of SCA, with a projected opening day in 2020. Due to the decline in ski participation, an increase in the number of ski resorts, the changes in global climate, and the failure in keeping up with the pace of change, the Korean ski industry has struggled to regain its growth path and overcome the managerial challenges.

For a better future prospect, the following approaches should be undertaken. Firstly, the existing ski resorts should be modified and adapted to provide a truly integrated resort experience, especially to attract domestic consumers now with a higher service expectation. The Korean ski industry can learn much from the literature on experience goods and the service-dominant logic (Schulze, 2013). Secondly, most of the ski resorts and the destination cities have failed to strategically build their brands and destination image to both the domestic and international skiers and tourists. It is because the developers had employed the 'if we build it, they will come' approach. Thus, the ski industry has not played a proactive role in branding and marketing for the survival of the industry. On a positive note, the ski industry has seen some successes in attracting visitors and tourists. For example, the MCST, in collaboration with the Korean Tourism Organization (KTO) and ski resorts, ran a winter travel week campaign in 2017 and witnessed an 11.3 % increase in ski visitors if comparing to the same week in the previous year (MCST, 2017). This example illustrates the need for more concerted efforts and better working relationships with the related agencies and organizations by the ski industry. Thirdly, in conjunction with the branding and destination image issues, the ski industry should work on developing the legacy of the 2018 Winter Olympics to better build the destination image and maximize the (possibly) long-lasting benefits of the Olympics. However, the nation has failed to tap on the potential of the hallmark event by not having proper strategies and efforts. For example, follow-up annual winter sport and cultural events should be considered to keep the momentum going. Lastly, the globalization of sport and tourism markets should be well embraced within the industry's strategic plans given the shrinking populations in Korea. For Korea, China and South East Asian countries have been the top contributors to tourism in Korea. Thus, the Korean ski industry must find better ways to attract Chinese and East Asian tourists and also continuously expand its marketing territories internationally.

References

Bae, Sangwoo. 2009. "Factors affecting ski destination choices". *Korean Journal of Physical Education* 48(2): 297–307.

Carter, Jamie. 2017. "Five of the best Asian countries for skiing, recommended resorts and our top pick for this winter". Retrieved from: https://www.scmp.com/lifestyle/travel-leisure/article/2120179/five-best-asian-countries-skiing-recommended-resorts-and February 14 2020.

Chung, Jaeyong and Won, Doyeon. 2011. "The authoritarian policy in South Korean sport: A critical perspective". *European Journal of Social Sciences* 20(1): 146–157.

Exane. 2017. "Cost of luxury goods relative to global average". Retrieved from: http://www.theatlas.com.

Gallup, Korea. 2013. "Preferred winter sports in Korea". Retrieved from: http://www.gallup.co.kr.

Hanrahan, Mark. 2018. "As South Korea counts down to Olympics, a ghost ski resort festers". Retrieved from: https://www.reuters.com.

Hwang, Sunhwan and Won, Doyeon. 2009. "Factors influencing the college skiers and snowboarders' choice of a ski destination in Korea: A conjoint study". *Managing Leisure* 14(1): 17–27.

International Telecommunication Union (ITU). 2017. *Measuring the information society report 2017.* Geneva: ITU.

Korea Institute of Sport Science. 2018. *2016 Sport white paper.* Seoul: Korea Institute of Sport Science.

Korea Institute of Sport Science. 2018. *2016 Sport industry white paper.* Seoul: Korea Institute of Sport Science.

Makens, James. 2001. "A ski-industry challenge". *Cornell Hotel and Restaurant Administration Quarterly* 42(3): 74–79.

Ministry of Culture, Sport and Tourism (MCST). 2017. "Increased visitors to ski resorts and indoor attractions during the winter travel week". Retried from: www.mcst.go.kr.

Park, Hye-Jung, Rabolt, Nancy and Jeon, Kyung-Sook. 2008. "Purchasing global luxury brands among young Korean consumers". *Journal of Fashion Marketing and Management* 12(2): 244–259.

Schulze, Gerhard. 2013. "The experience market". In Handbook on the Experience Economy, Sundbo, Jon and Serensen, Flemming (eds.). Northampton: Edward Elgar Publishing, 98–121.

Shin, Nari, Won, Youngshin and Choi, Younglae. 2013. "The subcultural characteristics of freestyle snowboarding participants". *Korean Journal of Sociology of Sport* 26(2): 19–41.

Siomkos, George, Vasiliadis, Chris and Lathiras, Petros. 2016. "Measuring customer preferences in the winter sports market: the case of Greece". *Journal of Targeting, Measurement and Analysis for Marketing* 14(2): 129–140.

Ski Resort Business Association of Korea (SBAK). 2016. "List of ski areas and resorts in Korea". Retrieved from: http://www.skiresort.or.kr.

Statistics Korea. 2016. "Projections of population growth: Ages 0 to 14". Retrieved from: http://www.index.go.kr.

Statistics Korea. 2019. "Total number of registered motor vehicles in Korea". Retrieved from: http://www.index.go.kr.

Won, Doyeon and Chiu, Weisheng. 2018. "Politics, place and nation: comparing the hosting of sport events in Korea and Taiwan". *Sport in Society* 23:1, 142–158.

Won, Doyeon, Bang, Hyejin and Shonk, David. 2008. "Relative importance of factors involved in choosing a regional ski destination: influence of consumption situation and recreation specialization". *Journal of Sport & Tourism* 13(4): 249–271.

Won, Doyeon and Hwang, Sunhwan. 2008. "Ski destination choice and recreation specialization". *Korea Journal of Sport Science* 19(3): 96–105.

Won, Doyeon and Hwang, Sunhwan. 2009. "Factors influencing the college skiers and snowboarders' choice of a ski destination in Korea: a conjoint study". *Managing Leisure* 14: 17–27.

1.4

An Evolutionary Path of Resort Governance: The Case of British Columbia, Canada

Peter W. Williams, Chris Sheppard & Alison M. Gill

Introduction

A growing body of literature highlights the importance of building and employing effective governance systems to increase the resilience, sustainability and competitiveness of resort community destinations (Bramwell and Lane, 2011). In a tourism destination context, governance concerns the overriding values, policies, rules, and processes employed by government institutions to influence the planning, development, and management of such places (Bramwell and Lane, 2011; Beritelli, Bieger and Laesser 2007). Research suggests that the importance of these various factors in shaping governance approaches and focus depends on a combination of path dependent and path creation factors (Beritelli et al. 2007; Gill and Williams 2011, 2014). Resort destination governance systems do not emerge in vacuums and they are not static. They evolve in response to global and local forces that are leveraged by agents to either change or maintain the status quo.

This chapter reports findings emanating from a case study exploring the driving forces and mechanisms shaping the development and evolution of a unique resort municipality focused governance system in

British Columbia (BC), Canada (Sheppard, 2017). From a theoretical perspective, it provides insight into the intertwining of path dependent and path creation forces shaping BC's resort area governance system. From an applied viewpoint, it identifies key components of that system that have been adopted and refined over time in order to accommodate BC's expanding network of resort municipalities. There is a long history of destinations emulating (and seeking to surpass) one another in efforts to maintain competitiveness. Dolowitz and Marsh (1996, p. 344) define policy transfer broadly as, "a process in which knowledge about policies, administrative arrangements, institutions etc. in one time and/or place is used in the development of policies, administrative arrangements and institutions in another time and/or place. However, as Boschma (2009, p.25) states, "instead of copying best practice models or selecting winners, policy should take the history of each region as a starting point". Reflecting this perspective, the term 'policy mobility' is increasingly employed by researchers to capture the more complex nature of the mutations that occur in policy transfer (McCann, 2011; Peck, 2011).

Approach

The chapter's case study examines the forces shaping the evolution of BC's resort municipality governance policy between 1960 and 2017. This 'made in BC' system was selected due to the significant role it has played in guiding and supporting the strategic development of BC's growing number of designated resort municipalities, as well as other international mountain resort destination developments in North America, and around the globe.

The data informing the case's findings were collected via a two-phased process. Initially, 13 personal interviews were conducted with a purposefully selected set of key informants directly involved in the creation and refinement of BC's mountain resort municipality policy at various points in its evolution. They included municipal and provincial government officials, professional destination planners, and resort community stakeholders. While few were present throughout the full fifty years of the policy's existence, individually each informant provided their personal perspectives on the forces and critical events shaping the policy during their periods of engagement. As such, they offered unparalleled and often nuanced qualitative insights into the evolution of the province's policy from its initial inception through to its current status.

The second stage of the data collection process involved an extensive review of existing public documents (e.g. government, institutional and media based) that elaborated on the narratives provided by the informants. It helped confirm, support and elaborate on the role and importance of many of the policy shaping critical moments and forces identified by the key informants. It also provided valuable insights reported by influential but now deceased stakeholders engaged at various temporal points in the policy's development and implementation. Building on the convergence of data collected in these two phases, a guiding framework for analyzing the forces shaping governance evolution was employed.

Guiding framework

Ensuing findings were organized and reported within the broader context of contemporary evolutionary economic geography (EEG) concepts. EEG focuses on interpreting and explaining how economic landscapes are transformed over time by specific mechanisms of change (Boschma and Martin, 2010). Several scholars have analyzed tourism destination development and evolution through EEG lenses (e.g. Brouder and Eriksson, 2013; Halkier and Therkelsen, 2013; Ma and Hassink, 2013). A few of them have focused their analyses on the evolution of tourism destination governance systems (e.g. Brouder, 2014; Gill and Williams, 2011, 2014). They suggest that it offers a useful framework for exploring processes of change within tourism systems.

Two complementary EEG models help focus the reporting of the findings. The first framework offers a "heuristic device to assist in understanding the complexities of governance change in tourist destinations" (Gill and Williams, 2017, p. 44). It identifies a generic set of path dependence and path creation factors to consider when exploring the constraints and catalysts of resort community governance evolution. From a path dependence perspective, the framework suggests that there are specific lock-in mechanisms that restrict change and promote path dependence. Martin (2009) emphasizes that such lock-in forces constrain the available choices of action compared to those at a previous juncture. Drawing from Grabher (1993), three specific lock-in path dependent mechanisms are incorporated into the Gill and Williams (2014; 2017) conceptualization: structural, political, and cognitive. Structural lock-in includes functional feedback processes such as sunk costs or transition costs that inhibit change. Cognitive lock-in relates to institutional embedding of 'how things are done'. Finally, political lock-in refers to dominant power structures that have a vested interest in the current path and are resistant to change. The path creation mechanisms highlighted in Gill and Williams' model (2014; 2017) are primarily drawn from the research of Garud and Karnøe (2001). These are factors that have the potential to change the status quo, un-lock existing pathways, and facilitate the creation of new ones. Collectively the path creation and path dependence factors provided the frame for exploring mechanisms shaping the evolution of BC's resort governance system.

Complementing the aforementioned framework, the cyclical process model developed by Sydow, Schreyîgg and Koch (2009) offered an additional organizational and temporal frame for examining the evolution of BC's resort governance system. Figure 9 provides a hybrid version of these two models. It highlights four sequential phases of resort governance evolution and the key mechanisms associated with them. Overall, it presents a useful lens for understanding the evolutionary fluidity of BC's resort governance system, as well as an appreciation of the historical context in which it transpired.

Phase 1 is a pre-path creation period in which contextual events collectively provide the «change-based momentum" required for "pursuing a new trajectory" (Sydow, Schreyîgg, and Koch, 2005, p. 28). Generating momentum is not an unbounded process, and is typically constrained by factors such as existing: policy frameworks, resource allocations, previous

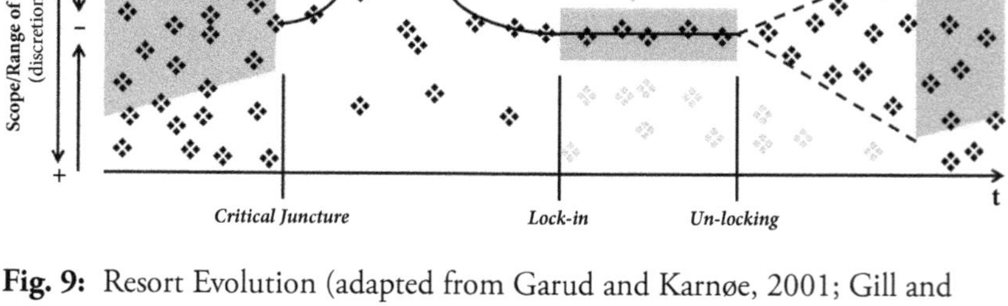

Fig. 9: Resort Evolution (adapted from Garud and Karnøe, 2001; Gill and Williams, 2017; and Sydow; Schreyîgg, and Koch; 2005, 2009)

decisions, and geographic constraints. However, Sydow, Schreyîgg and Koch (2005) suggest that a variety of critical forces (local or global) may emerge which create the momentum for change. This momentum may increase until a critical juncture or moment is reached when a decision is made that reduces available options, for action, as well as makes it significantly more difficult to return to a previous state (Mahoney, 2000).

Phase 2 is a path creation stage in which existing momentum is employed to identify a new pathway. In this phase a critical mass of support for a preferred pathway is developed. The implementation of the pathway is largely dependent on the capacity of a proponent to muster a 'critical mass or threshold' of political support for the new trajectory (Witt, 1997). Essentially, policy makers and proponents must perceive the newly created pathway to be easier to use than previous options (Ebbinghaus, 2005). In a Canadian context, public policy change typically occurs via hybrid public / political processes in which astute change leaders mindfully create, interpret and communicate proposed policy shifts in ways that align with the needs of their direct stakeholders, as well as the equally important priorities of the government's prevailing political party.

Phase 3 is a path dependence-building period in which self-reinforcing lock-in mechanisms are employed to support the chosen path. The lock-in mechanisms can be either structural, cognitive, political or a mixture. Sydow, Schreyîgg and Koch (2005) stress that lock-in mechanisms should not be viewed as unmovable fixed barriers, but represent social constructs that can be interpreted and acted upon in a variety of ways – along a preferred converging path.

Phase 4 is a path breaking stage when path dependent trajectories unlock due to previously unanticipated critical events, and the intervention of other path intervening change agents. Their actions can lead to another round of path creation or alternatively to a path switch or path cessation.

Resort municipality significance

This following case study employs the preceding framework to help explain and describe the evolution of British Columbia's (BC) unique resort municipality governance system. These designated communities (figure 10) have a relatively long and important history of contributing to the economic and social fabric of the province. In 2015, the tourism industry contributed approximately C$7.4 billions of value added to the BC economy (Destination British Columbia, 2016). An estimated C$905 millions of this revenue came from operations associated with the province's uniquely designated resort municipalities. (Resort Municipality Initiative Resort Collaborative, 2016). Currently 9 of BC's 14 operating resort municipalities are alpine winter sport destinations with growing summer programming. Whistler, the province's largest winter and summer resort municipality, generates approximately C$1.37 million in tax revenues for government every day (Economic Partnership Initiative Committee, 2016).

The ability of BC's resort municipalities to contribute in such a significant way to the province's substantial tourism economy is in no small part due to the unique state-based governance system that guides their development and management. It provides a supportive set of legislative, policy, and management structures designed to help shape localized approaches to land-use planning, revenue generation and infrastructure development in ways that make these places more resilient and responsive to the forces of global change. For instance, one of the more innovative support programs in the provincially based resort governance portfolio is the Resort Municipality Initiative (RMI). Since its inauguration in 2007, it has the provided approximately C$108.6 million to these resort communities (Resort Municipality Initiative Resort Collaborative, 2016). These funds have been used to support municipal investments in local transportation corridor improvements (e.g. streetscape aesthetics, welcoming and way-forwarding signage, and nordic-skiing, hiking, and biking trail facilities); public space enhancements (e.g. town squares and green space amenities); cultural /heritage facility upgrades (e.g. arts, festival and event venues, heritage museums); and hosting facilities (e.g. information centers upgrades, convention and meeting facility upgrades). It is but one example of several mechanisms embedded in the ever-evolving governance system. The following sections depict the system's evolution.

Phase 1: path creation – generating momentum

A combination of external and internal events during the 1960 to 1975 period were particularly influential in generating the momentum needed to trigger the development of an overriding resort governance system for

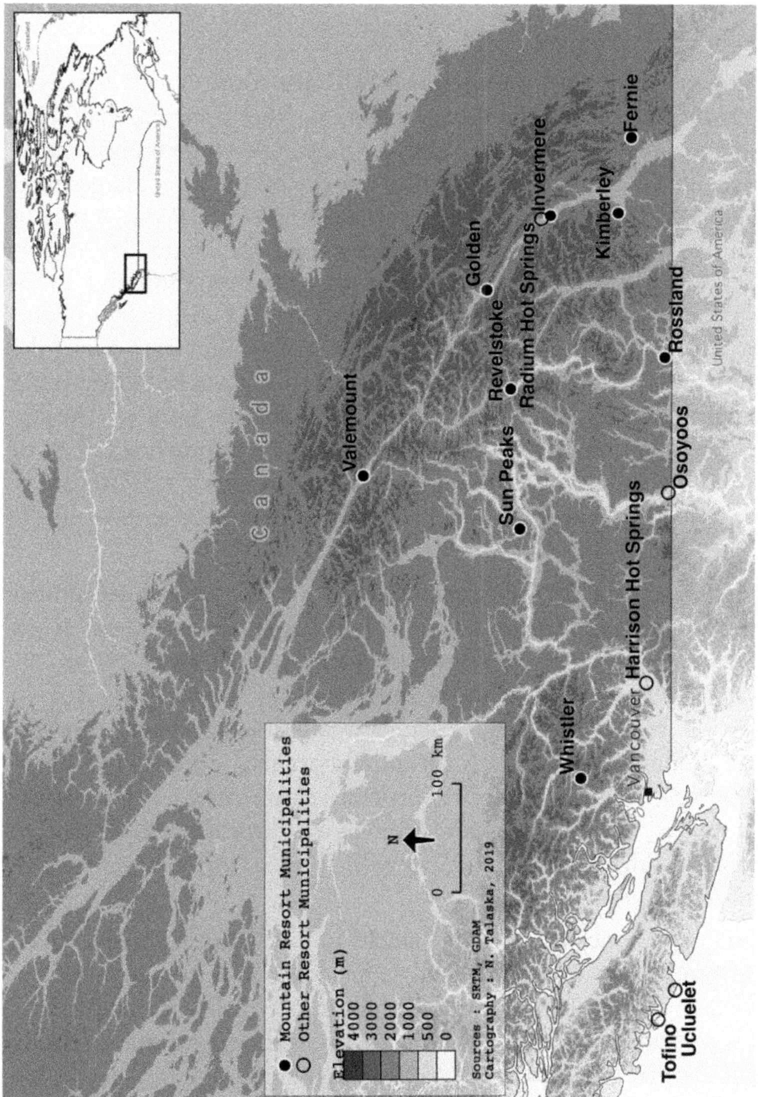

Fig. 10: Designated British Columbia Resort Municipalities, 2018.
Prepared by Flurina Beiger, 2019

BC. As elaborated below, particularly important in this regard were BC political and ski industry proponent responses to emerging population trends; US ski area development dynamics; US environmental activism; and an unsuccessful attempt to create a workable ski development approach in BC.

In response to a burgeoning post WWII growth in North America's young and increasingly mobile population, the Canadian and US governments pursued several strategies for meeting their growing outdoor recreation needs. This included placing priority on the use of large tracts of public wild lands for recreation purposes (Wolfe, 1964). Amongst other initiatives, this led to unprecedented private sector investment in U.S. ski area development on public lands between1960 and 1975. Approximately one new ski area per year was created during that period (Briggs, 2000). As this rapid growth proceeded, so did awareness of the potentially dramatic negative environmental implications of transforming a mountain from its natural areas into places for ski areas to develop. The seemingly unfettered growth and negative environmental effects of such developments led to a chilling and litigious response by civil society (Albrecht, 1976). It virtually halted all new ski area development in the US by the mid-1970s.

In response, the BC government and its ski area development proponents saw the slow-down in US ski area development as an opportunity to enhance investment in such facilities in the province. It touted to American developers the virtues of investing in BC. In particular, it highlighted that it had fewer barriers-to-entry, excellent potential products, and a willing government to help (Ski Policy Industry Informant #1). Not the least of its purported competitive advantages was a newly minted ski area governance system that provided a clear path to development – the Resort Municipality of Whistler Act (also known as, the Whistler model).

Elements of policy transfer are evident in the Whistler model, which was the product of lessons learned from other governance initiatives attempted in the 1960s and early 70s. BC's initial ski area governance model was a 'made in BC' public land use transfer process originally created for other regional resource development projects. An adapted version of its underlying "lease-develop-purchase" system was initially implemented at BC's Cypress Mountain in 1964. Referred to as the Cypress Bowl model, its fundamental lease requirements included the developer creating a critical threshold level of ski-lift capacity, in exchange for resource extraction (e.g. timber, minerals) rights designed to finance the area's ski capacity requirement. However, in the Cypress Bowl case the developer used resource extraction permits as a license to clear and harvest lands primarily for residential real estate purposes—almost to the complete exclusion of promised investments in ski development. In response to extensive outcry, the provincial government rescinded the lease and created a provincial park on the land on 1969. The same model was also tested in two other proposed ski areas and failed to achieve ski area development goals there as well. The critical lesson learned from the Cypress Bowl model's application was that the financial viability of ski resort development was dependent on massive upfront value creation investments in on-slope skiing infrastructure that could be recouped via the subsequent sale of higher valued off-slope real estate lands. This required a more nuanced and longer standing performance agreement between the developers and the government.

Subsequently, the Cypress Bowl model was replaced by the BC government's Resort Municipality of Whistler (RMOW) Act in 1975. It established Whistler (a previously unincorporated small settlement) as the province's first Resort Municipality and incubator for future resort municipality policy development. The RMOW Act's origins were rooted in the province's Instant Town Act (1965)—legislation designed to expedite the rapid development of permanent resource communities (e.g. mill towns, mine towns, rail towns, etc.). The subsequent Whistler model adopted the more entrepreneurial and private sector driven approach to governance legislation embedded in the Instant Town Act. However, reflecting the distinct character of resort towns, the RMOW Act (aka Whistler model) created its own governance pathway by unabashedly borrowing and adapting polices and lessons learned from the US Forest Service's extensive experience with ski area developments. This included adopting a novel perform-and-reward model, which allowed "the people who put the investment in the lifts…[to] control the land at the bottom" (Ski Resort Policy Informant #1, July 2015). This was facilitated via the BC Land Act of 1979 which charged the government with managing the allocation and use of about ~95 % of the province's vast land base by private, public and other institutions such as municipalities. In Whistler's case, this entailed: providing land tenures and permits for private sector investment in on-slope commercial recreation infrastructure and services; as well as transferring ownership of large tracts of off-slope land to the fledgling municipality for a complementary mix of off-slope commercial,

Fig. 11: Whistler Village 'Pedestrian Stroll'

institutional and residential land uses. This included lands for a proposed commercially focused and alpine-themed 'pedestrian' village, as well as adjacent acreage for residentially-oriented neighborhood developments. See figure 11.

In retrospect, the adaption of the Whistler model represented a critical moment in the province's resort governance evolutionary pathway. It strategically positioned 'the Whistler model' as an innovative and self-contained experiment—a ski policy incubator that provided protected space for testing and shaping governance policies and mechanisms that would eventually be exported across the province and around the globe.

Phase 2: developing critical mass

Critical mass is developed through the injection of events that reinforce the pathway— making each ensuing decision easier (Ebbinghaus, 2005). As Whistler's development progressed, so did government enthusiasm for many of the governance policies and mechanisms it was refining through practice. As elaborated on in this section, proponents of the Whistler model subsequently leveraged several exogenous forces and events—the rise of neoliberalism, a global recession, the interest of a large multi-national ski area developer, and a World Fair to build the critical mass of support needed for its application on a broader provincial basis.

Shortly after the enactment of the RMOW Act in 1975, a more neoliberal focused Social Credit Party displaced the New Democratic Party (NDP) led government. While the RMOW Act was originally positioned as a "signature project of the NDP" (Provincial Officer Informant #1, July 2015), it had to be repositioned and reframed to better reflect the political ideologies of the new government. Specifically, the Whistler model's governance structure was reconfigured to provide the municipality with more power over decisions concerning investments in municipal infrastructure, works and services, as well as the levying and collection of rates and charges for public land and improvements. This politically driven pivot allowed Whistler to reinforce its search for a more autonomous approach to governance.

Subsequent to the preceding shift in governance emphasis, Whistler solicited proposals in the late 1970s for the planning and development of alpine lands adjacent to Whistler Mountain and the Village's signature core. The successful proponent—Aspen Skiing Corporation—subsequently leveraged its expertise and influence to secure an amendment to the RMOW Act. The modification required the RMOW to install a single 'leading voice' resort marketing association. Its mandatory business membership and taxing powers became essential components of the Whistler model.

Further support for the Whistler model came in the form of public funding for much needed tourism oriented infrastructure. As awareness of the extraordinary costs associated with developing the infrastructure for resort municipalities emerged, so did the need for government funds and incentives that would meet these financial requirements. In response, Whistler sought and obtained significant public funding for several initiatives. Specifically, in 1979 a joint federal-provincial program Travel Industry Development Subsidiary Agreement (TIDSA) contributed C$10.5 million towards Whistler's development. Secondly, the Federal government launched a creative Multi Use Residential Building (MURB) program that creatively provided tax shelter incentives for private sector investments in primarily condominium style hotel accommodation investments in Whistler. Finally, in response to a crippling global recession in the early 1980s that essentially bankrupted the Municipality of Whistler, the BC government stepped in and secured a C$21 millions bail out of the development arm of the municipality owned Whistler Village Land Company. The level and timing of these funding initiatives were central to the survival of Whistler at a particularly fragile stage in its development. At critical moments, they helped develop key infrastructure and investor confidence in the long-term prospects for Whistler.

The Whistler model's innovative and unique combination of local governance autonomy and supportive public policy subsequently provided the critical mass of support for export of the approach to other places across the province. More specifically, it offered innovative insights and approaches critical to the creation of a province-wide Commercial Alpine Ski Areas Policy (CASP), supportive Municipal and Regional District Tax (MRDT), and broader Mountain Resort Act.

As Whistler's growth and reputation as an emerging world-class resort community grew, so did political and industry support for exporting its governance approach to other BC destinations. In response to this demand, the CASP was created in 1982. It provided the foundational development framework for all future ski area development initiatives in the province. It leaned heavily on the Whistler model for several of its key policy and programming elements.

Notwithstanding the Whistler model's many innovative governance achievements, it remained an evolving "experiment" well into the 1980s. A particularly challenging concern for Whistler was the unanticipated burden of financing, developing and maintaining public infrastructure (e.g. public transit, parks, streetscape beautification, parking, and way-forwarding installations) for a rapidly growing number of visitors. In most cases, these responsibilities were initiatives "that typical municipalities wouldn't even think of funding" (Municipal Officer #1, July, 2015). However, raising property taxes to fund these initiatives was increasingly perceived to be unfair to local residents and disastrous for politicians. In response, Whistler leveraged unprecedented levels of favorable political awareness of tourism's significant contributions to BC's economy during Vancouver's Expo '86 World Fair to acquire a more equitable approach to financing public infrastructure in resort municipalities. The ensuing Municipal and Regional District Tax (MRDT) of 1987 sanctioned an additional tax (above existing provincial hotel taxes) that resort municipalities could charge their hotel guests. Funds collected via the tax were earmarked to "assist local communities operate convention centers and other tourism amenities [...] necessary to become year-round destination resorts" (Couvelier, 1987, May 21).

Whistler's ensuing investments in public, commercial and private real-estate development, and market promotion through the 1990s led to unprecedented levels of rapid market growth. Confronted with a growing concern over the negative effects of this growth on quality of life and environmental quality, resident civil society organizations created the political urgency needed to shift Whistler governance from its long standing focus on growth, to one with more of 'sustainability fix' orientation (While et al. 2004). The particular approach adapted for this purpose employed principles and methods embedded in The Natural Step, an international model for sustainability governance used previously by a wide range of corporate and community entities (the Natural Step 2019). Its widespread international acclaim provided Whistler with the legitimacy and trust needed to readily adopt TNS as the guiding framework for its journey towards greater sustainability (Gill and Williams, 2011; 2018).

Collectively these policy initiatives provided the critical mass of tools and support needed to shape and lock-in resort governance approaches across BC. As one key informant noted: "We are fortunate to have Whistler, Whistler goes through the industry hiccups sooner than everyone else" (Provincial Officer Informant #2, July 2015).

Phase 3: path dependence

The path dependence phase involves employing reinforcing mechanisms to make existing approaches to governance resistant to change. Several critical events have reinforced the use of the Whistler model of across the province. These include favorable political influence, and the hosting of the Vancouver 2010 Winter Olympics.

The return to power of a New Democratic Party in the early 1990s led to a strategic push for stronger economic ties with Asia. One of the hallmark projects they established was a 1993 Master Development Agreement with Nippon Cable (a Japanese company) to develop Sun Peaks Resort (CitySpaces, 2005). This Agreement was quickly reinforced with a supportive Mountain Resort Act (1995) that essentially cemented the basic building blocks of the Whistler model within a broader context. The embedding of the Whistler model's governance approach within the new Act's purview was largely the product of a collective political agency and urgency to get the project up and running in a politically expedient fashion.

In 2003, Vancouver won the bid to host the 2010 Winter Olympic Games (the Olympics). As the official Mountain Host for the Olympics, Whistler used this position to secure a unique Resort Municipality Initiative (RMI) agreement with the BC government. While initially intended to help Whistler meet its Games-related financial and development responsibilities, it was eventually broadened to include other designated resort municipalities. The RMI's revenue sharing model provided designated resort communities with access to a formula based portion of local hotel tax revenues. These funds were specifically designated for municipality-based investments in tourism industry supporting infrastructure, amenities and events.

Complementing this initiative, the neoliberal leaning Liberal Party led government of the day committed to doubling the size of the BC tourism industry by 2010. Central to achieving this goal was creating a larger and more competitive set of resort communities. Building on the recommendations of an industry led Resort Task Force, the government adopted several significant policy directives including: establishing an All Season Resort Policy that extended beyond alpine skiing destinations to include other forms of four-season resort development: (e.g. marine, golf, casino, agri-tourism); expanding the existing roster of designated resort municipalities (from 2 to 14); and, (3) providing designated resort communities with access to existing resort municipality revenue sharing programs. Collectively all of the aforementioned policy initiatives helped reinforce lock-in to the Whistler model's approach to resort governance and management.

Phase 4: path breaking

Phase 4 entails anticipating and responding to potential path-breaking forces. Though BC's resort municipality governance system remains locked into the essential components of the Whistler model's governance system, several potential path-breaking forces may challenge its resilience. Relevant possibilities emanating from this study's research follow.

Climate change

The effects of global climate change have potentially dramatic impacts on the province's designated resort municipalities (Pidwirny, Bahbahani, and Pedersen, 2017). While some BC ski areas are relatively well positioned for on-going snow coverage, many are increasingly susceptible and are responding to issues of snow availability, flooding and fire hazard in a site specific as opposed to system wide fashion (Hagar, 2018). Currently, no definitive guideline requirement for proponents to include climate change assessments and response strategies in proposal, master plan, or environmental reviews exist. It is quite possible that site selection and climate change mitigation and adaptation issues will emerge as a critical resilience factors in future resort governance decision making and Resort Municipality Initiative funding requirements.

Aboriginal engagement

Global pressures on governments to reconcile the lost human rights of marginalized Indigenous populations extend to Canada and BC. In 2003 the BC government initiated a "New Relationship" policy that sought to create and enhance partnerships between the government, First Nations and the broader provincial population. These policies were fortified by several legal judgments reinforcing and clarifying Aboriginal rights to land title, as well as the need for consultation and accommodation in Crown land decisions. From a resort community development perspective, First Nations have subsequently used their legal authority to not only leverage significant land development opportunities in Whistler in the run-up to the 2010 Olympic Games, but also ensure their direct involvement in its future Official Community Plan creation processes (Williams and Gill, 2017). More recently they employed their significant political capital to slow the approval of a controversial proposed ski resort municipality adjacent to Jumbo Glacier in BC's interior (Harris, 2017). These and other engagements with existing and proposed resort developments across the province, clearly signal that existing resort governance systems need to more fully incorporate Aboriginal values and perspectives into BC resort governance systems.

Resort municipality identity

From a resort municipality governance perspective, the balance between resort business success and community quality of life can be precarious. While for the most part Whistler has experienced and accommodated continual and sometimes rapid growth since its inception, it is now challenged in its capacity retain to retain its community identity and lifestyle. This has created a growing dissonance between the community and its resort businesses, Whistler's aggressive quest to retain top international destination status has come at a cost to local residents who are increasingly challenged by issues such as housing affordability, cost of living, local business participation, and recreation site traffic congestion. Collectively these impacts have contributed to growing economic and social displacements of employees and residents to "down valley" locations. Aggravating these circumstances is the recent acquisition of many of Whistler's prime on-slope and commercial operations by Vail Corporation. It has promised extensive investments in infrastructure and international marketing designed to attract a more upscale and globally influential clientele.

At a more local level, in 2010, the innovative RMI was changed from a relatively stable revenue sharing initiative to a grant program, making it more susceptible to random political decisions (e.g. changing political mandates, budgetary cuts, etc.). Collectively these and other externally generated forces related to capital mobility and corporatization may lead resort communities and provincial government policy makers to consider pathways that are less rather than more like the Whistler model.

Discussion

This chapter's findings suggest that BC's unique resort governance system is essentially locked into a particular pathway—the Whistler model. Over the past four decades actors have strategically leveraged and mitigated global and local forces in a number of cases adopting and adapting policy and practices from elsewhere, in order to shape the evolution of the Whistler model. In turn, various components of the Whistler model have been transferred elsewhere within BC. A unique set of path creation and path dependent mechanisms have been purposively used to influence the governance trajectory. The following discussion highlights how these were employed.

Path creation

As Garud and Karnøe (2001) explain, during the path creation process actors leverage and/or mitigate forces by employing a specific set of tools. These are categorized as time, relevance structures, and objects.

Actors can leverage the urgency of events to gain the attention and commitment of policy makers. During the Whistler model's evolution, British Columbia hosted two mega-events: Expo '86 and the Vancouver 2010 Winter Olympic Games. In both instances, Whistler leveraged urgencies associated with delivering these events to gain new funding streams critical to its development. Drawing from the lessons learned from Expo '86, Whistler convinced the provincial government to create the Municipal Regional District Tax in 1987. Similarly, building on the rationale established via its Expo '86 experience, Whistler used the Olympics to persuade the provincial government to introduce the innovative RMI revenue sharing program. A key informant recalled: "We wanted additional financial tools [like] the RMI funds. We used the Olympics to get them" (Municipal Officer Informant #3, July, 2015).

Urgency was also a factor in the adoption of TNS to address how the municipal government could proceed quickly with respect to issues of growth and sustainability. Additionally, Whistler employed the urgency and momentum of the Olympics to obtain additional Legacy Funds supporting: on-going management of its Olympic sport venues; resident based affordable housing; and other community's sustainability goals. (Municipal Officer Informant #3, July 2015). Essentially, international mega-events were strategically leveraged to mindfully deviate funds towards projects key to the long-term benefit of the community (Gill and Williams, 2014).

Relevance structures refer to coherent and shared practices or meanings understood between actors (Koppers, 2015). An actor's ability to interpret existing structures (e.g. policies, programs and regulations) in ways that match with relevant goals can be a powerful tool in path creation processes. When the neoliberal Social Credit government moved into power after the socialist NDP government had established the RMOW Act in1975, the Act's existence was placed in political jeopardy because of its overdependence on government oversight. A local association of developers emboldened by the recent change in provincial government leadership, began to openly challenge the Whistler model. In response, Whistler's entire Council aggressively lobbied the government to retain the Act, but reframe its governance guidelines in a form reflecting prevailing neoliberal ideologies. Agreement on this strategy led to a more autonomous, local government governance approach. The ability to interpret relevance structures in ways that reflected prevailing socio-political priorities was critical to the Act's long-term resilience. It helped the actors shape and protect the core components of their preferred governance pathway.

Rather than 'reinventing the wheel', agents of change regularly employ governance strategies from elsewhere and adapt them to their needs. Many of the foundational polices shaping BC's resort governance system were borrowed from other places. The Cypress model employed a lease-develop-purchase method of public land development originally designed for the establishment of resource extraction towns. When it proved to be inappropriate in a ski town context, the Whistler model introduced a perform-and-reward approach model borrowed from the US Forest Service. It was subsequently embedded into BC's more pervasive CASP in 1982 and All-Season Resort Policy in 2005. As well, The Natural Step approach was also imported from Sweden as a solution to guiding sustainability planning and policy development.

Similarly, while neither TIDSA or MURB policies were created specifically for supporting Whistler's development, they were translated into programs that helped create the critical mass of infrastructure needed to bolster Whistler's development journey at a time when it was most needed. In all of these cases actors nimbly shaped the resort governance pathway by translating policies, expanding and reframing common understandings, and employing time as a resource. These tools were available for actors to either leverage or mitigate global and local forces affecting the governance system's evolution and trajectory.

Path dependence

While path creation emphasizes the role of agency in shaping evolutionary pathways; path dependence describes the stickiness associated with pathways. Through BC's resort governance evolution process, path dependent forces kept the Whistler model as the foundational path. For example, the development of Sun Peaks as BC's second resort municipality benefited from the "inexorable force" and political pressure of the government's Premier at that time. To get it done, policy makers decidedly chose to borrow from the Whistler model instead of creating something novel. The policy makers also leveraged political forces to ensure that the legislation had a wider mandate than just Sun Peaks. Similarly, the province's commitment to resort development was further driven home by another BC Premier whose Resort Task Force reinforced the cognitive lock-in of the

Whistler model and developed the structural components to help with its administration.

Since the emergence of Whistler as an internationally competitive, prosperous and increasingly resilient resort municipality in the early 1990's, there has been political will to expand the range and number other resort communities in BC. While the ideological intent behind this push has varied between political parties over the years, the strategic focus remains centered on encouraging development proponents to adapt core Whistler model principles and policies to the unique circumstances of subsequent developments. Interestingly, this transferring and sharing of mountain resort governance policy and practices increasingly extends via government, corporate and institutionally sponsored field visits, conferences and workshops, personnel secondments, and professional consultations, to an increasing number of proposed resort developments around the globe (UNWTO, 2013)

Conclusions

While the path dependent character of the current resort governance system remains in place, it continues to be buffered and retooled by internal and external forces. Its ability to endure and respond to these forces in a nimble and resilient way will determine its future path. Past decisions influence the choices made today, proving that history is woven into the present and therefore matters.

Drawing from a theoretical foundation in evolutionary economic geography, as well as path dependence and path creation concepts, this chapter explored how resort governance systems emerge and evolve when faced with mitigating global and local forces. From an applied perspective, the findings illustrated how broader socio-political, financial, environmental, and socio-cultural forces influenced the evolving pathway of resort governance in BC. It also identified how specific mechanisms were employed to respond to these broader global forces in a proactive and timely manner.

From a theoretical perspective, the research builds primarily on the EEG concepts and constructs proposed by Garud and Karnøe (2001), Gill and Williams (2017) and Sydow, Schreyîgg, and Koch (2009). As a hybrid of their perspectives, the framework employed provides a unique frame for examining the interplay between path dependence and path creation forces over time Despite Garud, Kumaraswamy, and Karnøe's (2010) concern about combining path dependence and path creation—by way of mixing ontologies, the evolutionary pathway framework used in this case study (Fig. 9) proved to be quite useful. The findings suggest that the role of agency was at the heart of every critical moment along the resort governance system pathway. Agents strategically used the urgency of time, reframed relevance structures, and translated objects to further their agendas. However, they remained subject to both the inertia and interplay of past decisions and emerging contextual realities. As evidenced in the preceding discussion, one emerging reality that affects the evolving dynamic nature of pathways of governance is the increasing global mobility of policies and best practices from elsewhere.

In conclusion, although in the empirical study presented, the transfer of best practices (mainly from Whistler's model) to other resorts has focused on the regional context of BC, it is relevant to draw attention to the influence of Whistler's governance practices to a wider global context. Policy transfer is now widespread in a globally connected world and the agents of mobility in this process are increasingly recognized (see e.g. Dolowitz & Marsh, 1996; McCann, 2011; Peck, 2011). McCann (2011, p. 114) observes that there are three broad categories of mobilizers of policy "local policy actors, the global policy consultocracy, and informational infrastructures". As the mountain operations in Whistler (known as Whistler Blackcomb) have merged with several large North American based corporations there has been considerable exchange of 'best practices' between corporate properties. Most evident were the effects of Intrawest, who extended the 'Whistler village model' of design to many of its resorts in North America and also to the French resort of "Les Arcs 1950" in France.

McCann (2011) also emphasizes that site visits are of considerable importance and value. Whistler's policy makers and planners have welcomed field visits by resort representatives from around the world drawn to its achievements in sustainable development and environmental management policy creation. In turn, Whistler's representatives have travelled globally to seek best practices elsewhere. Several highly respected, Whistler-based, international mountain resort consultants – including, *Ecosign* and *Brent Harley and Associates (BHA)* – have been intimately involved in the creation and refinement of BC resort policies for many years and in turn have provided their expertise in this field to existing and emerging mountain resorts around the globe. Currently, they are involved in mountain resort projects in Asia, where central governments have considerable control over land use and decisions affecting mountain resort development. As demonstrated in this chapter, developing resort governance policies for guiding and controlling land-use development, public and private sector infrastructure funding, market development, and community sustainability requires considerable focus on tried and tested core principles that can be readily adapted to individual circumstances. While such path dependent factors provide the fundamental frame for resort governance in the province, they are subject to on-going adjustment and reinterpretation as global and local forces emerge.

References

Albrecht, Stan L. 1976. "Legacy of the environmental movement". *Environment and Behavior* 8(2): 147–168.

Beritelli, Pietro, Bieger, Thomas, Laesser, Christian. 2007. "Destination governance: Using corporate governance theories as a foundation for effective destination management". *Journal of Travel Research* 46: 96–107.

Boschma, Ron. 2009. "Evolutionary economic geography and its implications for regional innovation policy". *Papers in Evolutionary Economic Geography* 9: 1–12.

Boschma, Ron and Martin, Ron. 2010. "The aims and scope of evolutionary economic geography". In *The Handbook of Evolutionary Economic Geography*, Boschma, Ron and Martin, Ron (eds.). Cheltenham (UK) and Northampton (USA): Edward Elgar.

Bramwell, Bill and Lane, Bernard. 2011. "Critical research on the governance of tourism and sustainability". *Journal of Sustainable Tourism* 19(4–5): 411–421.

Briggs, James. 2000. "Ski resorts and national forests: rethinking forest service management practices for recreational use". *Boston College Environmental Affairs Law Review* 28(1): 79–118.

Brouder, Patrick. 2014. "Evolutionary economic geography and tourism studies: extant studies and future research directions". *Tourism Geographies*, 16(4), 540–545.

Brouder, Patrick and Eriksson, Rikard. H. 2013. "Tourism evolution: on the synergies of tourism studies and evolutionary economic geography". *Annals of Tourism Research* 43: 370–389.

Chong, Ida. 2007. *British Columbia. Parliament. Legislative Assembly. Edited Hansard.*[38] *th Parliament,* [3] *rd session*. Retrieved from http://www.leg.bc.ca/hansard/38th3rd/h70328p.htm.

CitySpaces. 2005. *Harmonization Process: All-Seasons Resorts*. Retrieved from: https://www.for.gov.bc.ca/ftp/HDI/external/!publish/Web/resort-development/asr/harmonization_rpt.pdf.

Couvelier, Mel. 1987, May 21. British Columbia. Parliament. Legislative Assembly. Edited Hansard. 34th Parliament,1st session. Retrieved from: http://www.leg.bc.ca/hansard/34th1st/34p_01s_870521a.htm.

Destination British Columbia. 2016. *Value of Tourism 2015: A Snapshot of Tourism in BC*. Retrieved from: http://www.destinationbc.ca/getattachment/Research/Industry-Performance/Value-of-Tourism/Value-of-Tourism-in-British-Columbia-_-A-Snaps-(1)/Value-of-Tourism-2015-Snapshot.pdf.aspx.

Dolowitz, David and Marsh, David. 1996. "Who learns what from whom? A review of the policy transfer literature". *Political Studies* 44: 343–357.

Ebbinghaus, Bernhard. 2005. *Can Path Dependence Explain Institutional Change? Two Approaches Applied to Welfare State Reform*. Max Planck Institute for the Study of Societies Discussion Paper 05/2.

Economic Partnership Initiative Committee. 2016. *2016: Updated Summary of Key Findings & Economic Planning Report*. Retrieved from: https://www.whistler.ca/sites/default/files/related/2016_key_findings_and_economic_p lanning_report_-_final_sm.pdf.

Garud, Raghu and Karnøe, Peter. 2001. "Path creation as a process of mindful deviation". In *Path Dependence and Creation*. Mahwah, N.J. Lawrence Erlbaum Associates: 1–40.

Garud, Raghu, Kumaraswamy, Arun and Karnøe, Peter. 2010. "Path dependence or path creation?" *Journal of Management Studies* 47(4): 760–774.

Gill, Alison M. and Williams, Peter. W. 2011. "Rethinking resort growth: understanding evolving governance strategies in Whistler, British Columbia". *Journal of Sustainable Tourism* 19(4–5): 629–648.

Gill, Alison M. and Williams, Peter. W. 2014. "Mindful deviation in creating a governance path towards sustainability in resort destinations". *Tourism Geographies* 16(4): 546–562.

Gill, Alison M. and Williams, Peter. W. 2017. ""Contested pathways towards tourism-destination sustainability in Whistler, British Columbia: An evolutionary governance model". In *Tourism Destination Evolution.* Brouder, Patrick, Anton Clavé, Salvador, Gill, Alison, and Ioannides, Dimitri. (eds.). Abingdon, Oxon: Routledge, 43–64.

Gill, Alison. M. and Williams, Peter W. 2018. "Transitioning towards sustainability in the mountain resort community of Whistler British Columbia, Canada". *Tourism Recreation Research* 43(4): 528–539.

Grabher, Gemot. 1993. "The weakness of strong ties: the lock-in of regional development in the Ruhr area". In *The Embedded Firm: On the Socioeconomics of Industrial Networks.* Grabher, Gemot (eds.). London: Routledge, 254–277.

Hagar, Mike. 2018. "How BC's ski resorts are coping with global warming threat to their existence". *The Globe and Mail,* April. 10. Retrieved from: https://www.theglobeandmail.com/news/british-columbia/how-bcs-ski-resorts-are-coping-with-global-warming/article22731538/.

Halkier, Henrik and Therkelsen, Anette. 2013. "Exploring tourism destination path plasticity: The case of coastal tourism in North Jutland, Denmark". *Zeitschrift fÅr Wirtschaftsgeographie* 57(1–2): 39–51.

Harris, Kathleen. 2017 (Nov. 2). "Supreme court ruling removes barrier for year round ski resort on sacred First Nation land". *CBC News.* Retrieved from: https://www.cbc.ca/news/politics/indigenous-rights-ski-resort-1.4381902.

Koppers, Robin. R. A. 2015. *Implementing Living Labs in Regulations: A Case Study of the Realization of The Green Village* (Unpublished Master's thesis). Netherlands: Utrecht University.

Ma, Mulan and Hassink, Robert. 2013. "An evolutionary perspective on tourism area development". *Annals of Tourism Research* 41: 89–109.

Mahoney, James. 2000. "Strategies of Causal Inference in Small-N Analysis". *Sociological Methods & Research,* 28(4): 387–424.

Martin, Ronald. 2009. "Rethinking regional path dependence: beyond lock-in to evolution". *Papers in Evolutionary Economic Geography,* 09.10. Utrecht (Netherlands): Utrecht University.

McCann, Eugene. 2011. "Urban policy mobilities and global circuits of knowledge: Towards a research agenda". *Annals of the Association of American Geographers* 101(1): 107–130.

Peck, Jamie. 2011. "Geographies of policy: From transfer-diffusion to mobility-mutation". *Progress in Human Geography* 35: 773–797.

Pidwirny, Michael, Bahbahani, Kalim and Pedersen, Shane. 2017. *Climate Change Challenges for Alpine Ski Resorts in Western Canada*. Department of Geography, Okanagan Campus, University of British Columbia, Okanagan Campus. Retrieved from: sreso21https://blogs.ubc.ca/michaelpidwirny/.

Resort Municipality Initiative Resort Collaborative. 2016. *RMI Funding: Building on Success*. Retrieved from https://www.google.com/search?client=firefox-b-d&q=RMI+Funding+building+on+success https://www.google.com/search?client=firefox-b-d&q=RMI+Funding+building+on+success.

Sheppard, Chris. 2017. *The Evolutionary Paths of Resort Governance: A Case Study of British Columbia from 1975 to 2015*. Unpublished report no.665. Burnaby: School of Resource & Environmental Management, Simon Fraser University.

Sydow, Jörg, Schreyîgg, Georg and Koch, Jochen. 2005. *Organizational Paths:Path Dependency and Beyond*. Mimeo, Faculty of Economics and Business Administration, Free University of Berlin.

Sydow, Jörg, Schreyîgg, Georg and Koch, Jochen. 2009. "Organizational path dependence: Opening the black box". *Academy of Management Review* 34(4): 689–709.

The Natural Step. 2019. Whistler Blackcomb. Retrieved from: http://www.naturalstep.ca/whistler-blackcomb.

UNWTO. 2013. "Developing New Destinations for Snow Tourism". *Euro-Asian Ski Resorts Conference*, Almaty, Kazakhstan, October 8–9, 2019. United NationsWorld Tourism Organization, Madrid, Spain. http://www2.unwto.org/en/event/first-euro-asian-ski-resorts-conference.

While, Aidan, Jonas, Andrew E. G. and Gibbs, David. 2004. "The environment and the entrepreneurial city: Searching for the urban 'sustainability fix' in Manchester and Leeds". *International Journal of Urban and Regional Research* 28(3): 549–569.

Williams, Peter W. and Gill, Alison M. 2017. "Towards decolonization, Indigenizing resort Governance in Canada". In: *Tourism and Political Change*. Butler, Richard and Suntikul, Wantanee (eds.). 2nd Edition. Oxford: Goodfellow, 75–90.

Witt, Ulrich. 1997. "Lock-in vs. critical masses: Industrial change under network externalities". *International Journal of Industrial Organization* 15(6): 753–773.

Wolfe, Roy. 1964. "Perspective on outdoor recreation: a bibliographical survey". *Geographical Review* 54(2): 203–238.

Definitions of acronyms

BC: Province of British Columbia, Canada

EEG: Evolutionary Economic Geography – A sub-discipline of economic geography focuses on interpreting and explaining how economic landscapes are transformed over time by specific mechanisms of change. This extends to studies of the forces shaping the governance systems affecting economic landscapes.

MRDT: Provincially sanctioned Municipal and Regional District Tax (MRDT) of 1987 that permitted an additional tax (above existing provincial hotel taxes) that resort municipalities could charge their hotel guests. Funds collected were earmarked to help support local communities operate convention centers and other tourism amenities

MURB: Federal government sponsored Multi Use Residential Building (MURB) program that provided tax shelter incentives for private sector investments in primarily condominium style residential real estate developments.

NDP: New Democratic Party – a socialist leaning political party in British Columbia.

RMOW: Resort Municipality of Whistler – BC first officially designated resort municipality.

RMI: Resort Municipality Initiative – A government funded designated resort municipality grant program funding public infrastructure projects designed to enhance the attractiveness of the community for tourists and local community.

TIDSA: Tourism Industry Development Subsidiary Agreement – a federal government funded and provincial supporting regional economic development in BC during the late 1970s and early 1980. A particularly ambitious portion of the program focused identifying and nurturing emerging opportunities for internationally competitive ski area development.

UNWTO: United Nations World Tourism Organization –providing professional development support and statistical information for encouraging the sustainable development and management of tourism destinations across the glove.

1.5

Skiing in the Catalan Pyrénées: Between Private Ventures and Public Rescue

Beltran Oriol and Ismael Vaccaro

Introduction

The panorama visible from the Coll del Triador, a mountain pass at 2,083 meters of altitude located on the border between Vall d'Àssua and Vall Fosca, offers a symptomatic image of the history of the ski resorts of the Catalan Pyrenees, on the southern slope of the mountain range. On its eastern slope, the old municipality of Llessui (now integrated in the municipal territory of Sort) accommodated in the mid-1960s one of the first stations linked to the development of winter tourism in Catalonia. The town of Espui (belonging to the municipality of Torre de Capdella), located on the opposite slope, was, at the beginning of this century, the stage of the region's last construction project until today, dedicated to alpine skiing: the Vallfosca Mountain Resort. The image presented today by both places, only 25 km apart, in the landscape of the high mountain, is devastating. The ruins of the abandoned buildings, the accumulated scrap and the aligned pylons of the lifts still standing show the collapse of two ventures separated by four decades, and so seem to demonstrate, full of melancholy, the destiny of ski ventures in Catalonia.

The Llessui resort was built as a result of a business initiative with some degree of local involvement. Inaugurated in 1966, at a time when winter sports were still an elitist and minority practice, it was open on and off until it shut down definitively in 1987. With 30 km of ski trails, the station reached a capacity of 3,750 skiers per hour. Unlike previous initiatives, it was already connected to certain real estate interests, a feature that from then on would be shared by all the enterprises associated with the winter tourism in the Catalan Pyrenees. Two hotels and a few apartment buildings were constructed around the station.

The small scale of the first station built in the Pallars Sobirà differs to a high extent from the characteristics and dimensions of the project in Vall Fosca. Although the winter resort was designed to avoid overcrowding (42 km of trails, a maximum of 4,500 skiers per day) and limit its landscape impact (the station, 3 km from the urbanized area, would be accessed by cable-car), it had in fact a considerable growth potential: there were plans to establish a future connection with the neighbouring station of Boí-Taüll that would have added up to a ski area of 3,400 ha. The resort associated with the station would include 900 apartments and a hotel complex with 2,200 beds, together with a convention centre for a thousand people, a commercial area, a spa-wellness area and a 25 ha golf course. In the valley of Manyanet, there were plans to build a luxury urbanization (another 290 apartments, together with a hotel with a hundred rooms) as well as a zip train covering 12 km and close to a thousand meters of difference in altitude. Whilst construction work started in 2004, the Vallfosca Mountain Resort programme was seriously disrupted by the financial crisis, until 2008, when it was definitively paralysed.

Nevertheless, the history of initiatives related to alpine skiing in Catalonia is by no means a series of business failures. Of the thirteen resorts built since the pioneering stations were established in the 1940s, ten are still open. However, the views described above of the Coll del Triador indicate the great difficulties that the ski industry has had to go through to become established in this region.

This article describes the historical deployment of ski resorts in the Catalan Pyrenees from the perspective of the political ecology of tourism (Douglas 2014; Mostafanezhad et alt. 2016; Nepal and Saarinen 2016; Stoddart 2012). It intends to analyse the complex interconnection between the political, economic, and ecological processes intervening in the development of this tourism sector, by taking into account the actions of its different agents (Fletcher 2016).

As we will see, most of the resorts in Catalonia have suffered significant economic crises that have either lead to their closure or that have been resolved through the investment of public funds in their financing. The prominence acquired by the authorities in this sector can not be considered merely accidental in the process of modernisation of the region. From our perspective, the support of the public sector to skiing (authorising new ventures, while contributing to the rescue of failed initiatives) should be interpreted as part of an induced tertiarisation programme, which is the main government project for these mountain areas.

The historical development of skiing in the Catalan Pyrénées

Although the practice of skiing in Catalonia dates back to the early 20th century and was linked to hiking, which already had a long associative tradition in the area, it was not until the arrival of the railway to the districts of the Eastern Pyrenees in the 1920s that the first ski infrastructures would be established. Through the organization of training courses and the celebration of championships, the hiking entities favoured some level of promotion of winter sports among people of urban origin. The first mountain stations with stable installations would be created after the Civil War, in the 1940s, coinciding with the development of the railway line linking Barcelona with Puigcerdà: La Molina (1943) and Vall de Núria (1947).

The expansion and consolidation of skiing in Catalonia, however, was linked to its conversion into a mass sport, with an increasing number of ski facilities from the 1960s and onwards (Jiménez 1999). The Pyrenean regions, home of the areas with the most favourable orographic and climatic conditions for the practice of winter sports, had at that time already been fully integrated into the regional market. The implementation of the road- and rail network as well as hydroelectric installations played a prominent role in the process of economic change in the country's mountain areas. The places where the new tourist infrastructures were located came to experience an intense transformation.

The closure of agricultural holdings and the migration to metropolitan areas, subject to an important industrial growth, together with the aging of the local population and the abandonment of large areas of mountain territory, contributed significantly to the deployment of a still incipient industry. Paradoxically, the same snow that for centuries had hindered the execution of economic activities and the very development of social life, was to be exploited as the main resource of a sector with great potential. According to the model previously developed in the Alps (Barrère 1981; Godde, Price, and Zimmermann 2000), winter tourism, which could benefit from large spaces freed from the previous productive pressure, was perceived as a factor of modernisation (a creator of employment and an income generator; an alternative to the demographic collapse). The same argument would be used in all the ski related ventures that have been developed since then in the Catalan Pyrenees.

The local socioeconomic conditions were not the only factor to promote the construction of eight new stations between 1964 and 1975. This development was also favoured by the generalisation of access to vacation and to private cars, which took place during the same period of time. The locations chosen for ski stations were no longer dependent on the development of the railway lines, and their radius was therefore significantly broadened. The proximity to the metropolitan area of Barcelona, the origin of most of the potential clients of the new infrastructures, would even contribute to the selection of some pre-Pyrenean locations, although they couldn't guarantee optimal conditions for winter sports.

Nonetheless, the governmental action had a decisive weight in this deployment. In 1963 the Law of Centres and Areas of National Tourist Interest was approved, an instrument aimed at promoting tourism activities within the framework of the economic planning policies of the time (the so-called Economic and Social Development Plans sponsored by the technocratic Francoism) (Galiana and Barrado 2006). Tourism, thus, became a prioritised factor in the national economy. With the intervention in the regulation and promotion of the sector, the authorities sought to encourage the creation of new centres of activity, to promote its orderly development and to stimulate private initiative and investment (through fiscal measures, public credits and legal advantages). The first stations of the Catalan Pyrenees were declared as Centres of National Tourist Interest, contributing to the promotion of the tourism sector in the area.

The stations built at that time, which combined a business component with some local roots and support, were Baqueira-Beret (1964), Llessui (1966, closed today), Masella (1967), Espot Esquí (1967), Port del Comte (1973), La Tuca (1974, closed at the end of the following decade),

Rasos de Peguera (1974, active until 2004) and Vallter 2000 (1975). Subsequent initiatives, built over a longer period of time, resulted from more particular circumstances, although they also received explicit support from the government (regional or municipal): Port Ainé (1986), Boí-Taüll (1991) and Tavascan (1991). See figure 12.

The development of the sector has also been linked to the 'democratisation' of skiing: the improved facilities of access to the stations, together with the more affordable costs of the material or even of the ski passes, contributed to widening the social base of winter sports. The number of users in Catalonia tripled during the last decade of the 20th century, going from 850,000 in the season of 1989–90 to 2,250,000 in the 2001–02 season (Gili 2003a). Today the alpine ski resorts of the Catalan Pyrenees have a total of 390 ski trails (with a 25 % increase in the last fifteen years), with an overall length of 477 km. The transport capacity of its 128 ski lifts amounts to 175,000 skiers per hour.

The construction of facilities for cross-country skiing has been parallel to the development of alpine resorts. In the Catalan Pyrenees, seven facilities were established exclusively for this modality from the end of the 1960s to the early 2000s: Sant Joan de l'Erm (1967), Lles (1970), Tuixent-la Vansa (1978), Aransa (1986), Guils Fontanera (1993) and Virós Vallferrera (2001), in addition to Bonabé, closed in 2000. Some of the alpine resorts include cross-country ski circuits in their own domains (as is the case of Baqueira in Pla de Beret, Rasos de Peguera, Espot Esquí, Port Ainé and Tavascan). However, the dynamics generated by this type of stations, in aspects such as their volume of activity or their economic impact, is not comparable to the alpine facilities. It is estimated that they accumulate only 1.7 % of the expenditure by users in the whole sector.

Different scales, different models

The ski resorts of the Catalan Pyrenees present a significant diversity in terms of their dimensions, a factor associated both with their physical expansion possibilities and the success of their business activity. The three categories in which they can be grouped (very large, medium and small stations, obviously in relative terms) imply different dynamics in relation to their development, their socio-economic impact and even their own survival capacity.

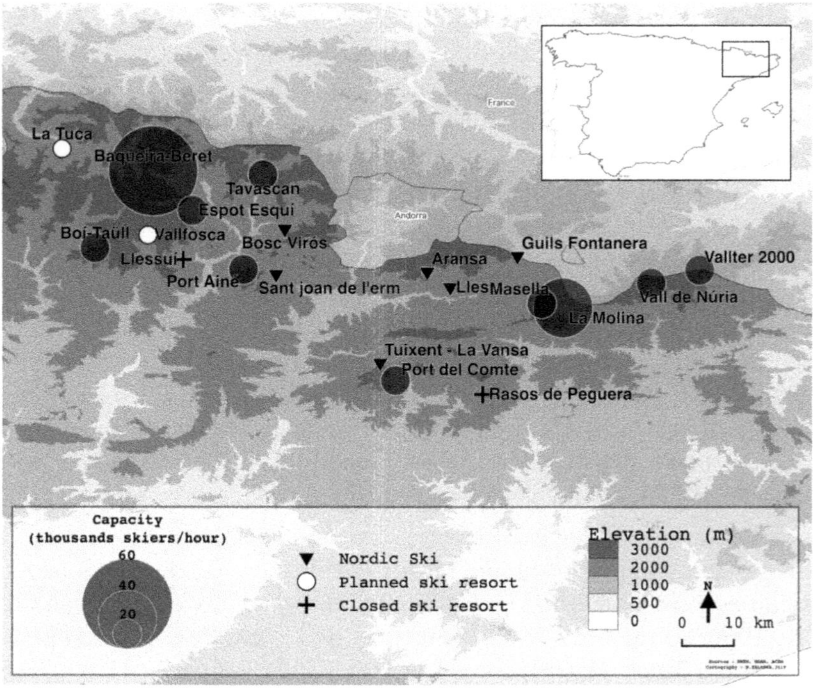

Fig. 12: Capacity and elevation of alpine ski resorts of the Catalan Pyrenees

Baqueira-Beret is the only Catalan ski station that can be considered as a large capacity station, which clearly differentiates it from the other resorts. As one of the first resorts created, its facilities have kept on expanding over the last 50 years. Both in terms of size and activity, this Val d'Aran station represents one third of the sector in Catalonia: a domain of 2,166 skiable hectares, 156 km of trails and a transport capacity of more than 60,000 skiers per hour. The fact that it has never needed direct public financing accounts for its business strength. Baqueira-Beret, however, has always received significant support from the Catalan public administration at many other levels, especially in the approval of its growth projects and its real estate initiatives (Gili, 2003a).

La Molina, Masella, Port del Comte and Boí-Taüll are stations that can be considered medium-sized: their ski areas include between 45 and 74 km of trails, and their ski lifts have a capacity to transport between 15 and 30,000 skiers per hour. Due to their dimensions, the activity of these resorts (in terms of labour, commerce and urban development) has effects beyond the municipalities where they are located, with an impact on the whole districts: Cerdanya, Solsonès and Alta Ribagorça.

The five remaining stations in the Catalan Pyrenees (plus the three closed stations, also to be included in this group), can be qualified as small: between 6 and 26 km of trails and a capacity to transport under 11,000 skiers per hour. Their direct impact is also more limited at the local level. Tavascan, due to its size, is a particularly paradigmatic case: it has the capacity to handle 2,600 skiers per hour and has only seven ski trails with a length of 6 km.

The economic impact of the stations located in the westernmost districts of the Catalan Pyrenees (Pallars Sobirà, Alta Ribagorça and Val d'Aran) is manifested above all through the hotel and construction sectors. Their distance from the metropolitan area of Barcelona (300 km or more), the home of most of its users, means that overnight stay is inevitable. By contrast, the stations of the Solsonès, the Berguedà and the Ripollès have from the beginning targeted a family audience, using the proximity to the city (less than half of the aforementioned distance) as an important commercial argument. It should be noted that, with the exception of Baqueira-Beret, which has traditionally hosted a large contingent of skiers from outside Catalonia (it has a permanent commercial office in Madrid and has implemented several campaigns to attract French and British skiers), the Catalan stations have been directed almost exclusively to the regional public (which also has other competitive offers in Aragon, southern France and, especially, Andorra). The expenditure made by skiers demonstrates the great difference between a ski day and ski with overnight stay: in 2003, the first modality accounted for 59 million euros, compared to 333 million euros for the latter. In the sector as a whole, the average expenditure per person per day was around 160–170 euros, a high budget compared to other forms of tourism (ACEM 2004).

The impact generated by the Baqueira-Beret station is evident not only in the Naut Aran, the municipality where it is located, but in the entire Eastern half of the district. Between 1975 and 2010 hotel rooms in Naut Aran multiplied by 4.7. At district level, the increase in this same period of time was 220 %, reaching 6,400 beds (the district had 10,200 inhabitants registered on 2010). An impressive 45 % of the Aranese hotel accommodations are concentrated around the station, an area where also approximately 48 % of the homes are secondary residences (5,450 in the entire Val d'Aran in 2011).

Tab. 2: Alpine Ski Resorts in the Catalan Pyrenees (2018).
Source: B (ACEM)

Ski resorts	District	Opening Year	Closure Year	Base (m)	Summit (m)	Lifts	capacity (skiers/ hour)	Runs	km	Snow Guns
Baqueira-Beret	Val d'Aran	1964	no	1 500	2 510	35	60 144	104	156	629
La Tuca	Val d'Aran	1974	1989	1 560	2 220	6	sd	21	28	0
Boí-Taüll	Alta Ribagorça	1991	no	2 020	2 751	11	19 690	45	45	285
Vallfosca	Pallars Jussà	(2008)		1 800	2 600	9	sd	13	42	sd
Port Ainé	Pallars Sobirà	1986	no	1 650	2 440	10	10 810	25	27	126
Espot Esquí	Pallars Sobirà	1967	no	1 500	2 500	6	8 220	22	25	163
Tavascan	Pallars Sobirà	1991	no	1 750	2 250	3	2 600	7	6	0
Llessuí	Pallars Sobirà	1966	1987	1 445	2 430	7	3 750	22	30	0
Port del Comte	Solsonès	1973	no	1 700	2 400	16	16 060	37	50	256
Rasos de Peguera	Berguedà	1974	2004	1 850	2 050	6	2 600	10	8	0
La Molina	Cerdanya	1943	no	1 700	2 445	15	28 380	61	68	480
Masella	Cerdanya	1967	no	1 600	2 535	16	14 560	65	74	558
Vall de Núria	Ripollès	1947	no	1 964	2 252	5	4 560	11	8	88
Vallter 2000	Ripollès	1975	no	1 959	2 535	11	9 722	13	19	86

Boí-Taüll has registered an extraordinary increase in hotel beds during its twenty-five years of history. Although the area where it is located already had a long tourism tradition (the Boí Valley has an old spa centre, one of the main entrances to the National Park of Aigüestortes and Estany de Sant Maurici, and a number of Romanesque churches that were declared World Heritage by UNESCO in the year 2000), during this period of time it doubled its accommodation offer, which represents about 80 % of the offer of the entire district. The number of secondary residences went from 186 to 1,327 between 1991 and 2011, concentrating 70 % of the district offer. The resort, located in the Pla de l'Ermita (2 km from Taüll), accommodates 1,200 beds distributed among three hotels, an apartment hotel, and two residential complexes with 60,000 m^2 of built-up land.

Compared with the greater and more expanded impact of Baqueira-Beret and the smaller and more concentrated one of Boí-Taüll, the small size of the Pallars Sobirà stations corresponds to a more limited footprint at the municipal level, however, with some repercussions in neighbouring towns. Its direct impact is very limited: the three municipalities that host the Pallars complexes (Espot, Lladorre and Rialp) account for 40 % of the hotel beds in the district (and only 15 % of secondary homes, but other larger towns with a wider range of services still benefit from the existing tourist flow). In comparative terms, however, its offer of tourist accommodation is much more limited than that of larger stations: there is an average of 3.8 hotel beds per km^2 in the three municipalities mentioned, 6.4 beds/km^2 in Vall de Boí and 11.3 beds/km^2 in Naut Aran; 1.7 secondary residences per km2 in the municipalities of the Pallars Sobirà, 6.0 in the Vall de Boí and more than ten in the municipality of Baqueira-Beret.

From snow to cement

In the sector of the so-called winter tourism, the number of beds in hotel accommodation and second-home housing does not constitute a mere indicator of the volume of activity associated with each venture or of the number of skiers that make use of its facilities. Studies on the political economy of skiing coincide in underlining the low profitability of the investments made in the construction of the infrastructures necessary for its practice, mainly the ski lifts and associated services that, due to their high number and the low return they generate, they are not amortized until after many years. In other words: the commercial exploitation of the ski area itself, that is, the strict transport of skiers, is not an economically profitable activity. In this context, the business of skiing lies in the indirect and induced impact of the activity, mainly the construction of houses destined to second residences and the exploitation of the associated tourist and commercial services: accommodation and restaurant businesses, fundamentally (Campillo and Font 2004). The promoters of the stations are therefore oriented towards expanding their field of action through the offer of complementary services and, especially, through the promotion and sale of real estate products (López Palomeque 1996).

For this reason, the development of mountain tourism, in particular that associated with the practice of alpine skiing, has been considered as an important factor in the urbanisation of the affected rural areas. Not only has this activity contributed to the articulation of marginal areas through the deployment of the road network or that these have specialised in leisure offers aimed basically at a public of urban origin; the increase in the number of tourist accommodation places as well as secondary residences can be considered literally as part of an intensive process of urbanisation of the rural territory.

The statistics on the increase of construction in the Catalan Pyrenees in recent decades reflect different patterns related to the size of the ski stations as well as other more specific circumstances in each case: such as the existence of urban areas near the resorts or the layout of the road network (Lasanta, Beltran, and Vaccaro 2013). Some resorts such as Baqueira-Beret or Boí-Taüll concentrate this tourist construction around the facilities themselves (and the comfort associated with the proximity between accommodation and slopes thus becomes an important commercial argument), while in other cases the new constructions cover a wider geographic radius.

The figures are eloquent: of the 14,129 dwellings that were registered at the beginning of this century in all the Western districts of the Catalan Pyrenees (Alta Ribagorça, Val d'Aran and Pallars Sobirà), 40 % corresponded to second residences built in the course of the last three decades. In the same period, three out of every five newly built homes were destined for this use. The growth of the rest of tourist accommodation tripled between 1975 and 2010. Thus, with a population of about 20,528 inhabitants in 2017, the three districts had a capacity of accommodation for 65,000 tourists.

The ownership of the companies that promote the facilities for skiing demonstrates the importance of the real estate business in the initiatives. The first ski stations were developed by a small number of investors who, in many cases, had personal links to the construction locations. In the mid-1960s, when the development of the sector had not yet set off, this performance was interpreted as an expression of the entrepreneurial character, and even of the personal charisma of its protagonists. In contrast, the ownership of the most recent initiatives, as well as the promotion of larger complexes, are now associated with investment groups or large real estate companies.

The project of Vallfosca Mountain Resort, for example, was promoted by Fadesa, one of the leading Spanish real estate groups (of Galician origin), which at the end of 2007 merged with Martinsa, another real estate company based in Madrid. An 80 % of the shares of the Boi-Taüll Resort, which belonged to an insurance company and an investment bank, was acquired by the real estate group Nozar (owned by a Mexican businessman) in 2004, at a time when 32,000 of the total 90,000 m^2 planned of this winter complex were still to be constructed. Most of the shares of Baqueira-Beret (74 %), on the other hand, are controlled by the Serra family through three companies, among which the insurance company Catalana Occidente stands out, where the aforementioned family is the reference partner. CaixaBank also counts with a share (11.7 %) in addition to further minority shareholders.

The ski crisis and the intervention of the public administration

Half of the ski stations that remain open today in the Catalan Pyrenees belong to the Generalitat de Catalunya following the intervention of the regional government to cope with the financial collapse of the stations and thus prevent their closure. The strong role of the Administration in ski tourism – to the point that it's considered a subsidised economic sector – can not be interpreted as a merely accidental result of its historical evolution. The different stages of this process, the actors that have intervened in it, the scenarios where it has taken place, the arguments put forward and, especially, the socio-economic and environmental results, offer sufficient elements of judgment to identify the nature, interests and intentions of its agents. The close relationship between the initiatives associated with alpine ski resorts and real estate interests indicates some essential aspects of this process. However, the characteristics of the process and the logic of the development of the sector deserve a deeper analysis.

The increasing presence of the public sector in ski tourism during the last decades (compared to the limited role it plays in the rest of tourism sectors) is not exclusive to the Catalan Pyrenees, but has rather become a common phenomenon in Spain. The station of Sierra Nevada, for example, belongs since 2004 to the Junta de Andalucía after twenty years of receiving injections of public capital, in the absence of private investment in what is considered an important economic engine of the province of Granada. The Government of Aragon, for its part, created at the beginning of the century a 50 % owned company with a regional financial entity to manage five of the seven stations of the Pyrenees of Huesca that had been declared bankrupt. The intervention included the creation of the largest ski area in the country as well as an important urban development project in the Castanesa valley (involving the construction of 2,530 dwellings and 3,000 hotel beds).

In the Catalan Pyrenees, the financial precariousness of a large part of the existing stations has been interpreted as the result of accumulated indebtedness as well as the need to face new investments (Nel·lo, 2006). In addition to the difficulty of increasing the number of users (due to competition from other destinations, among other reasons), the irregularity of snowfall has been an important argument in favour of modernising the infrastructures. In this context, the installation of snow cannons for the production of artificial snow, with the purpose of guaranteeing the season regularity and skiing conditions, has become a priority. The influence of climate change is an aspect that has rarely been taken into account in the diagnoses of the economic viability of the stations. However, the experts perceive it as a decisive factor for the future development of the sector: it does not only affect the instability or the lack of snow at lower and lower levels, but it is also detrimental to the conditions required for the production of artificial snow. The public subsidies destined for this purpose became, at the beginning of the 1990s, the main support policy for the sector (López Palomeque, 2009). The use of artificial snow has increased since then: whilst in 1994 the stations of the Catalan Pyrenees accounted for 560 snowmaking devices, today there are nearly 2,671 cannons.

The economic crisis that began in 2008 would further aggravate the financial situation of the sector. Apart from the investments necessary to modernise the facilities, the operating costs increased significantly (mainly due to the increase in energy costs), whilst also users faced higher costs. The practice of skiing is not a cheap activity: in addition to the price of the ski passes, there are also expenses in relation to travel, stay and material.

The Generalitat de Catalunya had already intervened in the stations of La Molina (1985) and Vall de Núria (1986), which had lacked capacity to attract, on their own, the investments required to update their infrastructures. In the first decade of this century, both Espot (2006) and Port Ainé (2007) declared bankruptcy, whilst Vallter 2000 went bankrupt in 2013. After years of receiving public loans from the Institut Català de Finances (ICF), the regional government (through the public company

FGC) acquired the property and assumed the management of these five resorts. Tavascan, on the other hand, was from the beginning a municipal initiative and is still owned by the town council of Lladorre. The Boí-Taüll station has received public credits at different times (to face debts with other private companies) and has a mixed ownership at present, although its management is still in private hands. Only Port del Comte, Masella and Baqueira-Beret remain with an entirely private capital.

The process by which the regional Administration has acquired its current relevance in the sector is, however, more complex. This can be exposed by looking into the case of the two Pallars Sobirà stations. At the end of 2004 the Espot station accumulated a debt of over 5 million euros with the ICF, while Port Ainé's debit amounted to 13 million euros. In Espot, the aforementioned public financial entity requested the concession of the property, to temporarily cede the management to a company formed by several town councils and tourism entrepreneurs of the district. The company was auctioned in 2005, leaving the rights in the hands of the Generalitat. The government declared that it did not intend to become an actor in the sector and that it wanted to limit itself to protecting the station, providing aid in order to avoid having to close the facilities. Its objective was to study formulas to favour the entry of new partners. After a brief period of management by a society formed by municipalities and entrepreneurs, the management of the station was finally assigned to the public company FGC. The intervention in the pioneering stations of La Molina and Vall de Núria, twenty years earlier, was also proposed as a temporary solution, with the aim of "cleaning up the stations to return them to the market" after a short time (and not to create a public holding, as had been done in Aragon). Today, the Generalitat has accepted that the role it exercises in the sector will not change in the short term.

The different financial capacity of the stations is indicative of the weight of the real estate activity in each case as well as the different size of the initiatives. In fact, only those that have not failed in the real estate business have been able to avoid public rescue. As we have seen, the scale of the undertakings associated with alpine skiing is important since the social, economic and ecological footprint of each station is proportional to its size. But it is also relevant to understand the economic survival capacity that each tourist complex has shown over time as well as to be able to foresee its future evolution. In this sense, the stations of bigger dimensions (medium and large) have had the best chance of profiting from their investments through products that are parallel to the practice of sports. By contrast, the smaller ones tend to end up in debt until they are rescued, by the Administration either with a one-off financial aid or even with the takeover of ownership and management of the facilities.

Although they formally respond to different company names, the urban promotions and hotel establishments are often closely associated with the same business and financial interests that once promoted the stations (or that acquired them, in some of the successive crises that have marked their path). Rescue and public intervention, however, only affects the companies that own the ski lifts, releasing private capital from the burden so that it can continue benefiting from the more profitable activities within the sector.

According to the Catalan Administration, in addition to the financial restructuring of companies and the injection of investments for the renewal of facilities, the promotion of synergies between stations, their specialisation according to user segments, actions to broaden the social base of skiing, or the organisation of sports competitions, should help the rescued stations to stop accumulating debt. De-seasonalisation has been a prioritised strategy, through the reconversion of stations into mountain centres, with a diversified offer of activities, contemplated not only for the winter season. In this sense, the stations managed by FGC are those that in recent years have made the greatest effort to generate and promote a broad programme of summer activities (active tourism and/or tourism aimed at a family audience: from BTT circuits to wildlife observation programs). The impact, however, is at present not very significant. In La Molina, for example, after reaching record figures in the last winter season (380,000 skiers), there was an aim to reach 40,000 visitors during the summer. In any case, technical and managerial measures have been prioritised to cope with the crisis as opposed to measures that could represent an effective transformation of the model (Gómez, Armesto, and Cors 2016).

Ultimately, although the public rescue interventions have reduced the volume of losses, the stations subjected to interventions by the Administration still maintain negative balances, whilst the more profitable complexes continue generating profits for their shareholders. According to the balance of the 2017–2018 season, the activity of the five stations of the Catalan Pyrenees managed by FGC implied losses valued at 7.6 million euros. Baqueira-Beret, on the other hand, closed this campaign with declared net returns of 4.5 million.

Ski tourism: economic driver or speculative operation?

Through the approval of new projects (either to build new stations and tourist complexes or to expand existing facilities) as well as the rescue of bankrupt stations, the Generalitat de Catalunya has, over the nearly forty years of regional management, shown an unconditional support to ski tourism. The repeated failures of the sector's business initiatives and the pessimistic forecasts of global warming would justify a more cautious attitude. The analysis of skiing as an economic activity also highlights the lack of profitability of most existing stations, which require financial assistance to maintain their operations. However, all the political parties that have governed the territorial policies of the region have not only coincided in supporting this sector, but have also shown a direct opposition to any questioning of it.

The Administration justifies its support to the ski resorts with their social benefit, as they make up "a fundamental sector for the mountain districts", and "the only hope of the area during the winter months", according to affirmations of different political leaders. In this sense, recognizing that the rescued facilities are not by themselves a profitable investment, it is considered that the activity generated by the ski stations as a whole could represent a high percentage of the labour market and the income in the area, especially during the winter season.

According to data released in 2017, the Generalitat estimated the direct territorial impact of the Catalan stations as a whole at 250 million euros, an amount to which has to be added another 500 million generated by the indirect and induced impact. At the same time, each job created at a mountain station was calculated to generate up to five jobs in the surrounding area. According to a research conducted by the UPF Sports Lab the five publicly owned facilities managed through FGC involve the maintenance of 2,194 jobs, a net economic impact of 121 million euros and a total impact on the production of economic activity valued at 223 million (Bosh et al., 2018). These figures would justify the public investment made in terms of its social profitability, aimed at preventing the closure of the winter complexes and the damages that this would entail in their respective areas of influence: for each additional euro of expenditure in the rescued stations, two euros would be gained, and each additional million invested by the public company FGC would generate up to 20 jobs.

The balance offered by the deployment of the ski tourism industry in the Catalan Pyrenees, half a century after the opening of the first modern stations, presents apparently contradictory characteristics that need to be analysed in their complexity. Throughout this time, the relationship between the interests of private initiative and the actions of public institutions has played a fundamental role. The examination of this relationship raises several questions regarding the future both of the sector and of the mountain districts that accommodate it.

The contrast between stations that require continued support of public funds to maintain their activity and others that continue to generate business benefits is a key feature of this situation. As previously highlighted, the commercial exploitation of the stations does not under any circumstances cover the costs generated by the facilities for the practice of alpine skiing. In this context, only the largest stations with the greatest potential for expansion and economies of scale have been able to justify the maintenance of their most profitable activities, those that take place in parallel with the sale of ski passes for the use of the slopes: the offer of complementary services and, especially, the sale of real estate products.

Baqueira-Beret is the main example of successful initiatives in the sector. The station has registered a continuous growth throughout its trajectory. Although its main expansion project, the expansion to the Pallars Sobirà, announced in 1998 (which was intended to incorporate 1,400 ha of ski area, consolidate a new access road from this district and build 450 tourist homes), ceased to be a priority in 2015, its current dimensions place it as the station with the most kilometres of tracks in Catalonia and the largest ski area in the Pyrenees. All in all, Baqueira-Beret continues to open slopes and lifts every year. Next season it will add another 8 skiable kilometres to its facilities that will allow it to raise its maximum level to 2,600 m. The company that owns the winter complex has set its mid-term goal to reach 200 km of trails instead of the current 164.

The continued growth of the ski area and its associated infrastructures not only allows for a more powerful and attractive offer for skiing, but also justifies the expansion of real estate, hotels and related service projects. The urbanised centres of Baqueira, Tanau and, Val de Ruda concentrate an important part of the 2,590 secondary residences (according to 2011 data) and the 2,900 hotel beds (in 2016) of the Naut Aran municipality. Val de Ruda is, for now, the last major real estate operation in Baqueira-Beret. The initial project included the construction of three luxury hotels, an urbanisation with 500 apartments, 46 single-family houses, an underground car park for 1,500 vehicles and a cable car up to 1,800 meters. Despite its high price (5,500 euros/m^2), work began in 2004 with 40 % of the apartments already sold.

In Boí-Taüll, the growth potential of the station and that of its associated real estate was the main argument for the acquisition of the majority of shares by the Nozar group in 2004. The expansion of the 550 ha ski area and the 43 km of slopes to reach 1,000 hectares and 100 km was considered "essential to make the complex competitive". The operation, valued at 68.5 million euros, was also aimed at the construction of new luxury hotels that would increase the number of rooms in the resort by 1,500. The construction of a station in Vall Fosca, as we have seen, was also intimately associated with the intensive urbanisation of the Espui and Manyanet areas. The impact of the financial crisis on the real estate sector was precisely the main reason for the interruption of the project in 2008.

In contrast to these large establishments, smaller stations tend to have a more fragile financial situation and more limited growth possibilities. The historic complexes of La Molina and Vall de Núria, conceived at the time of opening without a direct connection to real estate interests, have been managed by the public administration for more than three decades. The rescue of the stations of Espot, Port Ainé and Vallter 2000 by the Administration coincided with the fact that they reached their threshold of profitability. In all these cases, the need for continued public financing indicates that this problem does not correspond to a specific situation, but that it is rather of structural nature.

It is significant that the intervention of the Generalitat in the sector has not been directly criticised by the private capital stations, considering the alteration to the conditions of free competition that it implies. However, in addition to contributing to the promotion of winter sports and the general maintenance of the sector, it should be remembered that the rescued stations do not compete directly with the private stations, as most of them are located in different districts and are aimed at other market segments. Ultimately, their supposed disadvantage ends up being an argument in favour of demanding, from the Administration, fewer obstacles to their growth plans and other forms of indirect support (promotion, improvements to the road network, etc.).

However, the contribution of public resources in the ski industry is not limited to government support to the stations through financial rescue or terms to carry out hotel and real estate projects. An aspect that is not usually considered in the analysis of this specific sector is that ski areas tend to be located on public land. In the case of the Catalan Pyrenees, the places that present the best conditions (relief and climate) for the practice of winter sports are always located in high mountain areas. As a result of its historical use as pasture areas, and unlike the sectors of the territory that were heavily used for agriculture, the zones of greater altitude formed during centuries a patrimony of communal ownership. The State intervention in peripheral rural areas during the 19th century would end with the assignment of the old communal property to the municipalities (Beltran and Vaccaro, 2014a).

The large territorial areas required for the slopes and ski lifts are a particular feature of the ski industry that specifically conditions the social process necessary to implement this type of initiative. For this reason, the stations always end up in areas of public domain, while the construction of houses and hotels is developed on privately owned patches of land. The territorial needs of alpine skiing, therefore, links this big capital activity to the only territorial reserves of Western Europe, the old communal areas, currently managed as municipal land.

The promoter of Baqueira-Beret obtained in the past the administrative concession to develop the ski area of the future station in the old communal mountain of Vaquèira (López Palomeque and Majoral, 1982). The municipal officials alleged at that time that the planned activity was perfectly compatible with the continuity of the pastoral use of the area by the local population. At the same time, the farms of the sector where the urban part of the station was to be built at 1,500 meters were acquired from private individuals. The agreement of the company with the city council of Naut Aran was subsequently updated and currently includes the deposit of 5 % of the turnover of the season to the municipality.

The Vallfosca Mountain Resort intended to apply a similar formula in its expansion project towards the Manyanet valley. This company won the tender for the use of the Llevata mountain (1,300 ha, one thousand of which are skiable) for a period of 50 years and at a price of 30,500 euros per year (which would increase by 20 % once the ski lifts had been built), committing to not developing any activity during the summer, to allow the continuity of the livestock exploitation. The small value of this amount contrasts with the 100 million budgeted for the first phase of the project, which included the acquisition of land to be urbanised, among other items.

The ski industry is thus situated in a complex relationship between public goods and private interests. The State has facilitated ventures for the creation of ski resorts in the high mountain areas, aiming to create poles of tourist attraction in economically marginal areas, favouring the connection between the great private capital with the old communal assets. At a later stage, when commercial exploitation ceases to be profitable, private capital has no longer any interest in maintaining the stations and the Administration is obliged to rescue them to avoid the negative social impact of their closure. Where there is still room for business, especially in terms of urban growth, the stations will remain in the hands of private enterprises.

Conclusions. The role of skiing in the tertiarisation of the high mountain areas

In February 1943, the first ski lift in the entire Iberian Peninsula was opened in the Fontcanaleta area (Cerdanya). The perspective offered by the 75 years that have passed allows us to assess the results of the winter sports industry in the Catalan Pyrenees. Despite their relatively small dimensions and their localised impact, ski resorts have played an important role in the process of tertiarisation of the high mountain districts, a process that has led to an authentic redefinition of the rural world (Vaccaro and Beltran, 2007; Woods, 2007). The tourist use of cultural heritage (through museums, fairs, monuments and local-based food products, among others) and the values of the landscape (where the massive deployment of protected areas – 40 % of the land of the whole region – has been decisive) also constitute prominent aspects of this evolution (Beltran and Vaccaro, 2014a; Vaccaro and Beltran, 2008).

Today, the economic transformation of these districts is an unquestionable reality. In the 1990s, the demographic data show an inversion of the negative trend that had been worsening since the 1960s, but that had actually started more than a century ago. The incipient recovery was associated with a certain rejuvenation and dynamism of the population, although the statistics also pointed to a concentration of these improvements in the better communicated areas located at the bottom of the valleys, as opposed to the depopulation of large areas of the upper regions (Nel·lo, 1997). The service sector has become the main sector of activity in the region, with an important weight of tourism.

These changes seem to be at the source of the apparent demographic recovery.

The ski tourism has some specific features that determine its contribution in the framework of the economic transformation of the Catalan Pyrenees. First of all, the direct occupation that it generates is directly linked to the functioning of the stations, that is, it is concentrated during a few months every year, which in addition are unstable: the campaigns are relatively short and changeable. Secondly, the importance of real estate products associated with the economic exploitation of skiing has other implications. Thus, an important part of the activity is implemented in the short term, that is, in the time dedicated to the construction and sale of new property, and comes to an end once this is completed. At the same time, the effects produced by the impact of second homes on the market, destined to a population sector with high purchasing power, favour the rise in prices and harm especially young people, whilst it also makes the arrival of new permanent residents difficult. The apartment buildings and the sectors where tourist homes are consolidated, on the other hand, generate certain revenues for the municipal entities at first (through the construction licenses), but they end up giving way to ghost neighbourhoods and housing developments, occupied only a tiny portion of the time, making service provision difficult at the local level (Gili, 2003b).

Neither is the exploitation of complementary services at the stations favourable for the local population and their initiatives. According to Campillo and Font (2004), the low profitability of the commercial exploitation of ski areas favours the stations' dependency on the internalisation of the revenue of the business chain generated by skiing (accommodation, restaurants, businesses, rental of material, etc.), with short-term economic effects. As ski stations extend their field of action, the possibilities of the local population to obtain benefits from their own initiatives are limited, and consequently, so are their opportunities for progress and well-being.

Beyond its economic effects, the tendency of stations to expand, the increase in size of the ski areas and the real estate promotions comes increasingly into collision with the environmental awareness of the population and with the demarcation of nature conservation areas. Parks and ski slopes compete for space very close to one another, although they are oriented to completely different tourist uses: extensive in the first case, intensive in the latter (Clarimont and Vlès, 2007). Despite being considered by the Administration and specialists as complementary sectors (following the pattern of alpine tourism), which would favour an "integral development" and a continuous tourism activity throughout the year, they are not of benefit to the same actors. The environmental conflicts generated by the stations (due to their high spatial needs and, increasingly, water for the snow production) have become a common dynamic in the Catalan Pyrenees.

In short, through its contribution to the economic transformation of the Pyrenean districts the ski sector can be considered an important gentrification factor in these mountain areas, implementing an unsustainable model both in social and environmental terms (Lindberg, Anderson, and Dellaert, 2001; Perlik 2011). Despite the symptoms of unease that it causes, and the costs involved for public budgets, the Government and most of the political groups show unconditional support for its continuity. It would seem, as stated by a regional political leader (when asked about the continuity of funding for skiing, given the evidence of climate change), that the ultimate reason for this position is the absence of an alternative project to tertiarisation, which is the programme allocated to this mountain region.

References

ACEM. 2004. *Pla estratègic del turisme de neu a Catalunya*. Barcelona: Associació Catalana d'Estacions d'Esquí i Activitats de Muntanya.

Barrère, Pierre. 1981. "La haute montagne, bien de nature où objet de profit". In *Supervivencia de la montaña*. Madrid: Ministerio de Agricultura (eds.). Espana. 459–475.

Beltran, Oriol and Vaccaro, Ismaël. 2014a. "Between communal herding and state parcellation: The conflicting territorialities of the Spanish Pyrenees". In *Negotiating Territoriality: Spatial Dialogues between State and Tradition*, Dawson, Charles Allan, Zanotti, Laura and Vaccaro, Ismaël (eds.). New York: Routledge, 21–35.

Beltran, Oriol and Vaccaro, Ismaël. 2014b. *Parcs als comunals. La patrimonialització de la muntanya al Pallars Sobirà*. Barcelona: Generalitat de Catalunya.

Bosch, Julia, García, Jaume, Murillo, Carles and Raya, Joseph M. 2018. "*Anàlisi de l'impacte econòmic de les estacions de muntanya de FGC*" [no publicado]. Barcelona: UPF Sports Lab.

Campillo, Xavier and Font, Xavier. 2004. *Avaluació de la sostenibilitat del turisme a l'Alt Pirineu i Aran*. Barcelona: Generalitat de Catalunya.

Clarimont, Sylvie and Vlès, Vincent. 2007. "El turismo en los Pirineos. ¿Un instrumento de desarrollo sostenible". In *Ecología política de los Pirineos. Estado, historia y paisaje*, Vaccaro, Ismaël and Beltran, Oriol (eds.). Tremp: Garsineu, 77–99.

Douglas, Jason A. 2014. "What's political ecology got do with tourism?". *Tourism Geographies* 16(1): 8–13.

Fletcher, Robert. 2016. "*Tours caníbales* puesto al día: La ecología política del turismo". *Ecología Política. Cuadernos de Debate Internacional* 52: 26–34.

Galiana Martin, Luis and Barrado Timon, Diego. 2006. "Los Centros de Interés Turístico Nacional y el despegue del turismo de masas". *Investigaciones Geográficas* 39: 73–93.

Gili, Mercè. 2003a. "L'ampliació de l'estació d'esquí de Vaquèira. Espais protegits i model de desenvolupament a l'Alt Pirineu i l'Aran". In *Aquí, no! Els conflictes territorials a Catalunya*, Nel·lo Colom, Oriol (eds.). Barcelona: Empúries, 227–251.

Gili, Mercè. 2003b. "Las viviendas de segunda residencia. ¿Ocio o negocio?". *Scripta Nova. Revista Electrónica de Geografía y Ciencias Sociales* 7(146), http://www.ub.es/geocrit/sn/sn-146(052).htm (last access 2 march 2020)

Godde, Pamela, Price, Martin and Zimmermann, Friedrich. M. (eds.). 2000. *Tourism and development in mountain regions*. Oxon: CABI Publishing.

Gómez Martín, M. Belén, Armesto López, Xosé A. and Cors Iglesias, Martí. 2016. "Crisis económica y estaciones de esquí en las comarcas catalanas de la Val d'Aran, Pallars Sobirà y Alta Ribagorça". In *Turismo y crisis, turismo colaborativo y ecoturismo*, Blàzquez, Macià, Mir-Gual, Miquel, Murray, Ivan and Pons, Guillem X. (eds.). Palma: Societat d'Història Natural de les Balears, 213–223.

Jiménez, Sole. 1999. *Patrimoni i turisme de muntanya a Catalunya: El cas del Pallars Sobirà. Una aproximació des de l'Antropologia Social*. Lleida: Universitat de Lleida. [PhD thesis].

Lasanta, Teodoro, Beltran, Oriol and Vaccaro, Ismaël. 2013. "Socioeconomic and territorial impact of the ski industry in the Spanish Pyrenees: mountain development and leisure induced urbanization". *Pirineos. Revista de Ecología de Montaña* 168: 103–128.

Lindberg, Kreg, Anderson, Tommy D. and Dellaert, Benedict. 2001. "Tourism development. Assessing social gains and losses". *Annals of Tourism Research* 28: 1010–1030.

López Palomeque, Francisco. 1996. "Turismo de invierno y estaciones de esquí en el Pirineo catalán". *Investigaciones Geográficas* 15: 19–39.

López Palomeque, Francisco. 2009. "Neu i estacions d'esquí i muntanya". In *Atles del Turisme a Catalunya. Mapa Nacional de l'oferta i els productes turístics*. López Palomeque, Francisco (eds.). Barcelona: Generalitat de Catalunya.

López Palomeque, Francisco and Majoral, Roser. 1982. *La Vall d'Aran. Medi físic i transformació econòmica*. Barcelona: Caixa de Catalunya.

Mostafanezhad, Mary, Norum, Roger, Shelton, Eric J. and Thompson-Carr, Anna (eds.). 2016. *Political Ecology of Tourism. Community, power and the environment*. New York: Routledge.

Nel·lo Colom, Oriol (eds.). 1997. *Bases per al Pla Estratègic de l'Alt Pirineu Català*. La Seu d'Urgell.

Nel·lo Colom, Oriol (eds.). 2006. *Pla director de les estacions de muntanya*. Barcelona: Generalitat de Catalunya.

Nepal, Sanjay K. and Saarinen, Jarkko (eds.). 2016. *Political Ecology and Tourism*. New York: Routledge.

Perlik, Manfred. 2011. "Alpine gentrificatiTn: the mountain village as a metropolitan neighborhood". *Revue de Géographie Alpine* 99(1). DOI: 10.4000/rga.1370.

Stoddart, Mark C. J. 2012. *Making Meaning Out of Mountains: The Political Ecology of Skiing*. Vancouver: University of British Columbia Press.

Vaccaro, Ismaël and Beltran, Oriol. 2007. "Consuming space, nature and culture: patrimonial discussions in the hyper-modern era". *Journal of Tourism Geographies* 9 (3): 254–274.

Vaccaro, Ismaël and Beltran, Oriol. 2008. "The New Pyrenees: Contemporary conflicts around patrimony, resources and urbanization". *Journal of the Society for the Anthropology of Europe* 8(2): 4–5.

Woods, Michael. 2007. "Engaging the global countryside: globalization, hybridity and the reconstitution of the rural place". *Progress in Human Geography* 31(4): 485–507.

II

Strategies and Forms of Adaptation of Resorts to Global Changes

2.1

Small Mountain Resorts between Change and New Expectations of Demand

Andrea Macchiavelli

Introduction

Mountain tourism is certainly not the only, and perhaps not the most important, type of tourism involved in global changes and in particular in climate change: the erosion of the coasts and the great risks that the small islands are running as a result of the increase in sea level are significant examples. However, in the last few decades many factors of change of different origins have been concentrated in Alpine tourism; they initially produced a condition of uncertainty, after decades of stability, and then were followed by attempts of interpretation and by initiatives oriented to preserve the loyalty of tourist, but not always coherent with the specific resources of the mountain and with the conditions of the alpine tourist offer. Even if a coherent and organic project is not yet visible at the country level, or at the level of the most important tourist resorts, the last few years seem to offer positive signals to the alpine destinations, opening new perspectives also for small resorts, more affected by the scarcity of snow because usually located at low altitude.

This contribution intends primarily to explore these prospects, with particular reference to small resorts, indicating the conditions that could strengthen their role in the Alpine tourism market; the analysis cannot disregard the context conditions that affect the whole mountain tourist offer and therefore also the main destinations, which, in turn, are called to interpret the change starting from their own conditions and resources.

The aim is primarily to capture the evolution of the market in the main regions of the Alps (fig. 13): this analysis was made possible thanks to the data published periodically by ASTAT, the statistical service of South Tyrol, which however offers overall data on tourism at the regional level; some of these regions (e.g. Bavaria or Salzburg) are not only characterized by an Alpine offer and for this reason they must be evaluated in their context, while the data of the regions with total mountain prevalence are more clarifying (e.g. Tyrol, South Tyrol and Trentino).

Further information comes from the Italian Alpine regions where it was possible to acquire more detailed and more recent data; the emerging trend is fully consistent with international trends and therefore, although the reference is to Italian regions, where the offer is certainly different from that of other Alpine countries (France first), the data, accompanied by an empirical consideration of the context, are sufficiently reliable to interpret the scenario of change that reality offers.

Starting from this analysis and the resulting interpretation, the last part of the contribution offers planning hypotheses for small resorts which, differently from the past, could today find their own specificity in the Alpine tourism offer market: the prospect is however very demanding and requires in many places a strong strengthening of the tourist governance system.

Evidence from mountain tourism: positive trends

After the strong growth of the '70s and' 80s, Alpine tourism began to show a stagnant trend since the 1990s (Godde, Price, Zimmermann, 2000, Chap.1), so to induce, in 2000, the scientific Director of the 2nd Snow and Mountain Tourism Congress, promoted by UNWTO, to place mountain tourism in that phase of the life cycle of tourist destinations (Butler, 1980) that requires a reinterpretation of its function to avoid its decline (Keller, 2000). The reasons are many and well known, partly exogenous to the tourism market, partly endogenous (Bourdeau, 2009; Macchiavelli, 2006). Climate change has affected ski resorts more than other factors, but it is not the only one (Sonois 2000). The gradual aging of population has certainly changed the potential reference market of

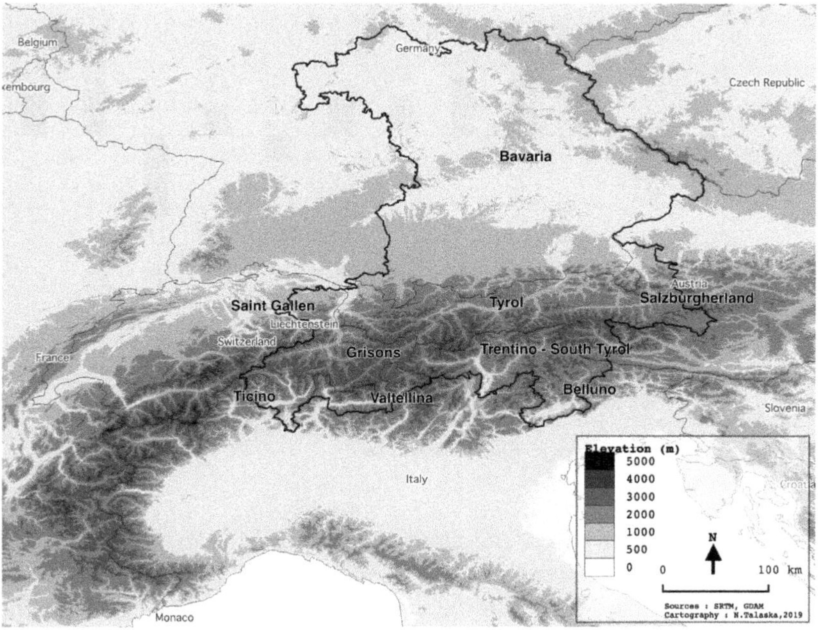

Fig. 13: Studied areas in alpine regions (Switzerland, Italy, Germany, Austria)

an offer that was strongly oriented towards sports tourism (Bourdeau, 2007); the new behaviors of population, especially young people, have also contributed to shift attention towards more playful and therefore less stable and ongoing expectations. The strong increase in tourist mobility, favored by the low cost flights with a progressive decrease in flight prices, has generated an increasing competition in the tourism market, so to shift the attention of the second holiday to destinations different than mountain; consequently the exceptional nature of the alpine landscape, traditional attraction of the mountain, has no longer been an exclusive factor, as other places, easily reached by tourists (e.g. Caribbean atolls or African deserts) offer at the same way exceptional landscapes, although different. Finally, the cultural and social change of population, who is less and less available for a challenging use of free time, begun to modify the expectations of the demand of snow tourism, requiring new services and new recreational opportunities.

The millennium has therefore opened with a long period of stagnation of tourism in most of the Alpine regions, well highlighted by the Fig. 14.

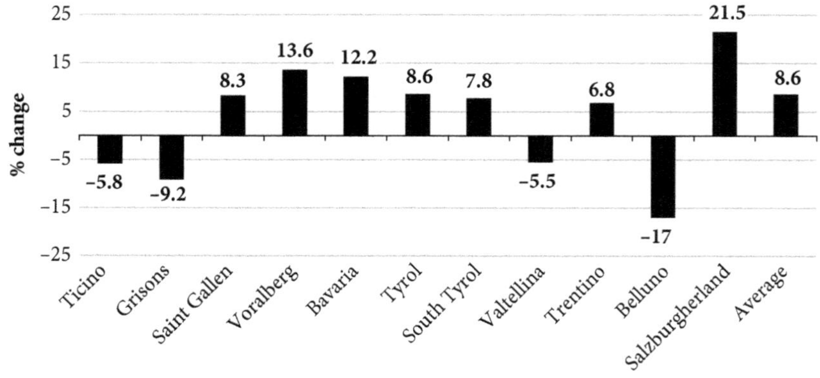

Fig. 14: Ten years of tourism in the Alpine Regions (%changes of hotels nights 2004–2014). *Source: ASTAT Bolzano*

Total nights in hotels and similar establishments in the considered regions on average rose only by 8.6 % in ten years, a very small increase if we consider that tourism in the same period had a strong development (international tourism in Europe, for example, increased by 36.7 %); the best organized Alpine regions (Tyrol, South Tyrol, Trentino) have recorded an average annual increase of less than 1 %, while others have suffered a decline, even strong (Province of Belluno, Grisons, Ticino). However, the last three years seem to provide new signals that would lead to a reversal of the trend. Table 3 shows the trend of hotel nights in the two-year period 2014–16 in the same Alpine regions considered previously (ASTAT, 2017).

As we can see, the negative trend concerns only Valtellina and the Swiss Cantons, penalized by the unfavorable exchange rate of the Franc, as well as by marketing problems. All the other Regions and in particular those with the most marked Alpine features (South Tyrol, Trentino and Tyrol, which together make up almost 60 % of the total nights), show significant growth rates, which bring the average growth to 6.5 % in two years, against 8.6 % of the previous decade. In Italy the phenomenon is clearly evident throughout the last three years, as shown in fig. 15.

The same phenomenon occurred in France, albeit with less strong accents; from 2005 to 2012 the change of hotel nights in Haute-Savoie was nil, in Rhône Alpes there was a 1 % decrease, while in Savoie (so at

Tab. 3: Change of hotel nights in Alpine Regions in the period 2014–2016. ***Source: ASTAT Bolzano***

Regions	**% Share 2016**	**% Total Change**	**Annual % Change**
Ticino	1,8	-1,4	- 0,7
Grisons	3,8	- 8,4	- 4,2
Saint Gallen	0,8	- 6,4	- 3,2
Voralberg	4,8	6,2	3,1
Bavaria	12,0	9,3	4,6
Salzburgland	15,8	8,1	4,0
Tyrol	27,6	5,1	2,6
South Tyrol	20,1	9,7	4,9
Valtellina	1,5	- 4,6	- 2,3
Trentino	10,2	9,3	4,6
Belluno Province	1,6	9,5	4,7
Total/Average Alpine Regions	**100,0**	**6,5**	**3.2**

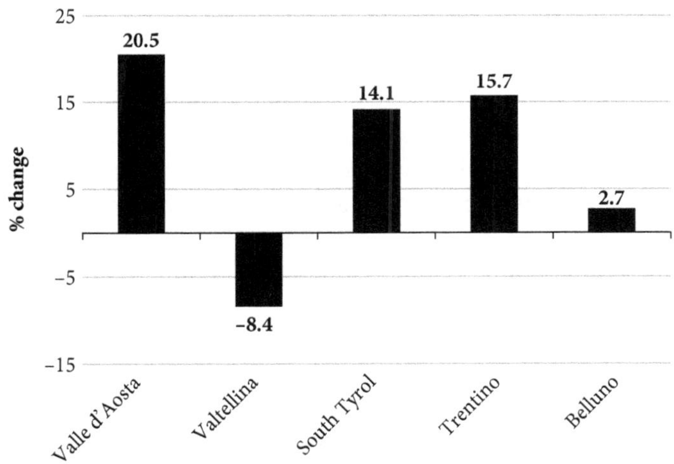

Fig. 15: % Change of total nights in the period 2014–2017 in the Alpine Italian Regions. *Source: Regional Tourism Observatories*

Tab. 4: Average stay in Alpine Regions (days). ***Source: ASTAT Bolzano***

Regions	Average stay 2004 (d)	Average stay 2016 (d.)	% Change
Voralberg	4	3,4	-15,0
Bavaria	3,5	2,7	-22,9
Salzburgland	4	3,4	-15,0
Tyrol	4,7	3,9	-17,0
South Tyrol	5,3	4,3	-18,9
Valtellina	4,1	3,3	-19,5
Trentino	4,8	4	-16,7
Belluno Province	3,6	3,4	-5,6
Average Alpine Regions	*4,5*	*3,5*	*-22,0*

lower altitudes) the decrease was even 9 %.[1] In the period 2010–2015 all these departments showed modest increases (1–2 % per year), while in 2016 compared to 2015, the number of tourists grew by 4–5 %[2].

In order to interpret the characteristics of Alpine tourism today, a second phenomenon concerns the sharp decrease in the average length of stay (see Tab. 4). As we know, it is not new and it is not just about alpine tourism, but in the mountains it is definitely more evident.

In the Alpine regions the average stay in hotel decreased in the last 12 years by one day, with a contraction of 22 %; for Italian tourism in the same period the decrease was about 15 %. This indicates that the traditional long-term stay is no longer the only or the most prevalent way of enjoying mountain.

Finally, there is a third phenomenon that should be noted. Generally, in recent years the "intermediate" months, those at the margins of the two tourist seasons (approximately May, June, September, October), have recorded rather high rates of increase, at least in areas that are able to offer services also out of season. In Savoie-Mont Blanc over a twenty-year period, with stability in total winter flows, there was a sharp increase

1 Cf. Observatoire Régional du Tourisme Rhône-Alpes, *Les indicateurs de l'economie du tourisme*, 2014.

2 Cfr. Insee, *Bilan économique 2016*. Auvergne- Rhône-Alpes, 2017.

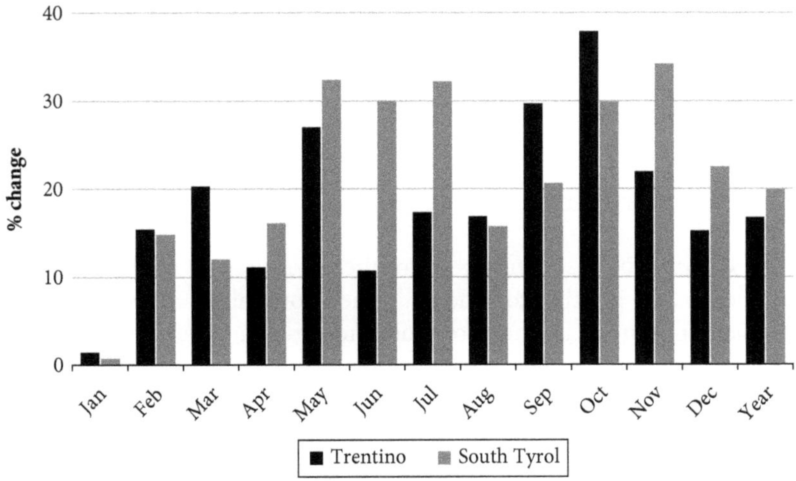

Fig. 16: Monthly % changes of tourist nights in South Tyrol and Trentino in the period 2005–2016. *Source: ASTAT Bolzano and Statistical Service of Trento Province*

in March (+32 %) and spring holidays (+8 %), compared to a decrease in January (-25 %) and Christmas holidays (-4 %)[3].

Fig.16 shows the monthly changes in tourist nights between 2005 and 2016 in the Italian South Tyrol and Trentino, which alone account for over 80 % of the number of visitors in the Italian Alpine region.

The months in which the increase is higher than the annual average (around 15–20 %) and where the monthly change is more than 30 % in both provinces, are those "non central", in particular May, October and November and, as regards South Tyrol also in June and July; on the other hand, the months of traditional winter holidays (January-April) are much less dynamic. In the same period the annual change has been 16.7 % in Trentino and 19.9 % in South Tyrol. But, generally also the pick summer months had a high level of increase, so the distribution of flows during the year doesn't change so much: only the months of May, September and October reinforce their weight in the year.

3 Cf. Savoie Mont Blanc Tourisme, *Les chiffres clés du tourisme en Savoie Mont blanc.* Edition 2018, page 17.

It is evident that these months are not determinant for the growth of the total nights, given the low weight in the total year, but the data still express an important signal of a tendency to seasonal enlargement, to which certainly climate change contributes, this time in a positive sense.

How to interpret these trends?

The three trends coming from the quantitative analysis become significant as jointly considered. The noticeable recovery in the number of tourists in the mountains recorded in the last three years does not suggest a cyclical trend; comes after years of stagnation and is accompanied by a further decrease in the average length of stay. Tourists seem to choose mountain again, but for shorter periods and therefore linked to motivations that induce short-term stays. Furthermore, there is a propensity to use less the central months, also favored by climatic conditions that allow to enjoy alpine landscapes even beyond traditional holiday periods (in autumn, for example, when the alpine landscapes are particularly evocative).

The interpretation repeated several times by the analysis of Atout France (*Carnet de route*, 2012, page 26 and *Tourisme festival*, 2016, page 3), which registers a difficulty of the French mountain to transform the attraction into frequentation, seems therefore here confirmed.

If this doesn't happen anymore, or rather occurs with times of attendance more contained than in the past, it is because the traditional factors of attraction of the mountain – those that generated longer periods of frequenting the Alpine resorts – are no longer the only reasons that guarantee overnight stays in mountain resorts, but are gradually joined by new and increasingly diversified motivations that lead to shorter, even sometimes repeated, times of stay.

The traditional factors that, until a few decades ago, guaranteed tourist stays in Alpine destinations can be summarized in two: landscape and nature in summer and alpine skiing in winter. On these two factors the tourist offer model of mountain resorts has been defined, the one that guaranteed flows, overnight stays and a good level of income for many Alpine populations.

In summer, since the end of the 19th century, when the imperial courts and the aristocracies began to attend the most beautiful Alpine locations to enjoy the landscape, nature and at that time also the consequent conditions of greater physical wellbeing (Battilani, 2009; Bartaletti, 1994); in winter, mainly after World War II, with the development of skiing in ever-wider sections of the population. On these two factors the model of tourist offer of the Alpine resorts has been substantially founded, enriched and qualified over time, but essentially characterized by components of the offer that supported and valued these two factors of attraction (Macchiavelli, 2017; Bartaletti 2006).

This offer model has been characterized for about 50 years by a tourist fruition with a long stay (summer holiday and winter "white week"), even if reducing over time, with few and well-defined activities (hiking and climbing in summer and Alpine skiing in winter), with a sport perspective and a severe relationship with the mountain that required a challenging continuity (SEATM 2000; Bourdeau, 2005).

Almost all the other resources of the mountain, which today we begin to consider potential attractions, have mostly been foreign to the tourist market.

We can think, for instance, to culture, both the immaterial and the material; only in relatively recent times the typical food products of the Alpine valleys, an expression of the culture and work of the territory, has been appreciated and they became a reason of complementarity with other resources or even as independent attraction. Even more, the material heritage (churches, castles, paintings, etc.) has never represented a real attraction for the mountain tourist, although there are valleys very rich of cultural monuments: "cultural tourism" has been always a prerogative of the "art cities", even if sometimes less rich in heritage of some Alpine valleys. The same can be said for torrents where today canoeing is practiced, for climbing walls today a terrain for free climbing practice, and in some places, even for spa, once a resource of many Alpine resorts, then passed into the background compared to the primary attraction of skiing.

In recent years, the behavior of tourists in the mountains has significantly changed (Hudson & Hudson, 2015, chap.3) and this is another reason that mostly contributes to question the model of supply of the past.

First of all, it changed about the open air activities carried out in the mountain; a survey conducted some years ago by the University of Bergamo on the practices carried out by mountain-goers revealed the following incidences in summer and winter practices (Garibaldi, 2014)[4]:

---

4 The sample concerned the population of Northern Italy.

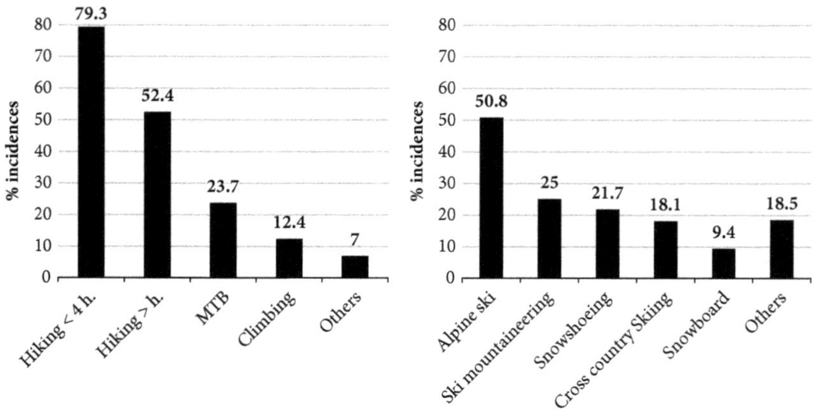

Fig. 17: Mountain practices. % Incidences. ***Source: Vetta Project – Regione Lombardia***

In summer, hiking is obviously the most widely practiced by visitors if the duration is less than 4 hours, (79.3 %), but if it exceeds, the incidence drops to 52.4 %. On the other hand, Mountain Bike represents almost a quarter of the activities practiced today. Similarly, in winter, downhill skiing involves about half of the activity practiced, but there are many activities carried out with incidences that revolve around 20 % (fig.17). In conclusion, referring only to the outdoor activities in mountain environment, we see that, despite the greater consistency of the traditionally practiced activities, many others, both in summer and winter, have now reached significant incidences. Similar conclusions come from parallel analysis made in France[5]. These data refer to physical activities that take place in the mountains; but there are many others that do not necessarily imply the use of slopes; for example, shopping, spas or quality restaurants are nowadays complements that qualify the alpine tourist offer and, in many cases, they become an attraction in itself.

The natural landscape and the snow therefore continue to be the great resources, but more and more are also enjoyed in ways that are different from the traditional ones and they need to be supported and integrated by other opportunities; while remaining the main attractions, they are no

[5] Cfr. Conso Montagne, *Premiers résultats de l'enquête de l'ANMSM sur les habitudes de consommation en station*, 2017.

longer enough to explain mountain tourism. The main condition that join a large part of the activities proposed to tourist is to place them at the center of the experience that the activity itself offers; in other words, to offer him the possibility of living an experience that can be remembered over time and therefore translate, as far as possible, into "emotion".

This is true for all tourist proposals, even if not all reach their goal. It's true for active experiences, where often the tendency to the extreme creates emotional intensity (i.e. mountain races, running, biking or others), but it is also true for sedentary activities (Hudson & Hudson, 2015, p. 257), where the context creates the conditions for the emotion (a thermal bath with landscape or a dinner in chalet in a snowy environment, for example).

On this topic, the tourist literature has today produced extensive reflections (Pine and Gilmore, 2004; Soscia, 2009; Gallucci, 2005) with reference to all tourist sectors; there are usually short-term activities, experienced on an occasional basis and with little effort, apart from some niche of enthusiasts for extreme activities.

These characteristics are usually accompanied by a good environmental sensitivity; tourist coming from the city goes to the mountains with an expectation of environmental respect far greater than in the past and generally higher than the average level expressed by the operators of the territory.

Considering the cases of new motivations of tourists for the mountains, two examples, among the many possible, are certainly interesting. The first is the *Maratona des Dolomites*[6], a cycling event on mountain roads that has been held for 30 years in the Dolomites valleys with departure and arrival in Val Badia and with routes that offer three levels of length and difficulty, ranging from 55 km with 1800 meters of difference in level to 138 km with 4300 meters of difference. If once this event was attended by few fans, today it has 9,000 participants, number set by the authorities for reasons of safety and environmental respect, since the applications now exceed 30,000 units coming from all countries in the world. Since it always takes place on the first Sunday of July, it has a strong tourist and economic impact on the territory, because in that period the numerous beds in the valley are still scarcely occupied by holidaymakers and therefore they find a wide demand, composed by the participants

---

6 It is in the Ladin language, spoken in Val Badia, where the event takes place.

and by accompanying persons, that stuff the accommodation offer for a few days. During the day of the event, some of important Alpine passes, usually very busy in summer, are closed to the traffic and this led the local community to consider further closure on other summer days to promote environmental respect, so that in summer of 2017 they were closed a weekly day during the months of July and August, with some entertainment initiatives[7].

The second case concerns *Bernina Express*, (or the Red Train of Bernina) which connects the town of Tirano, in Valtellina (I), with the well-known Saint Moritz in the Engadine (CH), through 2253 meters of Bernina Pass in a spectacular railway route with natural adherence among the steepest in the world. In 2008 the Rhaetian Railway officially entered the Unesco Heritage and this favored the wide tourist revival of the "Red train" that perfectly expresses a proposal of emotional tourism with full respect for the environment; the mountain city of Tirano, which, unlike Saint Moritz, from the tourist point of view does not offer important attractions, has increased the tourist bed places of about 50 % from 2010 to 2016, having however an average length of stay of only 1,5 day.

The multiplication of outdoor and indoor activities in mountain destinations therefore explains, at least in part, both the renewed interest in mountain tourism and the reduction in the period of stay; the practice of these activities, however, takes place generally for some short moment, as a test, often followed by other activities, through which to seek a further recreational or regenerating experience. The variability and dynamism practiced in the television *zapping* or in the use of social media returns therefore in the practice of recreational activities and often is fueled by the use of new technologies that allow sharing experiences with others, inducing new ones to make them known to friends. This propensity already offers an important indication to the operators: the level of satisfaction of the tourist at the moment of experimentation becomes a decisive factor for his liking and in a wide context of choices this can guarantee the success of the proposal: the quality of the offer can therefore only be high.

---

7 The debate on the partial or total closure of some Alpine passes in the Dolomites is still in progress and will have probably new development in the future.

The need to have high quality products is also one of the reasons that promote the rental formula in order to have the equipment suitable for the practice of the various activities.

The frequent change of activities and above all the irregularity with which they are practiced, from the economic point of view, support the dissemination of the rental of the equipment; if once this was offered for skis and related boots, today other type of equipment are more and more involved, both in winter and in summer (snowboards, snowshoes, mountain bikes, cameras, etc.).

The economic convenience is clearly the most incentive element of this propensity, but it is combined with the need to have high-range instruments, so that the experience is very intense and can be shared with others, showing quality. This trend to "not own" of equipment is also spreading in many areas of consumption of goods and services, as evidenced by the different forms of sharing economy (Maurer, 2015) and so everything suggests that in the future rental services in mountain go further spreading and above all qualifying, to become a sort of "assistance and sorting service" for the tourist to the different outdoor activities on mountain.

A final issue following the change in enjoying the mountain is suggested by the possibilities of seasonal expansion. If the diversification of activities takes place both in summer and in winter, as we have shown, it is true that it is especially in summer (and in the intermediate seasons) that there are more opportunities for innovating proposals, because in summer nature is the basic element for the activities and so sport, outdoor life, hiking and climbing; without the limit of snow there is more competition between the destinations and operators and the possibilities for innovation are greater. In its analysis of summer mountain tourism, *Atout France* notes that in summer, attendance takes place above all "outside the resort"[8]. Summer is therefore the strategic season for the alpine destination and becomes the guarantor of winter's successes[9]. The trend is therefore to innovate in summer, both through sports and recreational activities, and through the offer of events; in this context there is an ever-increasing tendency to favor the use of the lifts,

8 Cfr. Atout France, *Tourisme estival des destinations de Montagne, 2016*, page 88.

9 Cfr. Jean-Marc Silva, manager of France Montagne, interview on *Montagne Leaders*, n °265, Jan–Feb. 2018 pages 64–71.

once used almost exclusively in winter. The use of downhill MTB has been a great opportunity for many locations to increase the use of lifts also through a seasonal extension (Saint-Martin, Savre and Terret, 2013). In many places, high-altitude resorts and restaurants have been created, with recreational activities that support the daily visit; in others the offer of cultural events at high altitude (concerts, performance) that favor the use of the lifts, also outside the traditional timetables.

It must also be considered that this development of activities aimed at offering new opportunities to tourists can also have undesirable effects on the balance of the Alpine environment. If the Maratona dles Dolomites has increased sensitivity towards the reduction of traffic on Alpine passes, the implementation of initiatives at high altitude to favor the use of the cables often produces too many infrastructures (restaurants, entertainment facilities, etc.). A problematic situation also begins to arise due to the strong diffusion of MTB. In most cases, in fact, the paths used by bikers are the same as those used by hikers, who don't appreciate the sometimes invasive presence of bicycles, also because these on a steep slope are also less easily controllable. For this reason in the most organized destinations (where there are the conditions of space and routes) a diversification of the routes dedicated to bikers it is beginning. All this highlights the problem – which we will recall later – of the need for effective territorial governance, which is more difficult because the factors that must be governed are always greater.

The gap between big and small resorts

With the first effects generated by climate change, we began to feel a growing gap between the small and big destinations of alpine tourism. The former are, in fact, generally located at medium and low altitude and the decrease in snowfall has certainly favored the locations with a higher altitude, which, being also usually well organized, have actually increased their weight on the market. This gap is felt even more today face the new and ever-changing needs of tourists. Since, as we have seen, there is an increasing trend of the tourist to practice different activities and in any case to change them frequently, it means that the tourist resorts must equip themselves to meet these needs; tourist destination therefore should manage in order to offer a wide range of services and facilities (Kampf & Hunziker, 2014). This happens especially in large destinations where the amount of tourists is greater and where even the resources available are greater. Opening any website of important alpine resorts (Verbier, Saint Moritz, Les Arc, Val Gardena or Andorra, for instance) we will easily notice a very broad list of proposals, some of which involving a large number of people, other only some niches of few enthusiasts. The big tourist resorts can therefore meet the expectations of the demand only multiplying their offer and creating the most original and rare proposals. An example explains clearly the phenomenon: a few years ago, in Andorra, a destination /country with a great winter tourist offer, was presented and proposed an accompanying service to practice diving, in winter, in the alpine lakes of the territory; the local agency "*Diving Andorra*" accompanied the tourist in snow clothing in the alpine lakes and there, diving equipped, was guided in diving by specialized instructors.

Today this service no longer exists, probably because of little interest from potential customers, but the example remains useful to explain the trend in diversifying the offer and in imagining innovative proposals, even with little appeal. In this context, a relevant role is assumed by the destination-village, that is the tourist destination born around an original village, like the most of the Swiss, Austrian and Italian alpine places.

The theory defines this type of places referring to the *Community Model*, different from the integrated French or American stations, where the *Corporate Model* prevails, that is an artificially grown ski resort, whose management is entrusted to one or a few industrial size entities (Hope & Flagestad 2001). Just the French analysis explain that in summer "au-delà de la préférence pour les massifs domestiques, les montagnes suisses et autrichiennes attirent plus que les montagnes françaises"[10], because the tourist offer of places with an original community is an attraction itself, with its life, its culture, its expressions and the many events that the life of the community produces (Rigoni, 2006). However, the big resort is able to pursue economic conditions without renouncing a good level of quality of the offer; in spite of many different targets, the size of the offer favors the economies of scale, if there is a good capacity for segmentation of the market and management of the various targets. Quite different are the conditions of small tourist destinations. Their resources are definitively scarce: the economic ones are scarce, but the human ones as well, among which find ideas, know how, ability of innovation and environmental and

---

[10] Cf. Atout France, *Tourisme estival des destinations* de Montagne, 2016, page 95.

cultural sensibility (Pechlaner and Zeni, 2006). Every attempt to imitate the larger resorts offering a wide range of services, infrastructures and proposals for tourists, therefore ends up failing, as the quality level will be modest. This was actually the trend that has characterized the alpine tourism offer to date, in line with the main attraction factors: all the places have tried to sell the same product, substantially defined by the landscape and the possibility of alpine skiing, where possible. If the product tends to multiply and to articulate itself, this policy can no longer respond to the needs of the demand, which perceives the differences in quality more than before, as well as the variety of products; the small resort cannot be competitive, considering that the margins to contain the costs, and therefore of the prices, are modest (think about the costs related to the installations and the artificial snow, for example). The emergence of new attraction factors therefore imposes the offer to reorganize in relation to them.

It should however be stressed that if, from the point of view of economic management and market control, the condition of small ski resorts is objectively penalized by the impossibility of producing economies of scale, it is precisely in these destinations that it is sometimes expressed more creativity that leads to generate new initiatives, often followed, sometimes with greater success, by the more organized places. As we will see in the next point, many small localities, in an attempt to exploit their scarce resources as much as possible, have produced interesting examples of innovation in the tourist offer, initially oriented to small targets but available to spend more time in the deepening of the own relationship with the territory. The itineraries through the mountain pastures, the visit to cheese productions or the cultural itineraries, to mention some examples, were initially proposed in small or medium-sized locations, as well as certain areas of free style ski slopes in a guaranteed and safe area, as it happens in Livigno (Corti et al., 2015; Macchiavelli, 2017). Often these activities are also the origin of the birth of small or very small service companies.

The destination as a network of quality proposals

In light of the new market conditions, the pursuit of a specialization or a characterization can therefore represent today, a perspective for the smaller resorts that could thus acquire their own original specificity in the market, guaranteeing that Unique Selling Proposition (USP) that until now in the mountain resorts has been missed (fig. 18). It is certainly not a new process in the company business, even less in tourism, but it is certainly a novelty in the tourist offer of Alpine destinations that until today have substantially proposed the same product. The specialization or characterization will tend to enhance the characteristics of the destination itself, attracting the tourist by virtue of special resources that the locality has: it may be a sporting or recreational activity (MTB, running, climbing or cross-country skiing, for example), a cultural proposal (places with artistic value, or expressions of the immaterial culture) or a typical product to taste on site or to buy; the important is that numerous and various tools, both in management and communication, converge around them, in order to push the chosen proposal towards a high quality level, so as to be more competitive than other locations, even more known. But it could also be a characterization consequent to the conditions of the place. There are significant examples that confirm it; Alpine resorts that 15 years ago were considered finished, have found a new youth thanks to particular markets to which they have oriented and from which they have obtained recognition.

The most frequent case concerns resorts oriented to the family and to children, but there is no shortage of those most affirmed by the youth target, where spaces, activities, events, music are strictly oriented towards this target; many places are oriented to the well-being of person, with the provision of services that encourage *running* and *fitness*, as well as *wellness*.

The prospect of specialization for small towns can be an opportunity, but it presents risks and needs management conditions that must be carefully pursued.

The first risk is to let the choice of the USP be defined outside of an overall project for the whole territory in which the locality is inserted or, even worse, on the basis of pure intuitions of someone or of external investment proposals. There are cases of places characterized by large recreation infrastructures (e.g. theme parks), which have nothing to do with the natural environment and the surrounding landscape, with evident impoverishment of environmental and cultural sustainability (Area 47 in the Oztal Valley, in Austria is an example). Hence the need for the tourist resort, even if small and with fewer resources than others, to refer to an overall *Vision* on its role within the territory in which it is located and endowed with a Policy that allows to pursue their goals. We mention two positive examples in this direction. To the small town of Werfenweng in Austria (1000 inhabitants) we owe the birth of the Alpine

Pearls, a network of alpine locations characterized by a soft mobility; the origin lies in the fact that, already many years ago, Werfenweng, identified in giving up the car its own characterization among the alpine resorts, representing an example for many others. Similar situation is that of Funes Valley (Italy), one of the few Dolomites valleys without a skiing offer, initially due to poor favorable conditions and increasingly more and more to a choice of its community; today the Funes Valley is characterized by a high quality environmental offer, both summer and winter, and being less equipped with hotels, as it has not at the time enjoyed the development generated by skiing (such as the nearby Val Gardena), is now able to practice prices higher for an increasingly consistent environmental tourism demand. The need for this enlarged vision to the whole territory in which the location is located (usually it is a valley) requires a strong relationship of cooperation / integration with other destinations, for at least two reasons: a) because a unitary Policy in the territory can only be conceived and agreed with the other localities; b) because the tourist needs a structured offer and certainly could not be content with a single-subject offer.

A unitary Policy for the whole territory means that each locality expresses its own role, that means that it exercises a tourist function expressing its specificity at the highest possible quality level, in complementarity with the other places of the area. Consequently, the destination, in this perspective, will no longer be represented by singular tourist resorts, as it has been prevalently until now, but from the entire territory, which

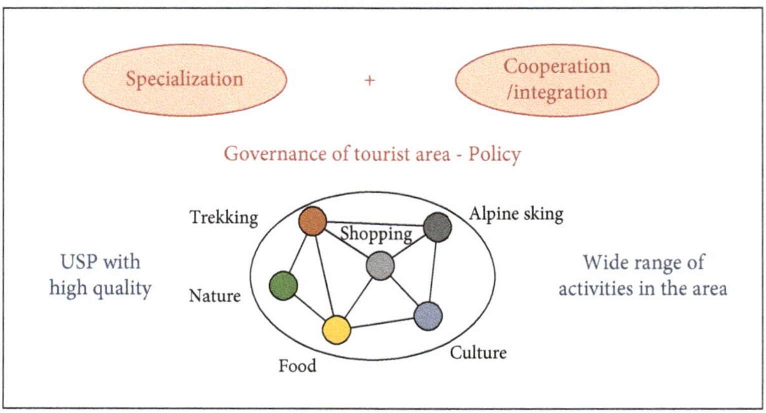

Fig. 18: A perspective for small mountain resorts

will be able to offer a variety of complementary opportunities, if all the locations will be oriented towards their USP; it will be a unique destination characterized by a network of locations with different and integrable proposals. Moreover, it is increasingly clear that the tourist demand could not be satisfied today with a monothematic offer; in other words, the identification of a prevalent resource in an Alpine place does not mean that it can fully satisfy the tourist's expectations. Hence the need to build an overall design and a policy that interests the entire territory. The tendency to make the specific resources the attraction for tourists is already underway, as we have underlined previously; the problem is that this almost never happens in a coordinated framework of choices and therefore it cannot give the desired results or, even worse, doing damage.

Some clarifications are appropriate:

a) Valuing one's specificities in order to define a USP does not mean abandoning all the other opportunities that the tourist resort has offered up to now; moreover, it is clear that no resort would be willing to choose this. It is a matter of implementing a process whereby the usually scarce resources of a small resort are progressively oriented towards the most favorable attraction factor, aiming to achieve the highest possible level of quality.

b) This means a strong convergence on the objective, first of all on the side of the local tourist operators. It is not enough that only those who have specific responsibilities and competence on the subject deal with it, but that the topic becomes a common heritage of the whole resort and that therefore the choices of other operators converge towards it. For example: if a locality opts for a specialization on running and physical well-being, Municipal Administration will set up paths and signs, but private operators should offer opportunities to measure physical conditions, traders sell equipment and clothing, restaurant offer appropriate menus, etc.

c) The essential condition, however, remains a strong cooperation among the resorts, searching the highest level of complementarity. First of all because only on this condition the tourist will be able to recognize the whole area as a destination and secondly because common policies on collective resources are necessary in order to relate the places. Always to give some example: if a resort offers a strong peculiarity in shopping, another in the cultural offer and a third in downhill skiing, it's essential that the mobility between the places allows an effective and quick use of the three opportunities with a transport system efficient and coordinated.

d) All this requires for this territory a solid governance of the tourism system (George-Marcelpoil and François, 2012), which should however be the goal of any tourist context. In this case, however, it is more necessary than elsewhere, but at the same time more difficult. More necessary, because the diversity of proposals requires a unitary direction, under which to bring out the complementarities and integrations; more difficult, because the coexistence of small localities tends to accentuate the parochialism and makes it easier to identify a leadership function, absolutely essential to manage a territory with different tourist functions looking for integration. The governance of the tourism system can therefore be only voluntary and pursued on the basis of shared objectives; therefore, a methodologically rigorous work will be necessary to build step by step the conditions for a unitary management of the whole territory. A serious governance that lasts over time is definitely inimical to short-term results, an objective often understandably pursued by public administrator with a mandate that expires.

e) Uniform communication and marketing is absolutely essential for the territory that presents itself as a single destination through a network of tourist destinations. This should be considered an obvious consequence, but experience says that difficulties are not lacking. First of all, in a marketing policy, which in this case will be more difficult, both because the target audience is different by reason of the proposals, and because each location has other markets that it does not give up and therefore must continue to cultivate. A unitary communication will have the advantage of favoring the application of economies of scale, very useful for places with few resources.

The two pillars (cooperation and specialization) of the proposed model require some deepening, above all because in this case they represent two complementary conditions: the added value for the area derives from the pursuit of both and not from just one, as is usually.

Cooperation has always been a condition of efficiency and often of cheapness in tourism; through cooperation a value chain (Pechlaner and Weiermair, 2000) is created that favors the creation of an Added Value compared to the production of a single subject, even if it should be said that this is not always distributed in a fair way among the components of the supply chain or of the working group; it may in fact occur that the advantage of one may not correspond to the same advantage to another. Precisely this is one of the reasons that make collaboration between companies more difficult, some of which could see little corporate interest in the face of the benefits achieved by the entire supply chain; this necessitates a solid leadership in the Governance, so that convincing benefits could be guaranteed to all, perhaps through other methods or in other times. In our case, the subjects called for collaboration and sometimes for integration within the area are the individual destinations, each of which must first find its own peculiarity in the area. There is therefore a first objective that is specific to the individual localities and which primarily calls the subjects (companies and institutions) operating in that destination, among which a good level of internal collaboration must however be developed to concentrate resources towards an objective and a unified strategy. But the choice of this objective, while being imposed in fact by the available resources, also requires effective coordination with the other localities of the area in order to generate synergies and not oppositions or parochialism. The marketing and promotional strategy of the area further imposes a strong coordination so that it is perceived as a single destination and the value deriving from its complementary offer is recognized.

If the difficulties are undoubtedly not lacking, being connatural with the small size of the localities and with the management of a plurality of subjects, it is however opportune to highlight with conviction also the advantages, both on the users' and on the of the localities themselves. Firstly, it would benefit the tourist who in this way could benefit from a variety of opportunities with a good quality service level; the concentration of resources and initiatives around the main resource should in fact lead to an increase in quality to the services and in this way increase the level of customer satisfaction. The resorts, on the other hand, would see their best resources valued and the gap between large and small resorts, which we have highlighted above, would tend to shrink because the lower-altitude locations would now be less exposed to snow dependence, and could therefore have a tourist function even in a winter setting with little snow. In the last few winters with little snow the places with a greater variety of alternative services held up more.

Three sensitive topics

Searching for and exploiting the peculiar resources of a small tourist resort and relating them to other specific areas of the territory requires analysis and knowledge that only a professional management of tourism can guarantee. As is known, this is basically poor in small towns and for this reason it is worth recalling some "sensitive" topics that most affect the tourist resorts and which require particular attention in this dynamic context.

Innovation in tourist resorts

In times of strong change, the capacity for innovation is almost always the variable that decrees the competitiveness of a tourist resort, as well as a company; particularly in this model of tourist offer because the enhancement of existing resources in a territory implies a capacity to reinterpret them and adapt them to the expectations of potential users, which stems from the capacity for innovation that the operators can express. Given that the ability to innovate should not be confused with creativity or geniality – which certainly can contribute to innovation, but which do not coincide – it should rather refer to a process of change that enriches and enhances a production process already in act, in a known context and in which consolidated professional figures operate (Flagestad, Hope, Svensson and Nordin, 2005). Innovation arises or compare therefore with subjects already operating and requires a "doing" that goes to introduce new elements or processes on an existing context. In the case of tourist resorts this requires two conditions: *the ability to read the changes in the market and the ability to interpret their resources differently*.

Tourism is a social phenomenon and is strongly influenced by the behavioral changes in society. Addressing tourism without carefully observing the social behavior risks to induce stakeholders to think that there are no significant changes in their customers and therefore to re-propose their offer by segments of demand that are no longer available to receive it in the ways of the past. It often happens that tourism operators – faced with a market crisis – await the resumption of behavior on the part of the market that probably will no longer occur in the future because they are the object of social change. The attention to social development is above all a cultural attitude and uses traditional channels of social observation, but in the specific of tourism activities it is now possible, with a minimum of competence, to observe the behavior of tourists using tools that do not involve high costs. The big companies (hotel chains, tour operators, etc.) operating on an industrial basis are well aware of this and are equipped with tools to observe the behavior of their customers; unfortunately, this does not happen with the same conviction among the small enterprises and consequently they struggle to readjust their offer to the new needs of the demand. The ability to capture customer changes is necessary but it may not be sufficient to introduce innovative components in its tourism proposal. The problem remains that of matching the resources of one's destination with the new expectations of demand. And if they have always been interpreted in the same way, according to the same point of view, they cannot represent an innovative offer. An example can help a better understanding of both expressed conditions. In a mountain valley of Trentino the stream that flows there has always been a resource for tourism, both because it is a fundamental component of the landscape, and because it is an opportunity for fishers; but only when some operators noticed a growing demand for canoeing tourism and realized that the stream could be a potential resource for that demand, the torrent has become one of the most important places to practice canoeing and rafting at the highest quality level, so that the Canoeing World Championships were held there. The recognition of a nascent (at that time) demand, as a new market trend, and the ability to reinterpret the tourist function of the river in relation to that social change were the conditions that favored innovative tourism in that territory.

Authenticity and technology

The smallest places in our Alps are usually those that most retain an authentic dimension, that is a visible testimony of their history, their culture, their productions; today the European tourist is more and more sensitive to this and he aspires to meet it, even when he is attracted to other factors. The available technology puts us in a position to share with others the experience we are living and it is not hazardous to say that this often becomes the main reason to do it: not only and not so much living an experience (sports, recreational or cultural) but can share it with other distant people and let people know where you are and what you are doing. The social networks and sharing tools offered by today's technology represent the new "word of mouth" that can decree the success or the indifference towards a tourist resort. Hence the need for a deep attention in expressing the authenticity of a place and in making it perceive to the visitor in a way that is more understandable, more pleasant and more attractive for him. All this puts into play the involvement of the tourist not only in the visit to the places, but in introducing him to the experience that those places allow to live and to transmit their origin and their history. The tasting of a typical dish that combines the knowledge of the authenticity and exclusivity of the ingredients, the visit to the places of production and the history of the products through the methods and tools of production is a very different experience from the traditional gastronomic offer of a territory (Flammer and Müller, 2012). Particularly interesting is the case of restorants in Val Badia (Dolomites), where within a few kilometers there are 3 restaurants Michelin star ratings, all with a strong characterization of the relationship with the territory. One of these in particular highlights how the starting point of its creative cuisine are the ingredients linked to the territory and this has produced a recovery of agricultural and food production that had been lost (Niederkofler, 2018). This is what could make the difference and determine the competitive success of the resort compared to others. The South Tyrolean development company (IDM) has constantly developed and updated a website (http://www.suedtirol.info/storiedavivere/) which has no direct tourist implications, but which tells stories of people who live and work in South Tyrol, because "We are convinced that stories are moving the world, so we give space to tales of dreams, rebirth, innovation and experience". The effect on the visitor is to recognize the value of that territory and to wish to meet it. This heritage of values and history is present in many Alpine villages and technology, which we all now have, allows us to easily make it known. Unlike the past, the costs of this communication and marketing operation are very limited and are therefore easily accessible even for small mountain resorts, which can therefore gain visibility and competitiveness. In this case, the poorest resource is not infrastructures, nor capital, but rather a cultural approach that recognizes, on the one hand, the value of the cultural heritage of which small mountain resorts are rich, on the other the potential marketing value that the new technology offers. In a context of globalization, the relationship between local productions and tourism is increasingly strengthened in all areas, through the most diverse forms of involvement of tourists in the places of production, also encouraging a more in-depth knowledge of the territory visited, whose characteristics are largely determined, yesterday and today, by local productions (Fernandes and Richards, 2017).

Productivity and de-seasonalization

Tourism is by definition less productive than other economic activities (Keller and Bieger, 2007, Introduction) and mountain tourism, in turn, tends to be less productive than other forms of tourism. The small size of the enterprises, the geographical marginality, the lowest managerial level of the operators are elements that affect the costs or the poor performance of the factors of production and that determine a lower overall productivity (Weiermair, 2000). Small mountain resorts are more exposed to these higher costs; but since there is little chance of lowering them further, the recovery of productivity can be faced up only by increasing revenues. We do not enter here on the management of traditional economic activities, which affect all tourism activities, but we limit ourselves to highlighting two problems that are common to almost all Alpine areas and which significantly affect the overall productivity of the destination's tourist offer.

The first concerns the phenomenon of cold beds in holiday homes; the low employment of a large number of beds means that these are very unproductive and this means a lower level of activity of the other accommodations and services (Colombo and Macchiavelli, 2007). The low average annual occupancy of beds in tourist homes means lower flows into the destination and consequently lower revenues for other activities: this is the condition in which many alpine resorts are located and where the prevalence of beds in the apartments is much higher than that of hotels, other economic activities (shops, restaurants, recreational activities) may have difficulty surviving, as happens in many locations close to large urban areas where tourist homes are prevalent. The increase in productivity of many small mountain resorts is therefore linked to a more productive use of beds in the apartments, through short rent formulas, services dedicated to second-home owners and an increase in the quality of accommodation.

The second phenomenon to which we intend to refer concerns the use of the lifts, which, as is known, have been made almost everywhere according to the ski slopes. In summer, many plants are closed and in general in the ski areas they are used on average about 20 % of their potential. It is clear that the opening of lifts in different seasons from the winter involves costs that are covered only where there are a good level of flows, generally fed by great panoramas; hence the difficulty to open out of ski season. The increase in the use of the lifts in the summer season therefore means an increase in productivity for the operators of the lifts, but also for the other operators who work at high altitude.

The possibility of greater use of the lifts in summer passes through three channels:

- Adopting flat-rate subscription formulas (card) which include other services, so as to favor the economic convenience of using lifts;
- Using the slope as recreational condition. The MTB offers today an important opportunity, but this involves the preparation of descent paths;
- The existence of attractions at high altitude. Apart from the panoramas, there is today in the most developed areas the tendency to create infrastructures and services at high altitude (adventure parks, fitness trails, restaurants) or to organize events (e.g. concerts) that attract constant flows

Climate change, as we noted at the beginning, does not have only a negative effect for the mountains (less snow availability), but also a positive one to allow greater use of the Alpine environment in the intermediate seasons. Interpreting alpine tourism today in the traditional dual reading of summer / winter, as two totally independent realities, is today misleading (Hudson and Hudson, 2015, chapter 10) because many activities are now usable in both seasons, or at least in those intermediate, since the conditions in offering the services are similar. In recent winters without snow, but with mild temperatures, there were some hikers who ventured on the trails in the most organized places in the Dolomites. Within certain limits, the problem of the seasonal enlargement of alpine tourism is not so much dependent on climatic conditions, but rather on the tourism system that in most of resorts (starting with the smaller ones) is not yet equipped to facing a potential (but increasingly real) demand off-season and the majority of services are not open. The possibility of a wider seasonal exploitation of mountain destinations would also contribute significantly to their consolidation, also in terms of quality, as tourism operators would find easily to supplement income, which could produce a more convinced professional qualification in tourism

Conclusions

The positive trends we have mentioned open new horizons for the tourism of the alpine resorts, but it's clear that the way to manage the tourist resorts cannot today be the same then in the past. The changes of the market that we have highlighted in this chapter are going to interface with many problems in the management of resources that makes all more complex – as it happens in the economic-management of each sector-and make it more difficult to achieve a balance sustainability, environmental, cultural, and – not least, economic. Complexity involves skills and tools that one could not do without; the smaller destinations are also the most fragile and have more difficulty in interpreting the changes in the market, but here too the generational change begins to produce its effects and brings new resources able to introduce management methods more suited to market expectations. In this framework of complexity, we must avoid to face reality according to preconceived schemes, which in most cases are trivial. So we make ours own this statement of Enrico Camanni, a deep connoisseur and passionate scholar of alpine culture:

> "The Alps should abandon any ancillary role and become (again) a place of proposal and innovation; in this perspective it is necessary above all to overcome trivialization and work on complexity. We need to recover and communicate the authentic reality of the Alps, which is that of an extraordinarily diversified environment from the natural and human point of view" (Camanni 2002).

References

ASTAT, *Il turismo in alcune regioni alpine*. In: Astatinfo, n° 51/2016 & n °45/2017.

Atout France. 2012. *Carnet de Route de la montagne*. Editions Atout France, Paris.

Atout France. 2016. *Tourisme estival des destinations de Montagne*. Editions Atout France, Paris.

Bartaletti, Fabrizio. 1994. *Le grandi stazioni turistiche nello sviluppo delle Alpi Italiane*. Patron editore.

Bartaletti, Fabrizio. 2006. "L'evoluzione del turismo invernale nelle Alpi italiane". In *Scritti in onore di Roberto Bernardi*, Salgaro, Silvino (eds.). Patron Editore.

Battilani, Patrizia. 2009. *Vacanze di pochi vacanze di tutti. L'evoluzione del turismo europeo*, Il Mulino.

Bourdeau, Philippe. 2005. "Imaginer et mettre en scène des produits touristiques de montagne". *Proceeedings of 4e "Congrès mondial du tourisme de neige et de la montagne*, Principality of Andorra.

Bourdeau, Philippe. 2007. "L'or blanc: de l'épopée à la complexité". In *Les sports d'hiver en mutation*, Philippe, Bourdeau (eds.). Lavoisier, 17–34.

Bourdeau, Philippe. 2009. "De l'après-ski à l'après-tourisme, une figure de transition pour les Alpes?". Revue de Géographie Alpine | Journal of Alpine Research 97-3 | 2009, mis en ligne le 09 décembre 2009, consulté le 14 février 2020. URL : http://journals.openedition.org/rga/1049

Butler, Richard W. 1980. "The concept of a tourism area cycle of evolution: implications for managements of resources". *Canadian Geographer*. XXIV(1): 5–12.

Camanni, Enrico. 2002. *La nuova vita delle Alpi*. Torino, Bollati Boringhieri.

Clivaz, Christophe. 2007. "L'immobilier en station de sports d'hiver: du lassaire-faire au savoir-faire?". In *Les sports d'hiver en mutation*. Bourdeau, Philippe (eds.). Lavoisier, 111–122.

Colombo, Emilio and Macchiavelli, Andrea. 2007. "Effective occupancy of the holiday houses: condition for the development of the destination". In *The problem of productivity in tourism*, Keller, Peter and Bieger, Thomas (eds). Erich Schmidt Verlag. Berlin, 123–134.

Corti, Michele. 2015. "Cibo, spazio, tempo, comunità". In *Cibo e identità locale, Centro Studi Valle Imagna*, Corti, Michele, De La Pierre, Sergio and Agostini, Stella (eds.). Sant'Omobono Terme (BG): Centro Studi Valle Imagna, 245–429.

Fernandes, Carlos and Richards, Greg. 2017. "La gastronomia tra tradizione e Innovazione. L'esempio della regione portoghese del Minho". In *In viaggio tra cibo e vino*, Garibaldi, Roberta (eds.). Aracne, 19–35.

Flagestad, Arvid and Hope, Christine A. 2001. "Strategic success in winter sports destinations: a sustainable value creation perspective". *Tourism Management* 22: 445–461.

Flagestad, Arvid and Hope, Christine A., Svensson, Bo and Nordin, Sara. 2005. "The tourist destination; a local innovation system? The creation of a model". In *Innovation in Tourism. Creating Customer Value*, Keller, Peter and Bieger, Thomas (eds). Atti del 55° Congresso AIEST, Brainerd, 245–259.

Flammer, Dominik and Müller, Sylvan. 2012. *L'eredità culinaria delle Alpi*, Casagrande, Bellinzona.

Gallucci, Francesco. 2005. *Marketing emozionale*, Egea, Milano.

Garibaldi, Roberta. 2014. "Immagine e fruizione della montagna nella popolazione del nord Italia". In *L'escursionismo in montagna. Fatti, comportamenti, prospettive*, Macchiavelli, Andrea (eds.). Edizioni ambiente, 33–53.

George-Marcelpoil, Emmanuelle and François Hugues. 2012. "De la construction à la gestion des stations". *Revue de Géographie Alpine | Journal of Alpine Research* [En ligne]. 100–103.

Godde, Pamela M., Price, Martin F. and Zimmermann, Friedrich M. 2000. *Tourism and Development in Mountain Regions*, CABI Publishing, Wallingford.

Flagestad, Arvid, Hope, C.-A. 2001. "Strategic Success in Winter Sports Destinations: a Sustainable Value Creation Perspective", *Tourism Management* 22, 445–461.

Hudson, Simon and Hudson, Louise. 2015. *Winter Sports Tourism*. Goodfellow Publisher, Oxford.

Kampf, Richard and Hunziker, Christian. 2014. *Succès et competitivitè du tourism alpin*, BAK Basel Economics.

Keller, Peter. 6-8 aprile 2000. "Les facteurs-clés du succès des stations de sport d'hiver. Rapport de base". In *Proceedings of 2e Congres Mondial du tourisme de neige et des sports d'hiver*. Principality of Andorra.

Keller, Peter and Bieger, Thomas (eds.). 2007. "Introduction". *Productivity in Tourism*, ESV, 1–12.

Macchiavelli, Andrea. 2006. "*Le stazioni alpine invernali: tendenze e questioni in gioco*". In *Il turismo della neve*, Macchiavelli, Andrea (eds.). Milano, 13–66.

Macchiavelli, Andrea. 2017. "Il turismo della neve nelle Alpi italiane: nascita, sviluppo e cambiamento". In *Histoire des Alpes – Storia delle Alpi – Geschichte 79der Alpen*. Università della Svizzera Italiana, Chronos Verlag, 49–72.

Manente, Mara and Cerato, Mirella. 2000. "Destination management per creare valore". In *Destination Management*, Pechlaner, Harald and Weiermair, K. (eds). Milano, Touring University Press, 55–72.

Maurer, Cristian. 2015. "Sharing economy and its impact on the competitiveness of destinations". In *Proceedings of Singidunum International Tourism Conference 2015*. Belgrad.

Niederkofler, Norbert. 2018. *Il progetto "Cook the Mountain"*. In *Destination Greenitaly. Modelli di Governance turistica dalle Alpi al Mediterraneo*, Pechlaner, Harald, Angelini, Paolo and Scuttari, Anna (eds.). Aracne, Roma.

Pechlaner, Harald and Zeni, Alice. 2006. "La gestione della destinazione: contesti, metodi, problemi". In *Il turismo montano tra continuità e cambiamento*. Macchiavelli, Andrea (eds.). Milano, 49–68.

Pechlaner, Harald and Weiermair, Klaus. 2000. Destination management: fondamenti di marketing e gestione delle destinazioni touristiche. TCI, Milano.

Pine, B. Joseph and Gilmore, James. 2004. *L'economia delle esperienze*, Etas, Milano.

Rigoni, Ernesto. 2006. "Le destinazioni turistiche invernali e le nuove sfide del mercato". In *Il turismo montano tra continuità e cambiamento*, Macchiavelli, Andrea (eds.). Milano, 95–122.

Rossanigo, Alessia. 2018. *Il turismo invernale nelle destinazioni alpine tra cambiamento climatico ed evoluzione del mercato*, Master Thesis discussed in June 2018 at University of Bergamo.

SEATM, 2000. Etude COFREMCA. *Carnet de Route de la montagne*. Editions AFIT, Paris.

Saint-Martin, Jean, Savre, Frédéric and Terret, Thierry. 2013. "Les premiers rassemblements alpins de vélos tout terrain". *Revue de géographie alpine/ Journal of Alpine Research* [En ligne], 100–103.

Scoscia, Isabella. 2009. *Emozioni & Consumo*, Egea, Milano.

Sonois, Jean-Pierre. 2000. "Investir dans l'avenir des stations de sports d'hiver: La vision de la compagnie des Alpes". In *Proceedings of 2e Congres Mondial du tourisme de neige et des sports d'hiver*. Principality of Andorra, 6–8 aprile 2000.

Weiermair, Klaus. 2000. "Know how and qualification gaps in the tourism industry: the case of alpine tourism in Austria". *Tourism Review* 2: 45–53.

2.2

How Ski Resorts in Quebec are Transforming to Deal with Climate Change

Michel Archambault

Context

In its most recent report, the Intergovernmental Panel on Climate Change (IPCC) sounded the alarm to limit the rise in temperatures to 1.5°C by 2050 (Pörtner, 2018). In an interview with Montreal's daily newspaper *La Presse*, the chairman of the insurance market Lloyd's of London, Bruce Carnegie-Brown, stated that "the number of natural disasters linked to climate change will only continue to increase [...] they are happening more and more, with losses estimated at some USD 2.9 trillion over the past 20 years [...] All industries, all regions of the globe, are affected by climate change" (Décarie, 2018). Even a northern region like Quebec can't escape these consequences.

According to a study published in the scientific journal *Nature Climate Change*, tourism is responsible for 8 % of global greenhouse gas (GHG) emissions (Lenzen et al., 2018). Researchers at the University of Sydney, the University of Queensland and National Cheng Kung University analysed data from 160 countries between 2009 and 2013. The vectors they took into account were connected to pollution and included travellers' transport, food, accommodation and purchasing habits. In 2013, CO_2 emissions reached 4.5 billion tonnes. Depending on different projections and on the rise continuing, tourism will produce between 5 billion and 6.5 billion tonnes of CO_2 per year by 2025. Scientists have proposed putting a carbon tax on tourist travel. As the sector's annual global growth approaches 5 %, and given the heavy dependence on (air and car) transport, players in the tourism industry are urged to take every possible measure to reduce emissions in the sector and contribute to efforts to decarbonise the economy in order to meet the goals of the Paris Agreement (Becken and Bobes, 2018).

Interdependence between the climate, land protection and tourism has been well-established. In Quebec, as is the case elsewhere, the alpine ski industry is suffering from the effects of climate change. Several studies in the field of adaptation science are seeking to improve knowledge about the effects on and vulnerability of this sector as a result of climate change. The gradual change in climate conditions and the natural environment could affect the profitability of activities, for both the resorts and the regions themselves.

For mountain resorts, there is an undeniable connection between climate change and tourist numbers on the one hand and their profitability on the other hand. According to the Regional Group of Experts on Climate in Provence-Alpes-Côte d'Azur (Antoine, 2018), global warming is affecting the Alps twice as fast as anywhere else. In the United States, the same warming of ski resorts means a loss of USD 1 billion in revenue and 17,400 jobs in the case of a light snow season (Hagenstad, Burakowski and Hill, 2018). This produces about a 10 % decrease in the number of visits. Without sustainable development strategies to curb the effects of climate change or actions to embrace opportunities, these results indicate that the viability of many businesses will be compromised.

This chapter begins with a brief description of Quebec's skiable territory. Based on an analysis of the observations made by a team of researchers in four regions in the province, a portrait is sketched of the ski industry's approach to adapt to climate change. The issues to be tackled and the actions to be taken will be laid out in further detail. The chapter also provides a few examples of resorts that have adopted integration and diversification strategies in the context of climate change and customer development.

The research team is comprised of experts, including the author. For the past several years, they have examined the impact of climate change on tourism in Quebec, especially at certain ski resorts. This group includes the Ouranos Consortium (a network of some 500 scientific researchers and experts, practitioners and decision-makers from Quebec who develop knowledge about climate change: https://www.ouranos.ca/ouranos), representatives and operators of outdoor activities, including the ski sector, Quebec's Ministry of Tourism and regional tourism associations.

Brief description of the ski industry in Quebec

With a population of more than 8 million and covering an area over three times the size of France, Quebec shares its border to the south with the north-eastern United States and to the west with Ontario, which, with some 14 million inhabitants, is the most populous province in Canada. Vast forests cover about half of Quebec's total area.

Skiing is the main activity at mountain resorts. Quebec has more ski resorts than Ontario (75 vs. 58), and they are located at a higher altitude, but there are more ski resorts in the north-eastern United States (135), some of which have higher peaks than in Quebec. Quebec's 75 resorts can be divided up based on altitude, ski lift capacity (number of people per hour), skier-days and ticketing revenue: major (8) and large (10), which make up 75 % of the total number of visits, and intermediate (20) and small (37). The resorts on the territory are mostly concentrated within a 100- to 150-km radius from Montreal and Quebec City, where more than 75 % of Quebec's population resides. The map below (fig 19) shows the location and the number of resorts by tourist region.

Quebec is also at an advantage thanks to its geographic location. It is better positioned than the Province of Ontario and states in the north-eastern United States, all of which are farther south than Quebec. These two regions, which are more severely threatened by global warming, draw the largest pool of skiers in North America. Thus, Quebec could exploit this advantage by attracting the skiers to its resorts.

Most of the ski resorts in North America are privately owned. They usually own and manage all of the ski-related services: ski lifts, grooming, ski school, catering, snowmaking etc. They own the mountain area or are renters on public land. There are also other private investors operating in the same area or close to the resort who offer a range of services, including hotels, restaurants, spas and games. This model is different

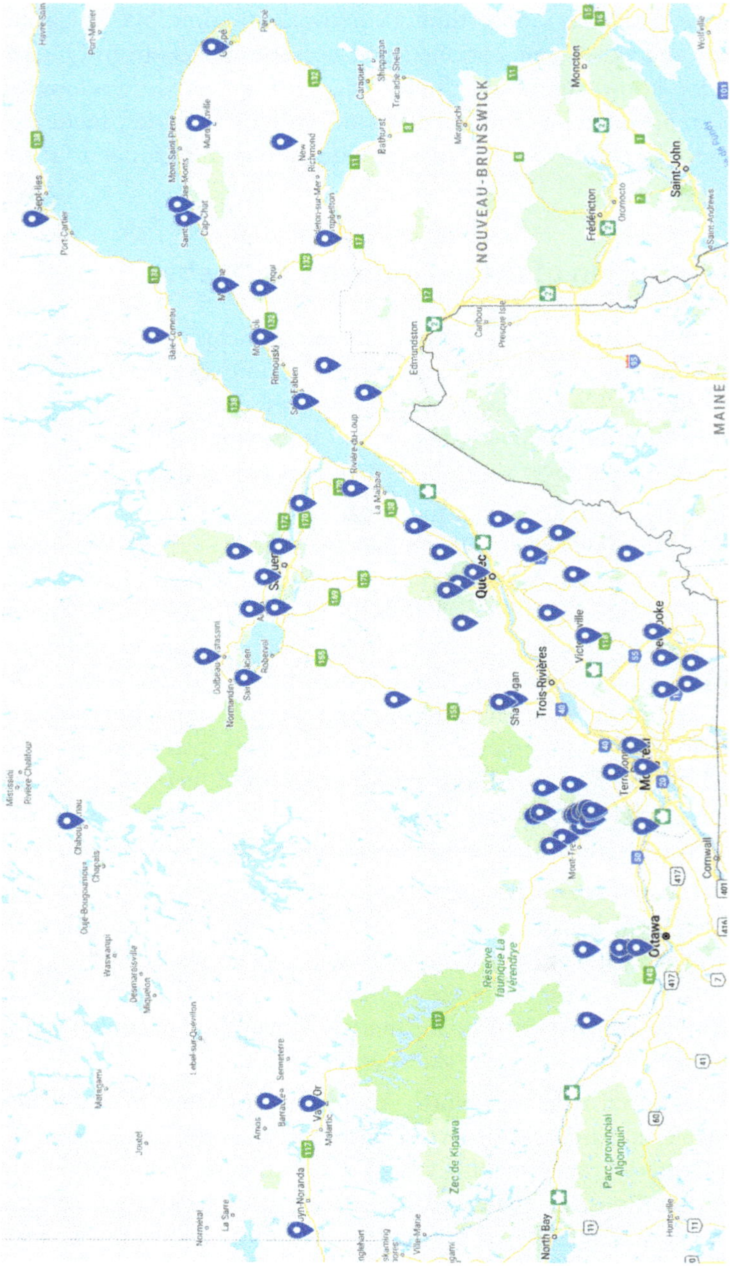

Fig. 19: Ski resorts in Quebec

from that of most resorts in Europe, where the ski lifts tend to be owned by municipalities. These municipalities mostly operate them in small and medium-sized mountain areas and often outsource the larger areas, by delegated management, to private owners over long periods of time (sometimes too long, some have argued: Cour des comptes, 2018). In addition, as is the case in France, the ski school is independent of the resort (the instructors are part of the National Union of French Ski Instructors or the International Ski Instructors Union).

In the 1980s, there were 116 ski resorts in Quebec, but in 2018 there were only 75 (Archambault, 2017). The same has happened in Ontario and in the United States. In the latter, the number of resorts has fallen from 735 in 1983 to 518 in 1993 to 472 in 2017 (RRC Associates. 2018). Among the possible causes for the closures, one suggestion is the intense competition between ski resorts for customers and, especially, the challenges faced by small resorts to remain competitive and appealing. These small resorts, largely undercapitalised, have also suffered from the impact of climate change, which is manifested in a lack of snow at the beginning of the season and at critical busy periods. In general, economic viability requires a minimum opening time of 100 days (Steiger, 2010). These resorts, which are already financially weak and often located far from the large pools of skiers and do not have additional services to offer nearby, have been unable to deal with the many challenges they face, including renovating or adding ski lifts and making the necessary investments in artificial snow.

Climate change and ski resorts' plans to adapt

In its synthesis on climate change knowledge in Quebec (2015), Ouranos notes that, since 1950, the rise in the average annual temperatures in Quebec has varied between 1 and 3°C depending on the region, and much more during the winter (https://www.ouranos.ca/synthese-2015). According to a plausible GHG emission scenario, increases of 2 to 4°C between now and 2050 – and between 4 and 7°C by the end of the century – can be expected for southern Quebec, and up to between 5 and 10°C in northern Quebec. This increase will make the snow season shorter. To add to this scenario, we know that, on the global scale, "2017 was the warmest on record without an El Niño event, and one of the three warmest years behind the record-setting 2016. The world's nine warmest years have all occurred since 2005, and the five warmest since 2010 according to the World Meteorological Organization (2018).

The combination of a rise in temperatures and in the amount of rain, along with a decrease in the amount of snow, will have an impact on ski resorts' operating conditions. The significant increase in temperatures will result in shrinking cold periods for making snow, especially at the beginning of the season, which is a critical time to ensure the resorts are open before the Christmas holidays. In addition, the decrease in snowfall between now and 2050 will increase resorts' dependence on snowmaking, thereby increasing the pressure on the effectiveness of the snowmaking systems, water management and control over the costs of preparing the ski area.

Four regions under consideration

For the purpose of our research, we selected four of the leading regions for outdoor tourism, including alpine skiing: the Laurentides and the Eastern Townships, which are located within a radius of between 75 and 150 km from Montreal, as well as the region of the provincial capital (Quebec City Area) and nearby Charlevoix. An initial study (Archambault, Germain and Bleau, 2012) was conducted between 2009 and 2012 in the Laurentides and the Eastern Townships), and a second, focusing on the Quebec City Area and Charlevoix, has been under way since 2016. In both cases, the analytical framework has five parts: 1) development of climate scenarios in relation to future traffic by using historical data; 2) workshops; 3) questionnaire development and validation; 4) benchmarking with best practices; 5) brainstorming session. The results of each of these parts informed the subsequent steps.

The Laurentides (north of Montreal) and the Eastern Townships (south-east of Montreal) were both chosen for the study because they are among the most important tourist regions from the point of view of supply and demand. Besides a diversified and mature sector of which recreational and outdoor activities for tourists are a strong economic component, these regions also have different climate conditions. They contain a large number of four-season holiday resorts offering a variety of activities, including water sports, cycling and golf in the summer and skiing in the winter. The Laurentides (13 resorts) received 31 % of all visits to ski resorts in Quebec in 2017, while the Eastern Townships (six resorts) made up 21.2 % of the total (Archambault, Morin and Nguyen, 2017).

Vulnerability analyses were carried out with the participation of more than 100 people. Dialogue between the researchers and the operators was facilitated through a collaborative approach (www.participatoryactionresearch.net) to study the climate, risk management, research needs and the integration of adaptive initiatives in operational and decision-making processes (Bleau, Blangy and Archambault, 2011–2012). Next, using traffic scenarios generated from climate projections, the group examined how the number of visits to ski resorts in the study areas has been affected by unusually warm winters over the past few years. This "spatial analogue" method helped in assessing the impacts during real climate phenomena and in capturing a range of adaptation strategies used by ski area operators. Sectoral and regional workshops were held, during which the participants could share their perceptions on the physical changes associated with the summer and winter activities and identify potential actions that might be taken.

A similar approach was taken in 2016 and 2017 for the Quebec City Area and Charlevoix region, with the active participation of 200 people from the industry and the relevant regions. The goal was to provide two overviews (one connected to tourism, the other tourism-oriented) and to diagnose the vulnerability to climate change of tourist businesses in seven sectors, including snow sports and ski resorts. The workshops made it possible to gain a better understanding of the risks posed to different areas of activity in which SMEs are involved and to identify opportunities resulting from climate change (Paque et al., 2018).

Mapping tools that illustrate the climate trends predicted by Ouranos for the next 30 years by using historical data, were tested to measure the players' capacity to adapt and were adjusted according to suggestions from the participants. Among others, the exercise sought to make the tools easier to assimilate and apply to the multi-institutional decision-making process and to better understand the foreseeable changes in the event that nothing changes. The visual aid below (Fig. 20) facilitates discussion and encourages the participants to discuss adaptation issues and solutions in order to better manage climate change.

Issues and actions

The two reports on the four regions highlight a number of issues related to the impact of climate change on tourism. Understanding these issues helps to identify the actions that have to be taken by and for the stakeholders. The major issues include:

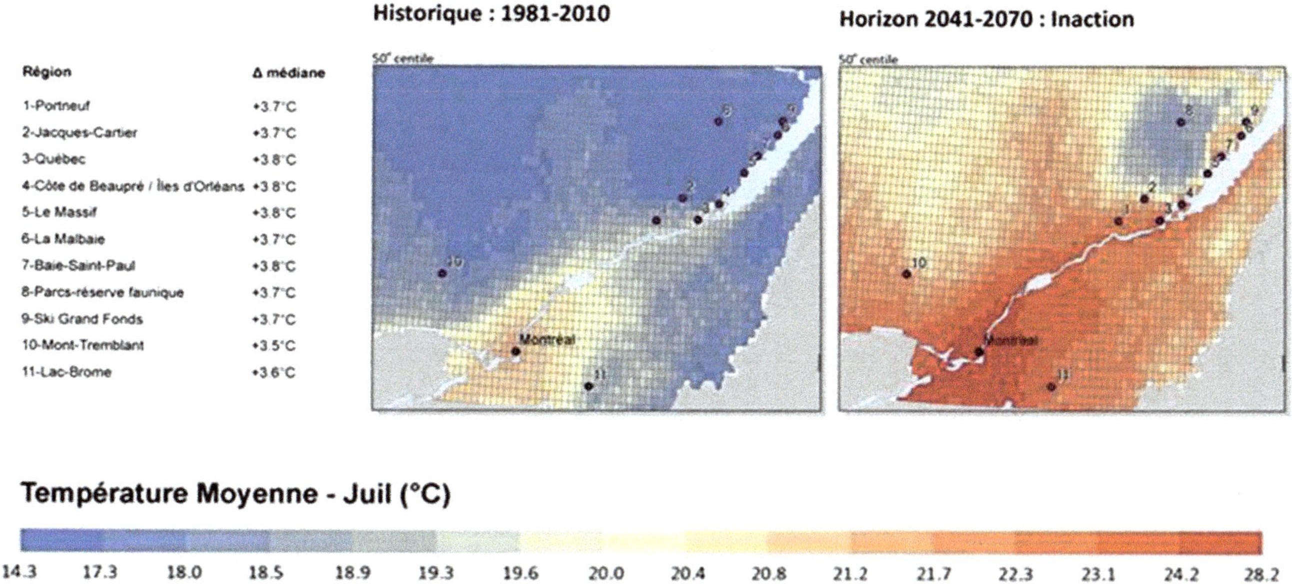

Fig. 20: Visual aid to facilitate discussion and encourage the participants.
Average temperature in July (°C). Historical data 1981–2010 and projections for 2041–2070 in case of inaction

- a lack of knowledge about the impact on and vulnerability of tourism and a shortage of strategic information about climate change to make informed decisions
- the need for more up-to-date and better-targeted information to facilitate decision-making regarding climate-related risks and the opportunities available to businesses, especially small and medium-sized (SMEs) ones
- a lack of expertise and research on adaptation, as well as a dearth of financial incentives to accelerate the implementation of profitable investments for the operator
- the advantage of better communication based on robust science for decision-makers in the tourism sector to promote the spread of climate research and knowledge transfer as part of their organisation's strategic planning.

In the studies that have been done, a series of actions that have been formulated would enable the operators to manage the risks arising from climate change more effectively:

- a rejuvenation of their infrastructure by installing better equipment for the mountain and reception
- an improvement of their energy and environmental report cards
- more modern and efficient manufacturing and snowmaking tools centred on automation
- an architecturally harmonious real estate development that takes the environment into consideration
- active participation by the municipalities to promote the mountain and its activities or, in some situations, even provide limited tax relief.

It was found that the presence of a ski resort drives the annual growth of the property wealth of municipalities near the mountain up by between 15 and 20 % (Archambault, 2017). This growth is mainly due to investments in the resorts, prompting developers or individuals to settle nearby or to modernise their real estate assets. All of this contributes to a quality of life and a living environment that people find appealing and may lead them to reside or acquire a second home there. This leverage effect then leads to the arrival of new businesses, leisure accommodation and road infrastructure in these same living environments that are now in demand. The municipalities of Mont-Tremblant (Laurentides) and

Bromont (Eastern Townships) have benefitted from the economic gains of their respective ski resorts: Over the span of 10 years (2005–2015), they saw their property value double or triple, which generated significant tax revenues.

In the second, more recent study conducted in the Quebec City Area and the Charlevoix region, it was also noted that the presence of a ski resort contributes to a municipality's property wealth. The report found an increase of more than fivefold in 10 years for the Petite-Rivière-St-François municipality (Le Massif ski resort) and more than threefold for nearby Baie-Saint-Paul.

- Diversification of activities to maximise the use of assets and human resources and to better respond to the target customer.

In Quebec, staying in the mountains, at certain ski resorts, does not necessarily mean that a large number of visitors are only doing alpine skiing. By offering, whenever possible and in partnership with other operators located on or close to the site, a wide range of winter activities (sleighing, snowshoeing, curling, dog sleighing, ice skating, cross-country skiing, snowmobile, spas and swimming pools, shops and restaurants etc.), it is possible to meet the varied needs of a family seeking to experience the mountain in different ways all while growing the market.

- In certain situations, having a range of summer activities (water sports or water park, mini-golf and golf, hiking trails, bike paths, mountain biking, horseback riding, tennis, festivals and events, meetings and conventions, weddings, triathlons, children's rides, outdoor obstacle courses etc.) helps to breathe life into the area and diversify sources of income by focusing on the seasonal adjustment of activities.

The changing mountain area

Mountain resorts have to reinvent themselves not only to respond to climate change but also to attract visitors with a variety of needs. In Quebec, those aged between 12 and 17 and between 18 and 24 ski less than their age groups did 20 years ago: Almost half of young people skied back in 1995; in 2015, this figure was down 58 % to just a quarter for the two cohorts (Archambault, 2017)! In the United States, there is a similar trend: Those between the ages of 13 and 17 and between 18 and 24 made up 17 % and 12.3 %, respectively, of all visits in 2008, compared with 13.2 % and 9.4 % in 2018. Among US millennials, also known as Generation Y (22–37 years), the figure decreased from 36.8 % to 25.2 % over the same period (Belin and Becher, 2018). The traditional development model is no longer adequate and has to be adapted to appeal to the visitors of tomorrow, who are more interested in moments of sharing and innovative experiences. The mountain no longer just means skiing; it is a place for games and responsible development that take the form of magical and authentic activities – a place that "sparks the imagination, speaks to our emotions… makes us dream, where it feels good to just chill" (http://www.poprock-agence.com/lindustrie-sports-dhiver-menace-t-lavenir-de-montagne/).

Many a ski resort is now defining itself more as a place of entertainment, pleasure, discovery and replenishment than merely as a snow park. Monoactivity has given way to multiactivity!

A resolutely green development

The leading resorts, such as Mont-Tremblant in Quebec, Whistler Blackcomb (Canadian Rockies), Vail and Aspen (Colorado), quickly understood that their future development would have to be done while respecting their environment, which is the core asset of their operations. These resorts are an example of the involvement of and collaboration between their municipalities and strategic actors, including the air transport sector. All of them are committed to protecting the environment. The president and CEO of Vail Resorts CEO, Bob Katz, has stated that his business has the following three goals for 2030: "Zero net emissions by 2030. Zero waste to landfill by 2030. Zero net operating impact to forests and habitat" (http://www.epicpromise.com/environment). This industry leader, which has more than 30,000 employees across its resorts and hotels (under the RockResorts brand), especially in the United States but also in Canada and Australia, is taking concrete steps to achieve each of these goals. The same is true of Aspen SnowMass, another leader taking action to protect the environment and reduce its own ecological footprint (https://www.aspensnowmass.com/we-are-different/our-environmental-commitment), which has received several prizes in recognition of the excellence of its programmes in this regard.

In 2000, Quebec's Mont-Tremblant resort endorsed the NSAA's (National Ski Areas Association: http://www.nsaa.org/environment/sustainable-slopes/) "Sustainable Slopes" environmental charter. In 2017, it signed on to the Sustainable Industries Council's Ecoresponsible Program to promote, accelerate and facilitate the integration of sustainable development principles and the implementation of environmentally responsible practices in all of its activities. The goal is to enable future generations to have experiences in nature that are just as enriching as the ones we have today (http://www.ecoresponsable.net/conseil-des-industries-durables). Along with 37 other resorts in North America, it has also been part of the NSAA's Climate Challenge since 2017 (http://www.nsaa.org/environment/climate-change-climate-challenge). Spurred on by its involvement in this programme, Mont-Tremblant has put in place a number of measures to better manage climate change and limit its impact on the environment (https://www.tremblant.ca/fr/a-propos/ecoresponsabilite).

Resorts and destinations: a four-season development

Over the past two decades, several ski resorts in Quebec have added a range of services and products to satisfy their customers while making greater use of their assets, which had previously been limited to skiing. New activities have been included in their strategies for periods that present opportunities to increase revenues throughout the year. They also attract potential investors looking for promising locations. This trend was confirmed by the arrival of Club Med at Le Massif ski resort, lying between the 1,200-kilometre-long St Lawrence River and the mountains in the Charlevoix region, whose first four-season village is slated to open in 2020 (https://www.lesoleil.com/affaires/un-club-med-aux-accents-de-charlevoix-348e17b952385ed5e88d47750a). The region is well-known for its stunning landscape and is also recognised as a UNESCO biosphere reserve.

A resort's chances of success increase insofar as it is located in a town or a municipality with a reputation for charm and whose many services, attractions and experiences complement those of the resort. Success will be all the more likely if it is easy to reach the destination by road or by air.

About 10 ski resorts in Quebec also offer activities outside the winter season. These activities make up between 20 and 35 % of their total business for the year and the number of visits. Thanks to this strategy, these resorts keep a good proportion of their employees, thereby reducing the cost of high staff turnover. Visitors passing through need to have access to accommodation, food and shops. They also engage in activities offered by other operators in the immediate area. Tourism is no longer seasonal and contributes to the development of an area that is revitalised by the mountain activities on offer throughout the year. A few examples will help to better visualise the impact of a traditional ski resort's metamorphosis into a four-season centre whose appeal stands to benefit the entire region and that has a diverse array of recreational activities.

Mont-Tremblant – Laurentides tourist region

The Mont-Tremblant resort stagnated for many years until the early 1990s. Its acquisition by Intrawest in 1991 marked a new beginning: community involvement, establishment of the Tremblant Resort Association, massive investments in the mountain, at the base of which a pedestrian village was built that took inspiration from the historical district of Old Quebec, along with a variety of hotels, restaurants, bars and businesses. These investments positioned Mont-Tremblant as the most dynamic winter resort in Quebec, and Ski Magazine has named it the "#1 ski village in eastern North America" on multiple occasions.

Subsequently, the nearby airport was re-opened, and golf courses, bike paths, tennis courts and an astronomical observatory were constructed, which started summer activities that have helped to revitalise the area and make its businesses more viable. In order to increase the number of visitors throughout the year, the Tremblant Resort Association has focused on a continuous event structure: a blues music festival, a world-class IronMan with around 3,000 competitors from more than 50 countries and a comparable number of volunteers, a 3-D sound and light show on a mountain circuit route (Tonga Lumina), the wellness-oriented Wanderlust Festival (https://wanderlust.com/fr/festivals/tremblant) with 7,000 participants and 40,000 visitors over four days, activities during the colourful autumn season (Indian summer), etc.

This "it" resort quickly appeared on the radar of not only ski enthusiasts but also fans of numerous activities offered both in winter and in summer. As a result, the numbers climbed from 400,000 visitors in 1991 (25,000 in the summer and 375,000 in the winter) to 2.5 million in 2017 (1.3 million in the summer and 1.2 million in the winter).

Bromont – Eastern Townships tourist region

Located less than an hour's drive south-east of Montreal, next to a highway exit, the resort is the municipality's biggest employer. It is certainly the main attraction in the area. The town's reputation and brand image are directly tied to the history and development of the ski resort, which also contributes to the residents' quality of life and the region's appeal among businesses.

Because of the jobs that are offered and the range of activities, the whole community benefits from having the resort there. The purchase of second or permanent homes is often linked to the mountain's reputation, which has made it possible to develop a diverse tourism offering over the years, as well as a pleasant living environment. We have already cited the resort's leverage effect on the municipality. Between 2006 and 2016, the municipality recorded a major increase in its population (75 %) and in the value of its property wealth (+158 %, that is, from 735 million dollars to 1.9 billion dollars). The number of residents on holiday increased by 3.5 times.

This resort is known for its large water park but also for its mountain bike trails, which fans can reach by ski lift and hurtle down to their hearts' content. Major events have been added to make the resort livelier: Summer of Chefs, Bromont Blues, BBQ and Nuit blanche. Accommodation and food can also be found at the base of the mountain. Other tourist partners in the region (vineyards, spas, golf courses, restaurants, museums etc.) benefit from these initiatives and enrich the tourist experience.

Sommet Saint-Sauveur – Laurentides tourist region

The Sommet Saint-Sauveur is located next to a highway exit less than 50 minutes north of Montreal. The Saint-Sauveur resort is part of the Les Sommets group, which operates five ski hills, and is the strongest in terms of turnover and the number of visits (more than 1 million). Thanks to a large snowmaking capacity that allows skiing to start at the beginning of November, the resort has been part of the landscape of the Saint-Sauveur Valley for more than 80 years.

A pioneer in the summer-oriented diversification of activities, the resort offers a variety of experiences that are well integrated with its host municipality. Like Bromont, one of its many initiatives includes a renowned water park. After the park reached its full potential, the

"Parc F.U.N." amusement park was added, which focuses specifically on families. It is described as follows on its website:

> "Enter [the] magical world of the Parc F.U.N.'s Amusement Park! Feel the thrill of moving at more than 35 km/h aboard the Viking, Quebec's only alpine roller coaster, and glide through the air on the Dragon, a double zip line. Continue your adventure on our five rides [and] end your day by testing your putting prowess at the 18-hole Minigolf course" (http://www.sommets.com/en/amusement-park/).

At the resort itself, there are also mountain bike trails and a mini-golf course. For entertainment stretching into the night, an open-air cinema has been added for screenings at night. In addition, the municipality offers quality accommodation and a wide variety of restaurants and speciality shops. Sommet Saint-Sauve is a leading driver of economic vitality in the valley and contributes to its property wealth by attracting both local and foreign investors.

Whistler Blackcomb – province of British Columbia

In Canada, **Whistler Blackcomb** (some 130 km from Vancouver in British Columbia) is undoubtedly the pioneer in this metamorphosis that ski resorts are undergoing. This particular resort has been ranked as one of the 12 most impressive resorts in the world (Hodgetts, 2017). It was one of the sites of the 2010 Winter Olympic Games and receives more than 3.5 million visitors a year, generating around 2 million skier-days in the winter. The town at its base offers more than 20,000 beds (https://www.whistler.ca/municipal-gov/community-monitoring/whistler-facts-and-figures/whistler-visitation). Over the summer, Whistler receives more than 1.9 million visitors. In collaboration with the municipality of Whistler, which is on the leading edge of environmental protection, development and resort planning (http://nrt-trn.ca/whistler-blackcomb-holdings-inc-case-study), the resort is committed to the territory's integrated sustainable development strategy. Year-round activities and services are offered for the benefit of residents and visitors alike (https://www.whistlerblackcomb.com/explore-the-resort/about-the-resort/environment.aspx; https://www.whistler.ca/municipal-gov/community-monitoring), including one that is very popular, the *Peak 2 Peak 360* – a gondola with 28 cabins, each with a capacity of 28 people, and a total length of 4.4 km. It links the summits of Whistler and Blackcomb and, at more than 3 km, has the longest span of any lift in the world between two pylons (https://www.whistlerblackcomb.com/explore-the-resort/activities-and-events/summer-activities/peak-2-peak-360-experience.aspx).

Conclusion

In the heart of nature, mountain resorts have focused on cold seasons with skiing and good amounts of natural snow. They take advantage of the climate and the beauty of their natural environment. Over the past few decades, climate conditions have changed because of global warming: The cold season is becoming shorter, while the amount of snow varies and is often not enough to start the ski season on time. High traffic in certain periods (Christmas, school holidays, Easter) is no longer guaranteed without making a significant investment in artificial snow.

Customer needs have also changed over time. Skiing is now but one activity among many. And yet, the mountain still seems to be a popular holiday destination, even if the practice of skiing is stagnating. The appeal of nature and the opportunity to stay there and do other activities go some way towards explaining this interest. At the beginning of the autumn in 2018, an Ipsos-mmv poll was taken and confirmed the interest of French visitors in staying in the mountain the following winter. It was found that "23 % of the French respondents (18 years and older) intend to go to the mountain [...] a higher turnover than expected. 53 % of those who planned to go this winter were also there last winter, which means that 47 % did not go there last year. What is even more surprising is that 18 % of the French respondents who will go to the mountain this winter haven't gone in any of the past five years" (Ipsos, 2018). The Quebecois are no exception and often visit the mountains. Several resorts are located close to national parks or protected wildlife reserves, which further enriches the recreational tourism offer.

The resorts are facing some major challenges:

- If they do not adapt, the start of the season will gradually be delayed as the effects of climate change increase. By 2050, the season will start about seven to 10 days later than it does today (2020). The length of the season will shrink by 10 to 20 days. Traffic at ski resorts could decrease between 2 and 15 % in terms of ticketing and 2 to 7 % in terms of season ticket revenues over 30 years (Da Silva et al., 2019). However, it is known that effective manufacturing equipment can maintain adequate snow conditions and make it possible to take advantage of cold periods by quickly making snow and opening the ski area on time at critical moments. However, the results of the financial analysis show that it will be a challenge for resorts to take on the necessary investments all on their own (Da Silva et al., 2018). Indeed, there are limits because the decline in energy consumption spending and the expected uptick in ticketing revenues will not be enough to recover the amounts allocated for this infrastructure. Will the resorts know how to persuade the public authorities or other partners to shoulder the financial risks associated with these investments?

- They have to ensure the sustainability of their winter ski operations by investing in efficient, state-of-the-art equipment with the smallest possible environmental footprint and limited greenhouse gas emissions.
- Nature and the beauty of the landscape need to be protected. Having both young and old people committed to the values of protecting the planet should encourage organisations to initiate – if they haven't already done so – innovative strategies to combine development and profitability with saving their environment.
- They have to be creative and develop new customer channels in order to diversify sources of revenue and reduce the dependence on skiing and the winter. To this end, a collaborative approach with other stakeholders in the area is a better gauge of success than working in silos.
- Experts and research can support managers in their decision-making. The research group in the Quebec City Area was in favour of drawing up "a table of monthly climate variables and indicators and a matrix of risks and opportunities for seven sectors of seasonal activity, including skiing [...]. These tools were developed to inform and raise the level of sectoral awareness..." (Paque et al., 2018).

The analytical framework that has been developed is a sophisticated tool that now makes it possible to integrate climate trends and to examine a number of potential adaptation measures that could be useful in a strategic management process. In addition, an investment simulator created in collaboration with managers is a visual tool that can help to compare the different adaptation possibilities with each other in terms of benefit and cost. The sector is also in major need of improving its capacity to analyse and to put in place more diversified and innovative adaptation measures, which requires a better consideration of the uncertainty surrounding this type of measure and specific market analyses. We hope that the approach presented here can help to delve deeper into a tourism economy that is more resilient to dynamic climate conditions.

> "… The complexity and cross-cutting nature of the climate issue require collaboration between actors on several fronts. It is a question of planning, developing and managing the territory. Therefore, it becomes interesting for managers and decision-makers to integrate known risks and seize opportunities in their guidelines and planning strategies in order to remain competitive, make better investments in the long term and maintain the integrity and sustainability of tourism assets" (Paque et al., 2018).

Having a vision, adopting a mission that integrates environmental values and connecting the area to the development of a way of life and of meaningful experiences are the prelude to a winning strategy for everyone!

References

Antoine, Aurélien. 2018 (13 novembre). "Réchauffement climatique dans les Alpes: opportunités ou menace?". *Alti-MAG*. https://www.alti-mag.com/alti-actus/rechauffement-climatique-dans-les-alpes-opportunite-ou-menace.

Archambault, Michel. 2017. *La modernisation des stations de ski du Québec – Prendre le virage de 2020*. Maneige (ASSQ), Terrebone.

Archambault, Michel, Germain, Kate, Bleau, Stéphanie and Matte, Dominic (avec la coll. de Blangy, Sylvie, Scott, Daniel and Chevalier, Jacques). 2012. *Analyse socioéconomique des impacts et de l'adaptation aux changements climatiques de l'industrie touristique au Québec. Rapport final*. Ouranos et Chaire de tourisme Transat (ESG UQAM) https://www.ouranos.ca/publication-scientifique/RapportArchambault2012_FR.pdf.

Archambault, Michel, Morin, Jean et Nguyen, Doan. 2017. *Étude économique et financière des stations de ski du Québec*. Maneige (ASSQ), Terrebone.

Aspen Snowmass. https://www.aspensnowmass.com/we-are-different/our-environmental-commitment (last access 2 march 2020).

Becken, Susanne and Bobes, Lucas. 2018. *Providing the Case: Carbon Reporting in Travel and Tourism*. 56 pages: https://amadeus.com/documents/en/travel-industry/report/proving-the-case-carbon-reporting-in-travel-and-tourism.pdf.

Belin, Dave et Becher, David. 2018. "2017-2018 National Demographic Study". *NSAA Journal* 26/(6): 1–11.

Bleau, Stéphanie, Blangy, Sylvie et Archambault, Michel. 2011–2012. *Stratégie d'adaptation aux changements climatiques pour l'industrie touristique du Québec*. Synthèse des ateliers participatifs et sectoriels.

Conseil des industries durables. http://www.ecoresponsable.net/conseil-des-industries-durables/.

Cour des comptes. 2018. *Les stations de ski des Alpes du nord face au réchauffement climatique: une vulnérabilité croissante, le besoin d'un nouveau modèle de développement*. Rapport annuel 2018, 441–522: https://www.ccomptes.fr/sites/default/files/2018-02/RPA2018-Tome-2-integral.pdf.

Da Silva, Laurent, Desrochers, Félix-Antoine, Pineault, Katherine, Gosselin, Charles-Antoine et Grenier, Patrick. 2019. *Analyse économique des mesures d'adaptation aux changements climatiques appliquée au secteur du ski alpin au Québec*. Montréal: Ouranos, Montréal,133 pages.

Décarie, Jean-Philippe. 2018 (23 octobre). "Changements climatiques et cybersécurité à risques élevés". *La Presse*: https://www.lapresse.ca/affaires/economie/grande-entrevue/201810/23/01-5201301-changements-climatiques-et-cybersecurite-a-risques-eleves.php.

Hagenstad, Marca, Burakowski, Elizabeth and Hill, Rebecca. 2018. *The Economic Contributions of Winter Sports in a Changing Climate*. POW et REI Co-op, Boulder.

Hodgetts, Rob. 2017 (13 juin). "Alpine Edge. 12 of the World's biggest ski areas". *CNN Travel*. https://www.cnn.com/travel/article/worlds-biggest-ski-areas/index.html.

Ipsos. 2018. "Qui passera ses vacances à la montagne?": https://france3-regions.francetvinfo.fr/auvergne-rhone-alpes/que-recherchent-francais-qui-partent-montagne-cet-hiver-etude-inedite-deconstruit-idees-recues-1544556.html.

Lenzen, Manfred, Sun, Ya-Yen, Faturay, Futu, Ting, Yuan-Peng, Geschke, Arne and Arunima, Malik. 2018. "The carbon footprint of global tourism". *Nature Climate Change* 8: 522–528.

https://www.independent.co.uk/environment/tourism-climate-change-carbon-emissions-global-warming-flying-cars-transport-a8338946.html (7 May 2018).

Les Sommets: http://www.sommets.com/fr/parc-amusement/.

"L'industrie des sports d'hiver menance-t-elle l'avenir de la montagne?". *Poprock*: http://www.poprock-agence.com/lindustrie-sports-dhiver-menace-t-lavenir-de-montagne/.

National Round Table on the Environment and Economy. "Whistler Blackcomb Holding Inc. – Case Study". http://nrt-trn.ca/whistler-blackcomb-holdings-inc-case-study.

NSAA. Climate Change/Climate Challenge. http://www.nsaa.org/environment/climate-change-climate-challenge/.

NSAA. Sustainable Slopes. http://www.nsaa.org/environment/sustainable-slopes/.

Organisation météorologique mondiale. 2018. Déclaration de l'OMM sur l'état du climat mondial en 2017.

Ouranos: https://www.ouranos.ca/ouranos/.

Ouranos. 2015. *Sommaire de la synthèse des connaissances sur les changements climatiques au Québec*. Montréal: Québec, 13p. https://www.ouranos.ca/synthese-2015/.

Paque, Gwénaëlle, Bleau, Stéphanie, Lebon, Coralie, Germain, Kate et Vachon, Marc-André. 2018. "Diagnostic des risques et des opportunités liés aux changements climatiques pour le secteur touristique des régions de Québec et Charlevoix". *Rapport présenté à Ouranos*. 125 pages: https://www.ouranos.ca/publicationscientifique/RapportGermain2018.pdf.

Pörtner, Hans-Otto (co-président d'une rencontre du GIEC en Corée du Sud tenue en octobre 2018). 24 octobre 2018. l'Express:https://www.lexpress.fr/actualites/1/monde/climat-le-giec-appelle-a-des-transformations-sans-precedent_2038778.html.

Resort Municipality of Whistler: https://www.whistler.ca/municipal-gov/community-monitoring/whistler-facts-and-figures/whistler-visitation.

RRC Associates. 2018. Kottke National End of Season survey 2017/18, NSAA.

Steiger, Robert. 2010. "The impact of climate change on ski season length and snowmaking requirements in Tyrol, Austria". *Climate Research* 43(3): 251–262.

Tremblant. *Le Développement durable au coeur de Tremblant. Ecoresponsabilité*: https://www.tremblant.ca/fr/a-propos/ecoresponsabilite.

Vail Resorts / Epic Promise. http://www.epicpromise.com/environment/.

Wanderlust Tremblant. 2019: https://wanderlust.com/fr/festivals/tremblant/.

Whistler Backcomb. https://www.whistlerblackcomb.com/explore-the-resort/activities-and-events/summer-activities/peak-2-peak-360-experience.aspx.

https://www.lesoleil.com/affaires/un-club-med-aux-accents-de-charlevoix-348e17b952385ed5e88d47750a.

https://www.whistlerblackcomb.com/explore-the-resort/about-the-resort/environment.aspx; https://www.whistler.ca/municipal-gov/community-monitoring.

2.3

From Crisis to Renewal: Adaptative Waves as an Approach to Understand Systemic Resilience in Mountain Tourism Destinations

Romano Wyss & Tobias Luthe

Regaining resilience in mountain tourism destinations

Tourism is the backbone of the regional economy of many mountain regions around the world, especially in many valleys of the European Alps (Mathieu, 2015). Tourism activities have allowed the largely agro-pastoral societies of the Alps to move to alternative, often more lucrative ways of generating net incomes in remote and naturally harsh environments. This development model has lately come under increasing pressure, amongst others due to climate change, demographic change and changing consumer preferences (i.e. Bätzing, 2015). To regain resilience, mountain regions will need to innovate for circular, collaborative and synergistic economies beyond tourism.

With the current developments in both the socio-economic (austerity, economic crisis of alpine areas, de-population) and ecological sphere (climate change, biodiversity loss), the economic model of many alpine regions is challenged in fundamental ways (Löffler and al., 2014). Tourists in traditional markets such as Germany and France are extending their demand to other activities which are not necessarily linked to the mountain environment, such as low-cost city trips and long-haul adventure tourism, while alpine tourism destinations are struggling to invest in ageing, but critical infrastructure (e.g. Bischoff and al., 2017).

In the present chapter, we present two destinations that have followed different development paths after their very existence has been questioned, mainly by factors beyond the direct control of the destination. The two cases are Ostana in the Piedmontese Alps (Italy) and San Bernardino in Switzerland (Fig. 21). The first has undergone a fundamental turn-around, while the second has been stuck to its previous development-path. Building on the adaptive waves model (Luthe and Wyss, 2015), we give an explanation of this difference and show practical possibilities how to further develop the touristic model in the two destinations by resorting to ideas inspired by circular economy thinking.

Two destinations after crisis, two pathways: San Bernardino (CH) and Ostana (IT)

Ostana and San Bernardino have drawn up very different management approaches to cope with these interconnected challenges, with which most alpine destinations are confronted. The community of Ostana in the River Po Valley, has undergone a substantial crisis during the 20th century, when the formerly summer-pastoral driven community was quasi abandoned with only five permanent residents left by 1980. 90 years ago, there were up to 1200 people living and working in Ostana in summer. Demographic development, a move of young people to the cities, and the big regional employer, car producer Fiat, moving most of the production to Asia, led to this drastic change. The visionary mayor Giacomo Lombardo put out three pillars of community renewal, in full awareness that the original economic model could not be re-built. These pillars were local traditions, knowledge and skills of people, and the breath-taking nature. The re-building process has since then been organized in participative bottom-up initiatives, with tourism being one amongst other sectors.

The community of San Bernardino in the Swiss canton of Grison, located at the Southern entrance of the San Bernardino North-South tunnel and mountain pass connection, experienced a different crisis. The community economy used to depend almost entirely on winter tourism, with guests coming mostly from Ticino and Italy as far as Milano. With the necessary renewal of the family-owned ski lift infrastructure, the creeping crisis of unpredictable snow conditions and diminishing interest in skiing, the crisis came to an outbreak. The ski lifts remained closed for four consecutive seasons, while the municipality was struggling to find new investors for a renewal of the ski lifts. The strategy was to prolong the existing economic model of ski tourism's singular growth, leading to another quantitative growth spiral where new investments in ski lifts would require a growth in ski ticket sales, followed by new parking, new hotels, eventually more financial risks under uncertain conditions (climate change, diversifying winter tourism demand). To the moment, there was no participative bottom-up process for the development of a future vision in San Bernardino, and a generalized lack of trust and collaboration has been the outcome (Rottmann, 2015).

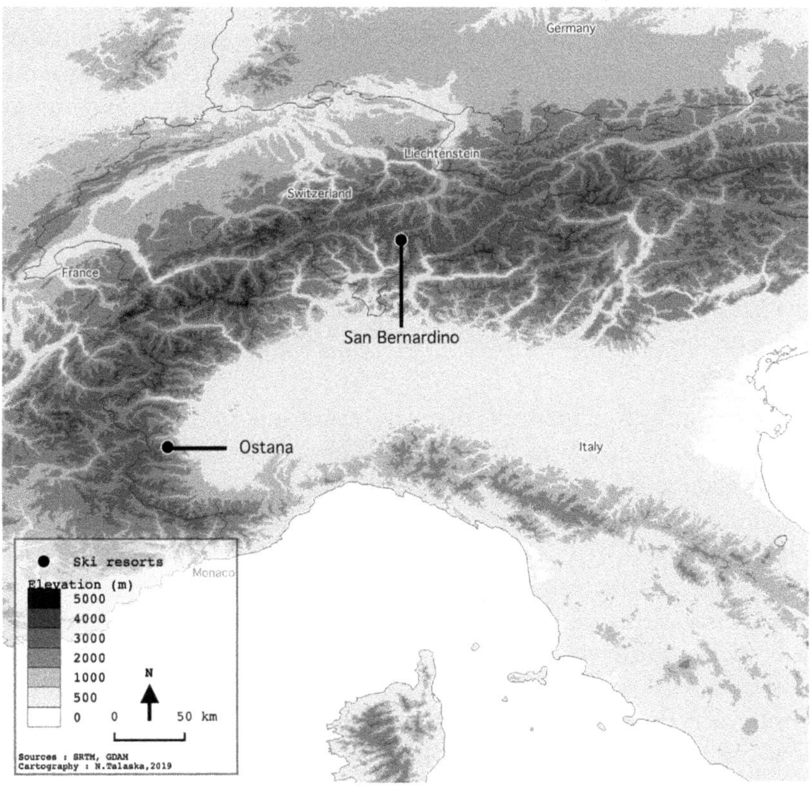

Fig. 21: Studied ski resorts in Italy

Deliberate transformation to higher resilience – the adaptive waves concept

In the context of systemic challenges and changes, such as those evoked for example by climate change in mountain tourism, the resilience concept provides an excellent starting point (Luthe and Wyss, 2014a). Resilience describes up to which level of external influence a system can maintain its basic functionality, when confronted with external shocks and stresses (Folke, 2006). External shocks can be very diverse in both quality and quantity. Certain shocks lead to quick and sudden changes, while others take a long time to manifest (Walker and al., 2004).

Luthe and Wyss (2015) have introduced the adaptive waves concept (see fig. 22), which provides a theory-driven explanation why oscillations in the composition and functional resistance of complex, socio-ecological systems can be found in a vast area of applications (see Luthe and Wyss for three distinct areas of application in tourism, fisheries and semi-nomadic pastoralism). Thereby, the resilience of the systems of interest can be strengthened, at least partially, by deliberate human actions, such as investments in physical and social capital, or changes in the governance systems which structure local tourism activities (Luthe and Wyss, 2016a; Luthe and Wyss, 2015). An additional important underlying factor which affects the resilience of a tourism system is the diversity of actors which make up the tourism sector, as well as the way actors are dependent on these resources and make use of them for their economic survival (Luthe and Wyss, 2017).

Overall, it is important that the transition process itself is also managed in such a way that the functioning of the (tourism) system is maintained along the transition path-way (Binder and al., 2017). In other words, tourism actors and governance bodies in tourism should make sure that there is no full collapse of system-relevant elements which necessitate a costly re-building process, as for example in the case of central infrastructure operators going bankrupt, but that the system is willfully transformed in view of an envisaged goal-system state, such as that of a zero-emission, 2000 Watt-destination. It is also important to ensure that the transition process is sustained over time, in other words that once the decision to fundamentally change the given system state is taken, the motivation and willingness of the population to support the change-process lasts while maintaining the openness to adapt the transition path-way if external conditions require this (Schilling et al., in preparation).

On a lower resilience level, oscillation is bigger than on a higher resilience level with dampened oscillating waves. San Bernardino and Ostana are positioned at their pre- and post-crises system state: San Bernardino with its closed ski area (first two pictures) re-built the status quo and has not gained higher resilience; Ostana (pictures 3 and 4) has deliberately transformed its economy, resulting in a higher resilience level.

When tourism destinations are confronted with external shocks leading to fundamentally new business environments, for example by unfavorable and highly volatile conversion rates or rapidly changing consumer preferences because of social media influence, destinations have the possibility to either reactively answer by optimizing their already existing offers, or set new goals and develop a transition agenda to attain a more long-term oriented business model that is based on the unique characteristics of the destination and can withstand these kind of shocks.

In order for such a transition to be successful, mental and physical barriers to change have to be addressed and a clear transition agenda – ideally supported by a majority of the local stakeholders – should to be set in place in an inclusive and participative way in order to prevent later set-backs due to unaddressed dissatisfactions of community members.

In practice, we can observe that transformative processes only really take off once a certain threshold of dissatisfaction with the status-quo

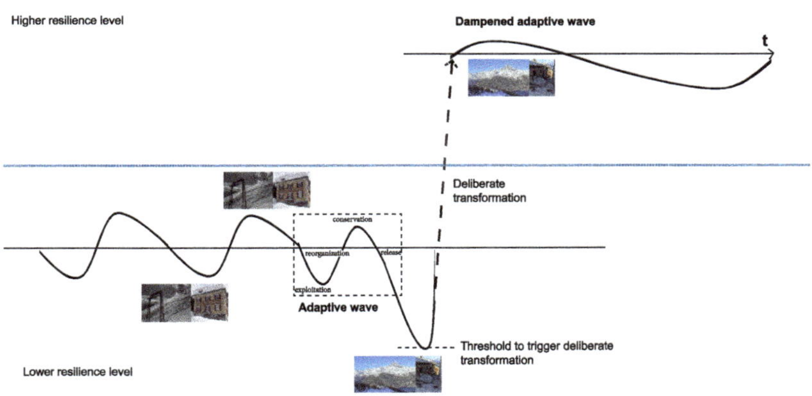

Fig. 22: The adaptive waves concept (adopted from Luthe and Wyss, 2015)

is reached and thereby resistances to change can be overcome (see the adaptive waves model). Practically, in mountain tourism, this means that the ski-based destination model, which has been challenged by changing climate patterns, altering consumer preferences and ageing infrastructure, must first be actively deconstructed by the same tourism actors that have depended on skiing activities in the past for their living, before new opportunities can be seized (Luthe, 2009).

While one option consists of riding the saturated peak of the wave until its end, when external pressures push the system to adapt (often with substantial losses) to the new situation, an alternative consists of defining a new system-state goal, in our tourism case for example that of a tourism model based on non-motorized summer activities such as hiking and biking, or the reorientation of winter activities towards ski touring and snowshoe hiking (Wyss and al., 2014b) – or even towards a more diverse economy where tourism is only one of other strong sectors. In a resilience context, this means that the actors in a tourism system define an alternative, desirable system-state before crossing the threshold when the resilience of the previous system is reduced to zero and the system passes into a new system-state in an uncontrolled and chaotic processes of ad-hoc adaptation. The actions that are required to be taken to achieve this new, more desirable system state can, for example, be defined in a design thinking and back-casting process (Fig. 23), where the envisaged system-state goal is taken as a starting point to define explicit measures which are to be implemented in order to reach a more desirable and sustainable functioning of the tourism system (Fischer, 2015).

The adaptive waves illustrated by means of San Bernardino and Ostana

In our two examples of alpine communities, we can illustrate how sudden, external shocks can accentuate slower, longer-lasting processes which lead to fundamental altering of how social systems function. Take our first village, Ostana in the Italian Piedmont region, shaped by agricultural activities and marked in its development by the geographical vicinity to the metropole specializing in car production. In Ostana, the traditional economic model, based on subsistence agriculture, was followed for a very long period of time, up to the point when the mismatch between the local systemic set-up and the general economic conditions

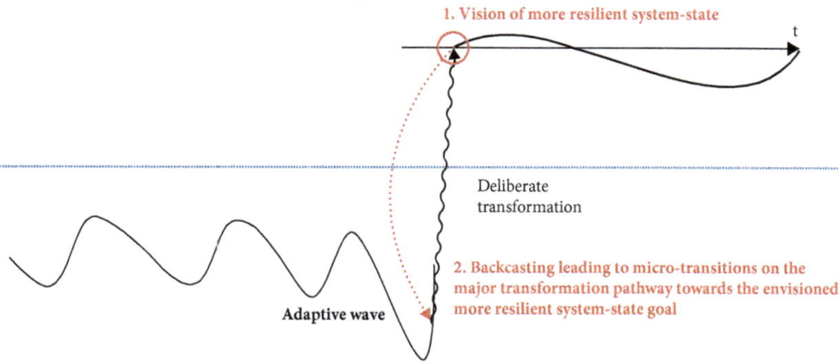

Fig. 23: The adaptive waves concept shown with the design thinking process for a desired, more resilient system-state (1), the following backcasting process (2) and the resulting micro-transitions on the major transformation path adopted from Luthe and Wyss, 2015)

and requirements became so evident that there was practically no way to support the local population. The traditional way of working the fields and living from subsistence agricultural activities lost much of its appeal in the early 20th century as new technologies spread to rural areas, new social norms were established and people were more and more aware of the living conditions in other, more urbanized areas. Then, with the two world wars taking place in the first half of the 20th century, large parts of the male population left the valley to fight for their country in the far-away plains of Western and Eastern Europe, and other mountain areas of the country. Of those who have survived the battles, only a minority returned to their families and their valley of origin, adding additional stress to the economic situation and leaving an important gap in the demographic constitution of the village.

San Bernardino, on the contrary, is just now at a bifurcation point, where the system actors are forced to choose their future development path, partially because of urgent investment requirements in the ageing skiing infrastructure. Therefore, inhabitants of San Bernardino have to decide (if receiving the chance and being asked) whether they want to continue with a socio-economic model that is heavily reliant on second-home owners from the big agglomerations to both the North and the South of the main Alpine ridge who fill the streets of the village for just two-to three weeks around Christmas and Easter. Or, as an alternative, whether their future should rather lay in an economic model that is based more on the resources and people that are (already) present in the region and would help to set up a (tourism) business model that is more balanced in terms of both the main guest segments that visit the destination and the time periods in the year when they do so. The economic model could even be more inclusive to other sectors, such as agriculture, forestry, transportation, services, or mobility, in order to diversify the economy, develop inter-sectoral synergies and become less dependent on classic ski tourism (Rottmann, 2015).

The inhabitants of the second village, San Bernardino in the Swiss Alps, in contrary, never participated in a war since the middle-ages. Their town has been left untouched by many influences of modern times after a project to build a railway-line through the valley and a long tunnel to the Northern versant of the Alps. It never reached completion due to financial difficulties in connection with the tourism crisis after the end of the Belle Époque. It was not until after World War II that external influences had a substantial effect on the community and the way it functioned. With increasing incomes of the middle-classes both to the North and the South of the alpine divide, and the amelioration of the road transport infrastructure, summer holidays became more and more common for middle-class families. At the same time, technological improvements in ski-lift design and the introduction of groomed ski-slopes lead to pursuing winter sport activities. The first ski lift was built in San Bernardino in 1949, the first all year-round connection to the North via the newly built road tunnel was established in 1967. The establishment of a secure connection to the source markets North of the Alps also attracted many investors to the village, which in 10 years' time dramatically changed its appearance. A big number of second homes were built, and a large area with a 200+ bed hotel was developed in the Northern part of the village.

Revelation of structure in stakeholder participation by social network analysis

An in-depth analysis of social networks allows to unravel the quantity and quality of the dependences between actors, as well as between actors and resources (van der Zee and Vanneste, 2015; Beritelli and Bieger, 2014; Baggio and al., 2010). By applying network metrics to destination structures, i.e. collaborative stakeholder networks and their interpretation in terms of resilience, key insights of the functioning of tourism systems can be derived. This with respect to both the capacity of the system to withstand external pressures and shocks – described in the literature as resistance – as well as to the capacity of actors to adapt system structures and system functions to external changes, sucsh as climate change– described in the literature as adaptive capacity (see Luthe and Wyss, 2015; Kelman and al., 2016 for empirical applications in the tourism context).

The social network of Ostana is today rather dense, with a clear core of actors who situate themselves around the central figure of the village, the long-time mayor and indefatigable promoter of the village, Giacomo Lombardo (MonViso Institute, 2017). At the same time, the economic networks have suffered by the sustained outmigration and the loss of most jobs in agriculture and forestry over the course of the last thirty years.

In San Bernardino, the social network is much more dispersed, with not clear core of actors that are in close contact to one another, complicating joint decision-making processes and quick collective action, e.g. in the case of natural disasters. On the other hand, the economic network is much more developed, even if thinning out over the course of the last years, with a great number of actors grouping around the destination organization, a handful of hotels, and a farm offering lama trekking activity (Rottmann, 2015).

Strengthening resilience by innovation in a circular economy

Mountain tourism regions will likely remain continuously and increasingly confronted with many challenges, reaching from climate and demographic change to currency instabilities and political turbulences. In these circumstances, destinations will have to adapt their current business models and undergo major transitions. The revival of mountain economies demands the development of a more resilient economic model that is more adaptive and innovative to prepare for and respond to change.

A promising option is that of a circular economy (e.g. Andersen, 2015; Murray and al., 2015) which may allow tourism dependent communities to thrive in the years to come. Such a more resilient economy is based on higher connectivity between different economic sectors, mimicking natural systems that function in circular ways where no waste exists, but outputs from one process are new inputs for another one. The design of a circular economy in mountain regions requires sophisticated tools and motivating illustrations to address complexity, and to overcome jealousy and lack of collaborative will.

What this means in a mountain destination context can be summarized as such:

- strengthening of network-thinking, by taking interlinkages between actors and sectors into account, in order to understand the dynamics, synergies and context specificities of destination development,
- taking into account the regional social, economic and natural context which allows place-specific resilience building (resistance and adaptive capacity),
- supporting regional connections within tourism and other economic sectors (circular thinking), allowing to build on the capacities that are present in a region, thereby potentially dampening the (negative) influence of external shocks and stress factors, while creating locally bound value.

Coming back to our both illustrative mountain destinations with different resilience pathways, Ostana and San Bernardino, we can exemplify possible applications of circular economy approaches:

In Ostana, Cannabis (hemp) has been historically cultivated as the "bread and butter" plant, displayed in the local museum. Hemp is amongst the oldest cultivated plants with a worldwide history of agricultural use (Herer 2008). In many mountain regions, the traditional mountain economy used to be based on industrial hemp. Unfortunately, hemp became largely misunderstood, and as a result was forbidden to grow, own and utilize. In the last years though, society has been re-discovering this plant and its genius capacity to power an entire economy. Hemp grows basically anywhere, produces fast biomass, improves the soil by loosening it with its deep root system, does not require pesticides, the fibers of the sheath can be used to produce fabric, clothing and paper, the stems can be mixed with lime stone as a building material, the seeds can be used to produce oils for the kitchen and nowadays for 3D printing of organic plastic.

The potential to use the hemp cycle for designing a circular economy in the mountain community of Ostana, Piedmont, Italy, has already started to connect agriculture, forestry, architecture, construction, gastronomy, tourism and textiles for building economic, social and ecological synergies as the trigger of a local, circular economy. At the real-world laboratory MonViso Institute (MVI) we grow the regional *Carmagnola* type of industrial hemp. Its fibers are used for locally building skis with a new hemp-composite, and for reinforcing a kind of natural cement-concrete (hempcrete) for building house foundations; the shives (the inner wooden part of the stem) are used as building insulation material mixed with lime; the male and female blossoms are used for cosmetics and medications, and for distilling a new local liquor. The locally produced amounts of hemp are still illustrative, but with the provided examples comes a new interest for selling the valuable seed's oil, making flour from the seeds in a newly developed bakery, and install bee hives in late autumn when the hemp still blossoms. The MVI hemp is now being advertised by the municipality as a tourist attraction, new hemp building courses are being offered to train architects, engineers and building companies; the hemp field in Ostana has already begun to re-connect existing sectors and develop new socio-economic-cultural synergies on the pathway towards a circular economy.

In San Bernardino, the location at the famous and heavily used mountain pass on a European North-South axis could as well be translated into a local circular economy. If the abundant road travelers were invited for a driving break in fresh mountain air, with child care for families to "release" parents for some time off to sleep or do sports, then this could be an incubator for a new type of economic system. "Travel through" experiences and services, adopted to the local topography with two lakes, open pastures on terraced land facing the Southern lower valley, could stimulate new collaborations and new economic potentials. Open pastures could be used for luxury car camping, "glamping" called, both for quick rests and one or two nights in between the long travel distances of e.g. Dutch guests driving to the Italian beaches. Solar panels could provide shade for the individual parking spots, the panels could even cover most of the Southern slopes, while being installed with sufficient space below to allow for mountain goats and sheep to graze. Their wool could be used to manufacture a local clothing/fabric brand and produce sense-making souvenirs. With the abundance of clear spring water, a local brewery could offer a new local beer, while mushrooms can be grown on the beer brewing residues. Beer and mushrooms can be offered in local "zero km" restaurants. The whole village could specialize on zero km resting in a responsible, healthy mountain environment, where resting becomes part of the vacation. Many more economic connections both on a matter and product flow, but as well on a service product and shared consumer flow basis are possible for building a local circular and more resilient economy.

Overcoming social resistance and moderating collaborative visions

A major danger for successful transition processes that allow to strengthen the resilience of mountain tourism destinations is that along the way the interests of particular groups and the existing structures are perpetuated, possibly leading to situations of socio-economic lock-ins, where the system-actors are no longer able or willing to adapt to changing external social and economic requirements. Such resistances have to be addressed very early on, and should ideally lead to geographically diversified development models, which take into account the specificities of both the social and natural properties of the region.

Our theoretical discussion and its illustration on the two destinations have shown, how a network-based understanding of mountain tourism destinations, a systemic resilience perspective and the circular understanding of regional value creation can conceptually be aligned. Those are pathways in which the economic and natural potential of mountain regions can better be put in value than it is currently the case.

As of next steps and practical suggestions on how to develop this systemic, circular economic potential, in order to help build and sustain resilient regional mountain tourism systems in the future, one suggestion is to start with mapping the existing human, social and economic capital (collaborative social networks and skillsets) of a place. By analyzing and visualizing the existing potentials, one can start a collaborative, illustrative and inclusive design thinking process for developing a commonly shared, place-based vision of the future. Through backcasting techniques, one then looks back from the future vision to the present and co-develops concrete steps how to bridge from the now to the future. External moderators with sufficient local understanding and acceptance, but as well with a healthy distant position, are critical to incubate and continuously stimulate such a process.

References

Andersen, Mikael S. 2015. "An introductory note on the environmental economics of the circular economy". *Sustainability Science* 2(1): 133–140, DOI:10.1007/s11625-006-0013-6.

Baggio, Rodolfo, Scott, Noel & Cooper, Chris. 2010. "Improving tourism destination governance: a complexity science approach". *Tourism Review* 65(4): 51–60.

Bätzing, Werner. 2015. *Die Alpen: Geschichte und Zukunft einer europäischen Kulturlandschaft*. CH Beck, München.

Beritelli, Pietro & Bieger, Thomas. 2014. "From destination governance to destination leadership–defining and exploring the significance with the help of a systemic perspective". *Tourism Review* 69(1): 25–46.

Binder, Claudia. R., Mühlemeier, Susan and Wyss, Romano. 2017. "An indicator-based approach for analyzing the resilience of transitions for energy regions. Part I: Theoretical and conceptual considerations". *Energies* 10(1): 36.

Bischof, Michael, Schmude, Jürgen and Bauer, Manuela. 2017. "Tourism and climate change-a demand-side analysis of perception and reaction using the example of the Alps". *Zeitschrift für Tourismuswissenschaft* 9(2): 221–248.

Fischer, Matthias. 2015. "Design it! Solving sustainability problems by applying design thinking". *GAIA* 24 (3): 174–178.

Folke, Carl. 2006. "Resilience: The emergence of a perspective for social–ecological systems analyses". *Global environmental change* 16(3): 253–267.

Gobiet, Andreas, Kotlarski, Sven, Beniston, Martin, Heinrich, Georg, Rajczak, Jan, and Stoffel, Markus. 2014. "21st century climate change in the European Alps–a review". *Science of the Total Environment* 493: 1138–1151.

Herer, Jack. 2008. *Die Wiederentdeckung der Nutzpflanze Hanf. Hemp & The Marijuana Conspiracy*. Nachtschatten Publisher, Solothurn. 41st edition. ISBN 978-3-03788-181–1.

Kelman, Ilan, Luthe, Tobias, Wyss, Romano, Tørnblad, Silje. H., Evers, Yvette, Curran, Marina M., Williams, Richard J. and Berlow, Eric. L. 2016. "Social network analysis and qualitative interviews for assessing geographic characteristics of tourism business networks". *PlosOne* 11(6), e0156028.

Löffler, Roland, Beismann, Michael, Walder, Judith and Steinicke, Ernst. 2014. "New highlanders in traditional out-migration areas in the Alps. The example of the Friulian Alps". *Journal of Alpine Research | Revue de géographie alpine, 102-3 | 2014, mis en ligne le 01 février 2015, consulté le 14 février 2020. URL :* http://journals.openedition.org/rga/2546

Luthe, Tobias. 2009. "Vulnerability to Global Change and Sustainable Adaptation of Ski Tourism: SkiSustain". PhD thesis, Deutsche Sporthochschule.

Luthe, Tobias. 2017. *The Resilience of Snow Sports: From Vision to Transformation. Zur Resilienz des Schneesports: Von der Vision zur Transformation.* FdSnow 51 2.2017. Fachzeitschrift des Skisports. Planegg, Germany. Deutscher Skiverband, Freunde des Skisports, Stiftung Sicherheit im Skisport (HrsG). ISSN 1864-5593.

Luthe, Tobias and Wyss, Romano. 2014a. "Assessing and planning resilience in tourism". *Tourism Management* 44:161–163. DOI:10.1016/j.tourman.2014.03.011 .

Luthe. Tobias and Wyss, Romano. 2014b. "Die Nutzung von sozialer Netzwerkanalyse zur Steuerung der Resilienz touristisch geprägter Räume". In *Zukunftsgestaltung im alpinen Tourismus – Schweizer Jahrbuch Tourismus 2013/2014*, C. Bieger, Thomas, Beritelli, Pietro and Laesser, Christian (eds.). Berlin: Erich Schmidt Verlag, 57–74.

Luthe, Tobias and Wyss, Romano. 2015. "Introducing adaptive waves as a concept to inform mental models of resilience". *Sustainability Science* 10(4): 673–685.

Luthe, Tobias and Wyss, Romano. 2016. "Resilience to climate change in a cross-scale tourism governance context: a combined quantitative-qualitative network analysis". *Ecology and Society* 21(1):27.

Luthe, Tobias and Wyss, Romano. 2017. "Eine sozial-ökologische Netzwerkanalyse zur Bedeutung von Ökosystemleistungen in der touristisch geprägten Schweizer Parc Adula-Gotthard Region. A social-ecological network analysis on the importance of ecosystem service sin the tourism dependent Swiss Parc Adula-Gotthard region". In *Wirtschaftsfaktor Sporttourismus. Band 19 der Schriftenreihe der Deutschen Gesellschaft für Tourismuswissenschaft e.V.*, Roth, Ralf and Schwark, Jürgen (Hrsg.). Eric Schmidt Verlag. ISSN 1612–8672.

Mathieu, Jon. 2015. *Die Alpen: Raum, Kultur, Geschichte [The Alps: Space, Culture, History].* Reclam, Stuttgart.

MonViso Insitute. 2017. "Social Network Analysis of the innovative community potential in Ostana". Course work of the USI (Universita Svizera Italiana) Master class Sustainable Tourism Colloquium. Presented in a public community meeting in Ostana.

Murray, Alan, Skene, Keith and Haynes, Kathryn. 2015. "The circular economy: an interdisciplinary exploration of the concept and application in a global context". *Journal of Business Ethics*. DOI: 10.1007/s10551-015-2693-2.

Neuhofer, Barbara, Buhalis, Dimitrios and Ladkin, Adele. 2015. "Technology as a catalyst of change: enablers and barriers of the tourist experience and their consequences". In *Information and communication technologies in tourism 2015*. Lugano, Switzerlang, Springer, 789–802.

Rottmann, Melanie. 2015. "Local participation and knowledge integration in the transition towards sustainable community development: Cooperative planning of a tourism-based Blue Economy destination model in San Bernardino". BSc thesis, HTW Chur.

Schilling, Thorsten, Wyss, Romano and Binder, Claudia R. 2018. "Going beyond the threshold – theoretical implications on a new way to look at societal transitions from a dynamic resilience perspective". *Sustainability* 10(12): 4593.

Van der Zee, Egbert and Vanneste, Dominique. 2015. Tourism networks unravelled; a review of the literature on networks in tourism management studies. *Tourism Management Perspective* 15: 46–56.

Walker, Brian H., Holling, Crawford S., Carpenter, Stephen R. and Kinzig, Ann. 2004. «"Resilience, adaptability and transformability in social–ecological systems"". *Ecology and Society* 9(2): 5.

Wyss, Romano, Abegg, Bruno and Luthe, Tobias. 2014. "Perceptions of climate change in a tourism governance context". *Tourism Management Perspectives* 11: 69–76.

Wyss, Romano, Luthe, Tobias and Abegg, Bruno. 2015. "Building resilience to climate change–The role of cooperation in alpine tourism networks". *Local Environment* 20(8): 908–922.

2.4

Critical Reflections on Projections of Climate Change Risk for the Ski Industry

Daniel Scott and Robert Steiger

Introduction

Ski tourism is a multi-billion dollars international market attracting over 400 million annual skier visits worldwide (Vanat, 2018; Research and Markets, 2018). With its direct dependence on climatic resources, the ski industry was the focus of the first climate change risk assessments in the tourism sector (Scott et al. 2012) and this literature has expanded to include 27 countries (Steiger et al., 2017). Studies that focus on the impacts of climate change on ski operations and season length have without exception concluded that climate change represents a significant risk to the profitability and sustainability of ski resorts around the globe, because of decreased and more variable natural snow, increased snowmaking requirements and reduced snowmaking opportunities and shortened and more variable ski seasons. Collectively, these impacts will alter the competitiveness of ski areas and reduce the number of operating ski areas in every regional market, with attendant implications for ski tourism employment and the value of vacation property real estate.

The need to understand the differential climate change risk within and among regional ski markets continues to increase. Over the past several years the authors have been contacted by increasingly diverse decision-makers (e.g., ski area owners, grooming and snowmaking equipment companies, insurance companies, institutional investors and pension fund managers and real estate buyers and developers) for climate risk information and expert advice on the implications of climate change for individual ski areas and regional ski tourism markets. Major real estate management companies are already advising clients on the climate resilience of ski destinations to inform vacation property investment decisions (Savills World Research, 2018). The Task Force on Climate-related Financial Disclosures, established by the Financial Stability Board of the G20 countries, has developed guidelines for companies to disclose their physical climate risks and opportunities (European Bank for Reconstruction and Development, 2018). The disclosure of physical climate risk will become increasingly formalized and is likely to be required by major stock markets in the early 2020s. When all publically traded ski conglomerates (e.g., Vail Resorts, Intrawest, Peak Resorts, American Ski Company, SkiStar AB) are required to assess and disclose their physical climate risk, the financial sector will demand the same information from all ski conglomerates and independent ski resorts. No longer will ski areas in any regional market be able to ignore their climate change risk in hopes of maintaining a perception of low credit risk.

The demand from diverse stakeholders for foresight on the implications of accelerating climate change will increase tremendously in the years ahead. Currently, many decision-makers are being influenced by media coverage of the ski industry's risk to climate change, which is dangerous because of the substantial misinformation in the media (Scott et al., 2012; Scott and Steiger, 2013). The tourism research community has an important role to provide science-based, decision-relevant climate change risk information on the ski industry to the above-mentioned range of stakeholders that will influence the future development of ski tourism.

Despite the consensus in the academic literature that climate change represents a serious risk to the ski industry, the range of methodological approaches (for a detailed review see Steiger et al., 2017) project very different severity and timing of impacts. An important reason for the uncertainty in the magnitude of impacts is the methodological limitations of many studies in the extant literature (Steiger et al., 2017), which can be summarized as: 1) inappropriate temporal and spatial resolution of climate data; 2) inappropriate ski industry performance indicators; and

Tab. 5: Characteristics of case study ski areas (Weather station data, provided by Zentralanstalt für Meteorologie und Geodynamik)

	Kitzbühel	Obergurgl
Climate station altitude	744 masl	1942 masl
Mean temperature Nov-Apr (annual)	1.31°C (7.92°C)	-2.37°C (3.29°C)
Precipitation Nov-Apr (annual)	484 mm (1310 mm)	360 mm (884 mm)
Days with snow cover	104	196
Annual snowfall	233cm	482 cm

3) the omission or proxy representation of snowmaking, a widespread and expanding climate adaptation. Studies with these limitations must be interpreted cautiously and Scott and al. (2012) and Abegg and Steiger (2017) have stressed the need for critical reflection on the state of knowledge and media speculation on ski tourism climate change risk. The remainder of this chapter discusses each of these important limitations and demonstrates how they adversely affect projections of ski seasons under future climate conditions using two case studies in Austria (Tab. 5): internationally renowned Kitzbühel (740–2000 masl) and the much higher elevation Obergurgl (1800–3080 masl).

Climate data resolution

Snow is a highly variable (temporally and spatially) resource. Using coarse monthly resolution climate data (e.g. monthly mean temperature) in physical snow models or to derive statistical relationships is insufficient to analyze potential changes in natural snow conditions (e.g., Breiling and Charamza, 1999; Galloway, 1988; Pickering and Buckley, 2010) or the potential for snowmaking (e.g., Bark and al., 2010). If the ski industry loses two or three weeks at financially critical times of the year, particularly around the Christmas-new year holiday or school holidays in February, their profitability can be put at risk. Monthly scale climate inputs simply do not capture this operating reality.

Using monthly resolution climate data to estimate periods of snowmaking potential are also problematic. Advanced snowmaking systems are highly automated so that snowmaking teams can take advantage of suitable snowmaking conditions that may last only a few

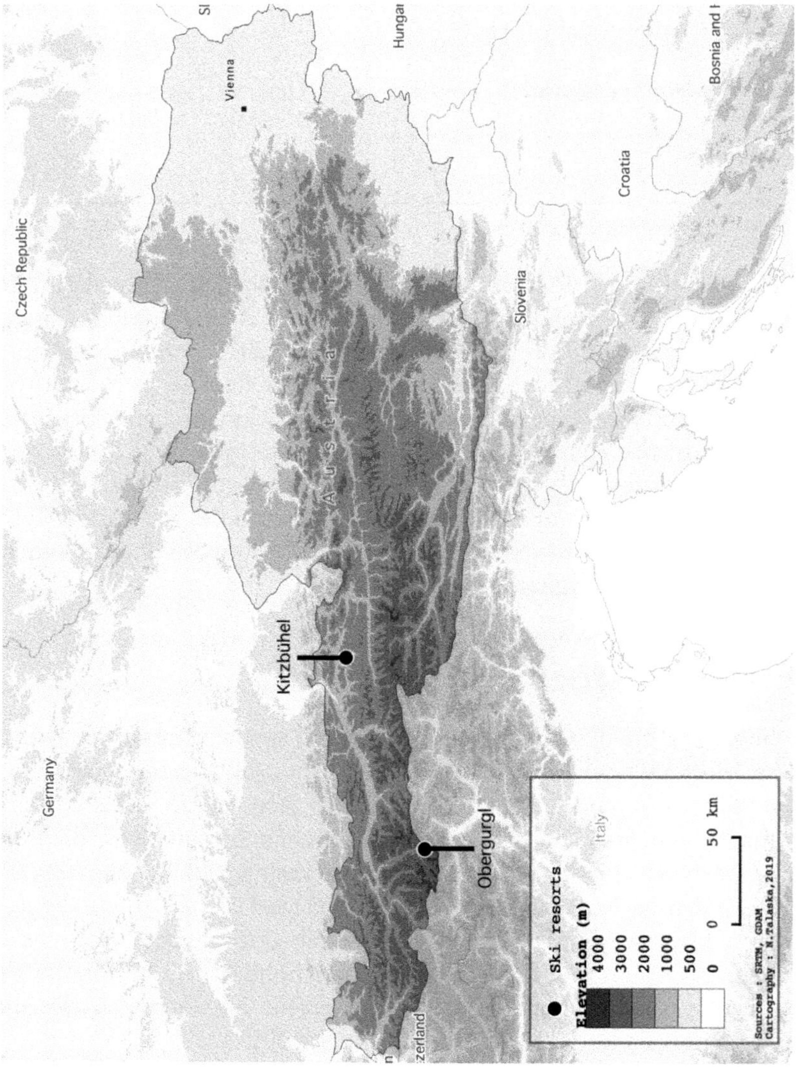

Fig. 24: Studied ski resorts in Austria

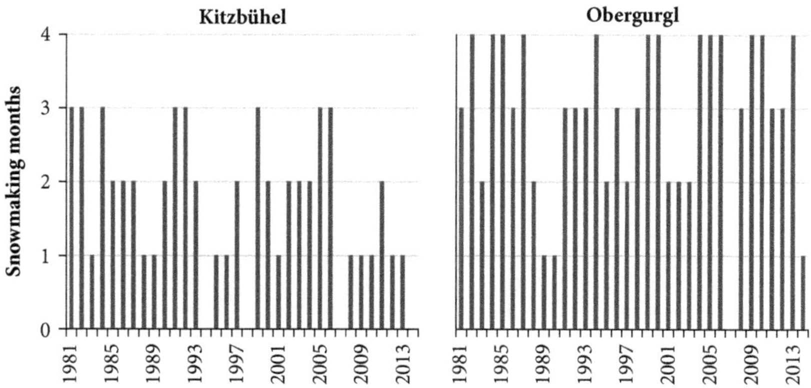

Fig. 25: Estimated snowmaking periods using monthly resolution temperature data

hours. Figure 25 shows that if monthly temperature data (monthly tmin <=-5°C) was used to calculate potential snowmaking periods in Kitzbühel, there would be many years with no snowmaking at the valley floor and in most years only one month of snowmaking. The results from these inappropriate climate data contradict the operating reality in the study area and many ski areas at similar elevations in the Alps. This approach indicates that Obergurgl would have much more climatically suitable snowmaking periods, but even at this high elevation there are many years with two or less months suitable for snowmaking. In contrast, Obergurgl has an average ski season from mid-November to May 1.

Even higher temporal resolution climate data must be considered carefully when estimating snowmaking potential. Figure 26 compares the number of days available for snowmaking at Kitzbühel over the 2003–07 seasons (using an operational temperature threshold of <=-5°C), using Tmean, Tmin and linear interpolation of temperature between daily Tmin and Tmax. Sub-hourly temperature data (15 minutes intervals) provides the most accurate estimate of potential snowmaking days (black bar). In comparison, using daily Tmean (e.g., as in Steiger and Mayer, 2008) underestimates potential snowmaking days substantially (-33 % over the period of analysis). Using daily Tmin (e.g., as in Scott et al. 2003 leads to a substantial overestimation of snowmaking potential (+74 %) in the study area. Steiger's (2010) method of linear interpolation between daily Tmin

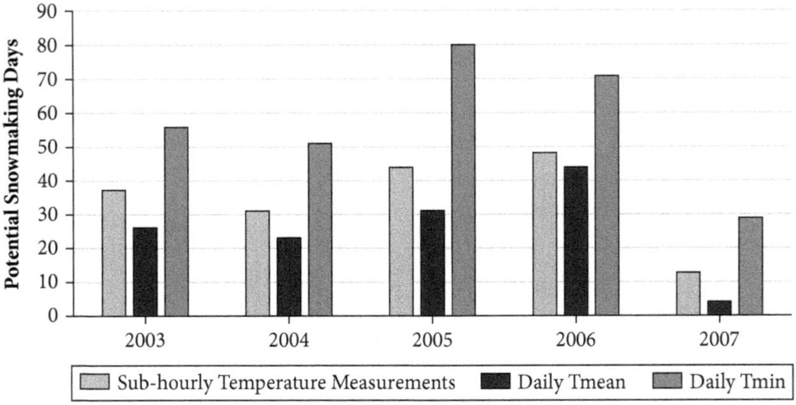

Fig. 26: Comparison of potential snowmaking days at Kitzbühel (Nov-Apr) using hourly and daily climate data

and Tmax to estimate snowmaking hours results in an improvement (-1 % over the period of analysis). Because hourly data is not widely available at climate stations (particularly among sparse mountain stations) and not at all in widely available gridded temperature data sets, this appears the most accurate approach to estimating the amount of time available for snowmaking at a location.

Another particular challenge for modelling climate change impacts on ski operations is the limited availability of climate data in close proximity to ski areas. Some studies use gridded model data to address this problem of data availability, but introduce potentially major biases between modeled and observed weather. Endler and Matzarakis (2011) for example report model (gridded data) biases of +4°C in annual mean temperature and +/- 20 % precipitation in the Black Forest region (Germany). Studies must investigate such potential biases in regional gridded data, as they lead to unrealistic snow depth values and uncertainties in the validity of climate change risk assessment. The biases introduced by using data from the nearest climate station (and most proximate elevation) to represent meteorological and snow conditions at a ski area remain uncertain. While many ski areas report onsite weather and snow conditions in real time, this data is not archived for potential comparisons to nearby National Meteorological Service stations and is often of uncertain quality.

Ski industry performance indicators

A second prevalent limitation in the literature has been the utilization of inappropriate measures of ski industry performance. The choice of relevant climate change risk indicators is very important for both the correct interpretation of climate change vulnerability and for credibility with ski industry stakeholders. Some studies have used modeled snow cover days (i.e., snow cover of defined threshold [often 2.5cm], but without known depth) near ski areas (e.g. Galloway, 1988; Harrison and al., 1999; Harrison and al., 2005; McBoyle and al., 1986; Sauter and al., 2009; Tranos and Davoudi, 2014). Although snow cover is defined differently in these studies (e.g., from 1 to 10 cm of snow), none of these depths are suitable indicators for ski operations that require a snow depth of 30 cm or more depending on the terrain (Witmer, 1986 and widespread consultations by the authors with ski area operators in Europe and North America).

Other hydrological indicators, such as the snow-water equivalent on the first day of April (a hydrologic indicator used to estimate summer water supply) have been used to project changes in ski season (e.g., Zimmerman and al., 2006), even though this indicator has no relevance to ski operations because it provides no information on the required operational depth throughout the ski season. Other studies have estimated ski season length solely on the basis of temperatures and precipitation (e.g., Heo and Lee, 2008), or on the basis of days with snowfall (e.g., Moen and Fredman, 2007), with no physical indicator of snow depth required for safe ski area operations. These inappropriate indicators have likely been utilized because they can be computed easily from climate station data (versus developing and validating a physical snow model capable of estimating snow depth). As Fig. 27 reveals there are only moderate correlation between snow depth on April 1 and the number of days with snow depth of 30 cm or more at the two study areas (r^2 of 0.65 Kitzbühel and 0.52 at Obergurgl). Figure 28 further illustrates the moderate to weak correlation between snowfall and days with snow depth greater than 30 cm in the study areas (r^2 of 0.56 Kitzbühel and 0.02 at Obergurgl). Correlations between snowfall and days with more than 30 cm of snow are likely to be weaker in regions with less precipitation, but which are sufficiently cold to conserve a snowpack (even more so at higher elevation as with Obergurgl). Furthermore, it is uncertain whether current correlations would remain unchanged in a warmer climate.

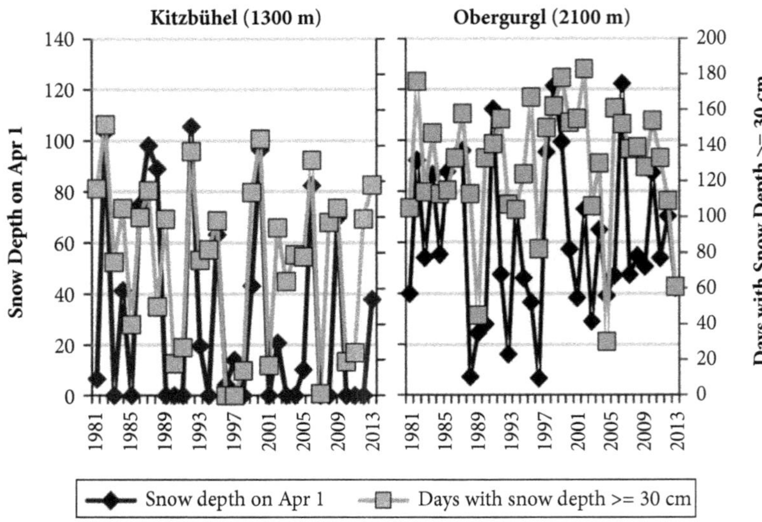

Fig. 27: Correlation between April 1 snow depth and days with >30cm snow depth

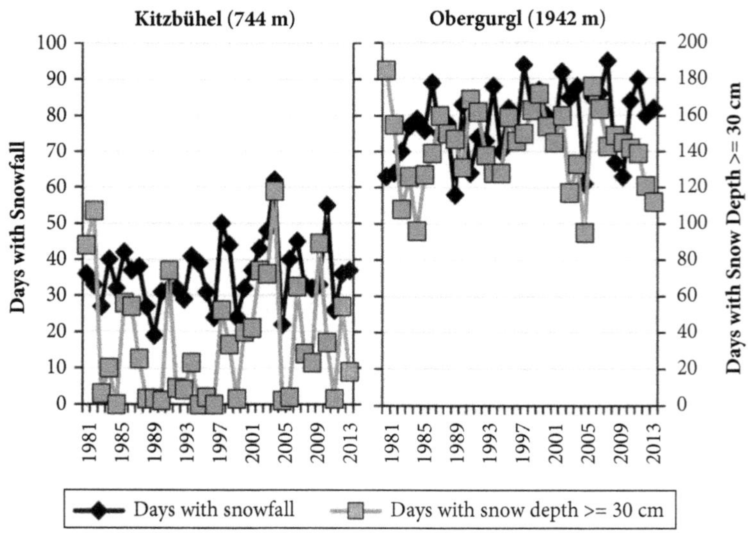

Fig. 28: Correlation between snowfall and days with >30cm snow depth

Snow reliability is another type of indicator that has been used in a number of studies. It is a general economic indicator of the financial viability of a ski area and has usually been defined as being able to achieve a 100-day long ski season (i.e., the '100 day rule' recommended by Witmer 1986) in seven out of ten seasons. Some studies have put forward altitudinal thresholds to define snow reliable ski areas, with those above all classified as reliable and those below unreliable (e.g., Abegg, 1996; Elsasser and Messerli, 2001; Elsasser and Burki, 2002; Abegg et al., 2007). Climate and snow conditions are highly variable in complex mountain terrain, so that establishing such a universal threshold for a regional ski market is problematic. For example, when the 1200m altitude of snow reliability, defined by a study for the OECD (Abegg and al., 2007), was examined at 20 weather stations that are above this altitude and near ski areas in the Austria province of Tyrol, only three were found to achieve 100 day or longer seasons 70 % of the time (between 1981–82 and 2000–01 (shaded area of Fig. 29). Due to macro- (e.g. maritime versus continental climate) and meso-scale (e.g. windward and leeward locations, prevailing weather patterns) climatic heterogeneity, it

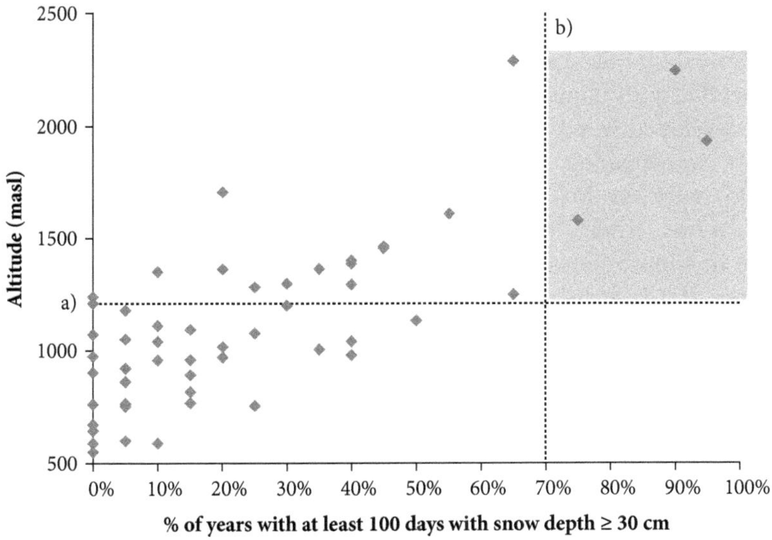

Fig. 29: Weather stations above 1200 masl in Tyrol (Austria) that are snow reliable
Adapted from Steiger (2010)

would be very difficult to generalize snow reliability or climate change impacts to certain elevations in almost any regional ski market (Steiger and, al. 2017).

Adaptation capacity of snowmaking

A third critical limitation of many climate change impact assessments has been the omission of snowmaking. Snowmaking was first used operationally at a ski area in New York State (USA) in the early 1950s and began to be used extensively in eastern North America in the 1970s and 1980s and gradually expanded to ski markets worldwide in the 1980s and 1990s. Early studies in what Steiger and al. (2017) defined as the 'pioneering phase' only considered natural snow and projected that vast majority of ski areas would lose almost all of their ski season by the 2050s in a high GHG emission scenario (e.g., Harrison and al., 1986; McBoyle and al., 1986; Lamothe and Periard Consultants, 1988; Galloway, 1988; McBoyle and Wall, 1987; Lipski and McBoyle, 1991; McBoyle and Wall, 1992; Abegg, 1996; König and Abegg, 1997; König, 1999; Bürki, 2000; Elsasser and Bürki, 2002; Fukushima and al., 2002). The ski industry was highly critical of these studies, arguing they did not represent their current operating reality with extensive snowmaking, let alone their future adaptive capacity and therefore misrepresented their vulnerability to climate change. In response a new modeling approach was developed in order to account for the rapidly evolving operational reality snowmaking provided to ski areas worldwide. Scott and al. (2003) were the first to fully incorporate snowmaking into ski season simulations. Their "SkiSim" model has been continuously improved upon in applications in Austria (Steiger, 2010; Steiger, 2012; Steiger and Stötter, 2013; Steiger and Abegg, 2017), Canada (Scott and al., 2007; Scott and al., 2017), China (Fang et al., 2018), Germany (Schmidt and al., 2012), Switzerland (Abegg and al., 2015), United States (Scott et al., 2008; Dawson and Scott, 2013; Scott and Steiger, 2013) and former Winter Olympic Games locations worldwide (Scott and al., 2014, 2018). Pons et al. (2015) integrated snowmaking in a climate risk assessment for the Pyrenees, but in their approach snowmaking was only activated on days when total snow depth was below a 30 cm threshold. This does not represent common snowmaking practices, when early season snowmaking is used to build a large and durable base (often double or triple the operational 30 cm threshold) in order to stockpile snow for the season. In France the meteorological downscaling system SAFRAN was combined with the snowpack model Crocus and recently modified by including snowmaking and slope grooming (Spandre et al. 2016). In this integrated model, the impacts of ski operations on snow depth and characteristics are simulated, but to our knowledge this model version referred to as "Crocus-resort" has not yet been used for assessing future climate change impacts on ski operations. Though model performance at four sites in France was very promising (Spandre and al., 2016) an application of this model for entire regional ski markets remains difficult, as Crocus requires hourly input data of numerous weather variables that are not available at the majority of climate stations. In every regional market, accounting for snowmaking significantly reduced climate change impacts on season length, revealing a complex set of differential impacts for individual ski areas that altered intra- and inter-regional market competitive positions.

Here we illustrate the difference in simulated ski seasons with only natural snow and with snowmaking at the two case studies (Fig. 30), using the SkiSim model (under RCP 4.5 emission scenario). The difference in season length when snowmaking is accounted for is massive in Kitzbühel, adding an additional 41 days at the top elevation and 95 days at base elevation in the baseline climate (1981–2010). Even at the high elevation Obergurgl site, snowmaking extends the season in the baseline climate 7 days at high elevation and 107 days at the base. At both locations, the length of the ski season with snowmaking is longer even under 2080s climate change scenarios than the current season with just natural snow. The difference between the ski season length with and without snowmaking is so large that not including snowmaking leads to a completely incorrect assessment of current operational reliability and future climate change risk.

Although Scott and al. (2003, 2006) have argued for over a decade that not including snowmaking is to ignore the obvious operating reality of most ski areas and misrepresent the current and future vulnerability to climate change, yet a number of recent studies still neglect snowmaking in their modelling approaches (e.g., Bark and al., 2010; Damm and al., 2016; Endler and Matzarakis, 2011; Hendrikx and al., 2013; Heo and Lee, 2008; Moen and Fredman, 2007; Tranos and Davoudi, 2014; Uhlmann and al., 2008). Such studies are considered invalid by ski industry stakeholders and contribute to the misinformation in the media.

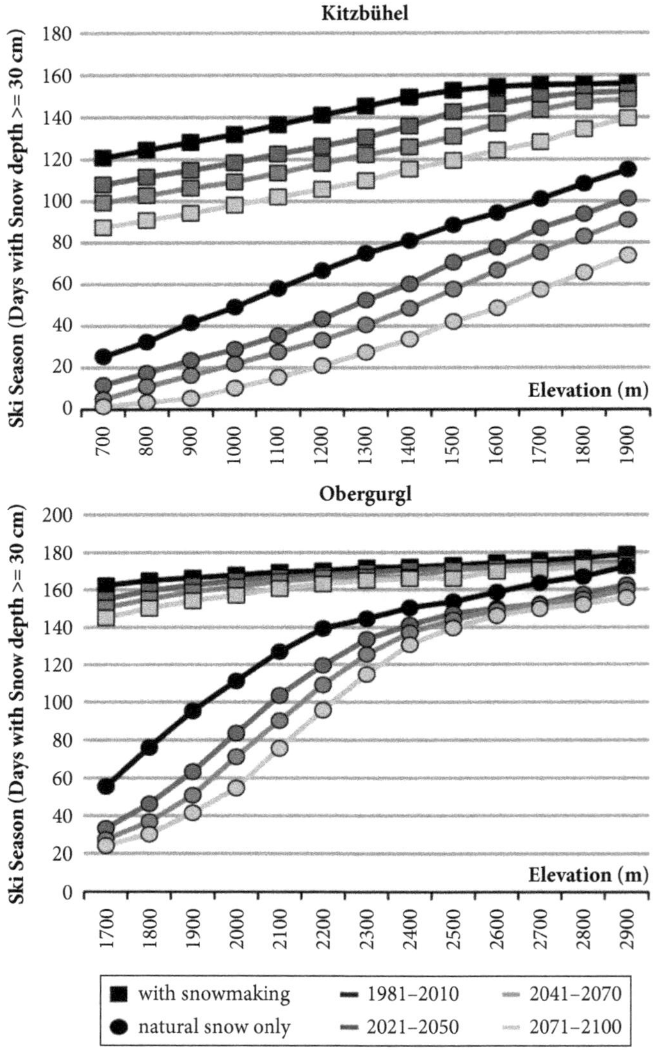

Fig. 30: A comparison of modelled ski seasons with and without snowmaking

A number of studies have attempted to account for snowmaking by using proxy indicators to represent snowmaking capacity, rather than physically incorporate machine-made snow into a modeled combined snowpack with natural snow. For example, in Australia Hennessy and al. (2008) modelled natural snow depth and separately analyzed potential snowmaking hours and whether snow production could meet the gap between the target snow depth and natural snow depth. The separate analysis of natural snow depth and snowmaking potential does not allow for an integrated calculation of snowmelt and total snow depth, nor does it allow the simulation of widespread snowmaking practices such as early season stockpiling of snow. This approach also does not allow for accurate representation of future snowmaking requirements under climate change scenarios. Similar non-integrated approaches were applied for Switzerland (Rixen and al., 2011), New Zealand (Hendrikx and Hreinsson, 2012), Bulgaria and Turkey (Demiroglu, 2016; Demiroglu and al., 2016) and the United States (Wobus and al., 2017).

In the case of Wobus et al. (2017) different indicators were used to define the start and end of ski seasons, rather than a consistent indicator of the presence of a defined operational snow depth. They used a natural snow depth threshold at the highest elevation of ski terrain to determine the season end, but either the same natural snow depth or a proxy of snowmaking hours (450 hours) at the base elevation to begin the season. Applying Wobus et al.'s (2017) proxy method to the two study areas (Tab. 6) over a seven years period reveals that it highly underestimates the ski season and leads to a completely wrong timing of the season, which is critical to understand potential risks to economically important holiday periods. At Kitzbühel the modeled season start was on average 60 days later than observed and the season ended on average five days early, resulting in a 65-day shorter season. At Obergurgl the modeled season similarly started 53 days late, but ran 25 day longer than observed, resulting in a 25-day shorter season. While better than the Kitzbühel result, the modeled season at Obergurgl started after the Christmas

Tab. 6: Comparison of Wobus et al. (2017) snowmaking proxy method to observe ski seasons (2006–07 to 2012–13)

	Kitzbühel	Obergurgl
Season Start[1]	60 days later	53 days later
Season End [2]	5 days earlier	28 days later
Total Season Length	-65 days	-25 days

1 – defined as the day when 10cm SWE natural snow or 450 hours of snowmaking occurs at base elevation

2 - defined as the day when SWE natural snow drops below 10cm at top station elevation

holiday season (which has never happened) and extended well past when the observed season ended due to lack of demand.

This model performance raises important questions about its exportability beyond the Rocky Mountain region of the USA, where it was developed, including lower elevation ski regions of the USA (Midwest, Northeast and Southeast).

Conclusion

The ski industry remains ill prepared for impending demands of institutional investors and the financial sector for disclosure on physical climate risks (Task Force on Climate-related Financial Disclosures 2017, European Bank for Reconstruction and Development 2018). One of reasons for this state of preparedness is the perception of many in the ski industry that climate change poses limited risk to their operations and long-term sustainability (Wolfsegger et al. 2008, Trawöger 2014, Steiger et al. 2017).

A wide gap exists between the climate change risk perception of the ski industry and what is portrayed in the media (Abegg and al., 2008; Saarinen and Tervo, 2006; Wolfsegger and al., 2008). Media coverage in many ski tourism regions have pronounced the impending doom of the ski industry (see Scott and al., 2012; Scott and Steiger, 2013; Rutty and al., 2017 for examples). Many of these media stories are based on the studies identified in this chapter as having important methodological limitations and may misrepresent the climate change risk of the ski industry. As the analysis in this chapter demonstrates, ski industry reservations and objections to many of the media stories are justified. Some in the ski industry have gone so far as to label all climate change research as "scaremongering". While criticisms of some studies is clearly warranted based on the results of this critical review, the obvious limitations of some studies has contributed skepticism and mistrust of even state of the art research by some segments of the ski industry. Our collective experience on two continents is that the collaboration necessary to facilitate improved information on ski tourism climate change risk and adaptation has been very limited. This is partly because the ski industry is so very image-sensitive and is very reluctant to engage in studies to understand its climate risk because they generate media coverage or appear in government reports that highlight the vulnerability of their businesses and destinations and will adversely influence financial institutions (e.g. lower credit rating, increase borrowing costs), investors (e.g. stock prices) and real estate developers and buyers,

Of course, the aforementioned limitations in the literature must be considered in the context of the development of the field of research. Scholars have over time developed methods to overcome the limitations of pioneering work and advance the field. New research must build on these advances and studies that do not reflect the current operational realities of ski areas, let alone future capacities, should not be acceptable. Our credibility within the ski industry, tourism development agencies, the financial sector (institutional and stock market investors) and real estate developers and buyer and our capacity to contribute to science-based adaptation demands this.

References

Abegg, Bruno. 1996. *Klimaänderung und Tourismus. Klimafolgenforschung am Beispiel des Wintertourismus in den Schweizer Alpen*. Zurich: vdf Zurich.

Abegg, Bruno, Agrawala, Shardul, Crick, Florence and de Montfalcon, Anne. 2007. "Climate change impacts and adaptation in winter tourism". In *Climate Change in the European Alps. Adapting Winter Tourism and Natural Hazards Management*, Agrawala, S. (eds.). Paris: OECD, 25–60.

Abegg, Bruno and Steiger, Robert. 2017. "Resilience and perceptions of problems in Alpine regions". In *Tourism and Resilience*. Butler, Richard W. (eds.). Wallingford: CABI Publications, 105–117.

Abegg, Bruno, Steiger, Robert and Walser, Roger. 2015. "Aktuelle und zukünftige Schneesicherheit der Skigebiete in Graubünden". In *St.Galler Schriften für Tourismus und Verkehr: Vol. 6, Schweizer Jahrbuch für Tourismus 2014-15*, Thomas, Bieger, Pietro, Beritelli and Laesser, Christian (eds.). Berlin: ESV, 1–16.

Bark, Rosalind H., Colby, Bonnie G. and Dominguez, Francina. 2010. "Snow days? Snowmaking adaptation and the future of low latitude, high elevation skiing in Arizona, USA". *Climatic Change* 102: 467–491.

Breiling, Meinhard and Charamza, Pavel. 1999. "The impact of global warming on winter tourism and skiing: a regionalised model for Austrian snow conditions". *Regional Environmental Change* 1(1): 4–14.

Bürki, Rolf. 2000. *Publikation der Ostschweizerischen Geographischen Gesellschaft: Vol. 6. KlimaänderungundAnpassungsprozesseimWintertourismus*. St. Gallen.

Damm, Andrea, Greuell, Wouter, Landgren, Oskar, Prettenthaler, Franz. 2016. Impacts of + 2 °C global warming on winter tourism demand in Europe. Climate Services.

Dawson, Jackie and Scott, Daniel. 2013. "Managing for climate change in the alpine ski sector". *Tourism Management* 35: 244–254. DOI:10.1016/j.tourman.2012.07.009.

Demiroglu, O. Cenk. 2016. "Technical climate change adaptation options of the major ski resorts in Bulgaria". In *Sustainable Mountain Regions: Challenges and Perspectives in Southeastern Europe*, Koulov, Boian and Zhelezov, Georgi (eds.). Cham: Springer, 77–88.

Demiroglu, O. Cenk, Turp, M. Tufan, Ozturk, Tugba and Kurnaz, M. Levent. 2016. "Impact of climate change on natural snow reliability, snowmaking capacities and wind conditions of ski resorts in Northeast Turkey: a dynamical downscaling approach". *Atmosphere* 7(4): DOI:10.3390/atmos7040052.

Elsasser, Hans and Bürki, Rolf. 2002. "Climate change as a threat to tourism in the Alps". *Climate Research* 20: 253–257.

Elsasser, Hans and Messerli, Pauk. 2001. "The vulnerability of the snow industry in the Swiss Alps". *Mountain Research and Development* 21(4): 335–339. DOI:10.1659/0276-4741(2001)021[0335:TVOTSI]2.0.CO;2.

Endler, Christina and Matzarakis, Aandreas. 2011. "Climatic potential for tourism in the Black Forest, Germany — winter season". *International Journal of Biometeorology* 55(3): 339–351. DOI:10.1007/s00484-010-0342-0.

European Bank for Reconstruction and Development. 2018. *Advancing Task Force on Climate-related Financial Disclosures Guidance on Physical Climate Risks and Opportunities*. https://climatecentre.org/downloads/files/EBRD-GCECA%20report.compressed.pdf.

Fang, Yan, Scott, Daniel and Steiger, Robert. 2018. "Ski tourism in China: historical development and challenges in an era of climate change". In: chapter in this volume.

Fukishima, Takehiko, Kureha, Masaaki, Ozaki, Noriatsu, Fujimori, Y and Harasawa, Hideo 2002. "Influences of air temperature change on leisure industries: case study on ski activities". *Mitigation and Adaptation Strategies for Climate Change* 7: 173–189.

Galloway, Robert. 1988. "The potential impact of climate changes on Australian ski fields". In *Greenhouse Planning for Climate Change*, Pearman, Graeme Ivan (eds.). Melbourne: CSIRO publications, 428–437.

Harrison, Raymond, Kinnaird, Vivian, McBoyle, Geoff, Quinlan, Christina. and Wall, Geoff. 1986. "Climate change and downhill skiing in Ontario". *Ontario Geographer* 28: 51–68.

Harrison, Sarah J., Winterbottom, Sandra J. and Johnson, Raymond C. 2005. "Changing snow cover and winter tourism and recreation in the Scottish Highlands". In *Aspects of Tourism: Vol. 22. Tourism, Recreation and Climate Change*, Hall, C. Michael and Higham, James (eds.). Clevedon: Channel View Publications, 143–154.

Harrison, Sarah J., Winterbottom, Sandra J. and Sheppard, Chris. 1999. "The potential effects of climate change on the Scottish tourist industry". *Tourism Management* 20(2): 203–211. DOI:10.1016/S0261-5177(98)00072-7.

Hendrikx, Jordy and Hreinsson, Einar Ö. 2012. "The potential impact of climate change on seasonal snow in New Zealand: part II—industry vulnerability and future snowmaking potential". *Theoretical and Applied Climatology* 110(4): 619–630. DOI:10.1007/s00704-012-0713-z.

Hendrikx, Jordy, Zammit, Christian, Hreinsson, Einar Ö. and Becken, Susanne. 2013. "A comparative assessment of the potential impact of climate change on the ski industry in New Zealand and Australia". *Climatic Change*: 1–14. DOI:10.1007/s10584-013-0741-4.

Hennessy, Kevin and al. 2008. Climate change effects on snow conditions in main-land Australia and adaptation at ski resorts through snowmaking, Climate Research, 35: 255–270.

Heo, Inhye Lee, Seungho. 2008. "The impact of climate changes on ski industries in South Korea – in the case of the Yongpyong Ski Resort". *Journal of the Korean Geographical Society* 43(5): 715–727.

König, Urs. 1999. "Climate change and snow tourism in Australia". *Geographica Helvetica* 54(3): 147–157.

König, Urs and Abegg, Bruno. 1997. "Impacts of climate change on tourism in the Swiss Alps". *Journal of Sustainable Tourism* 5(1): 46–58.

Lamothe & Periard Consultants. 1988. *Implications of Climate Change for Downhill Skiing in Québec* (88-03). Downsview: Meteorological Service of Canada.

Lipski, Steven and McBoyle, Geoff. 1991. "The impact of global warming on downhill skiing in Michigan". *East Lakes Geographer* 26: 37–51.

McBoyle, Geoff and Wall, Geoffrey. 1987. "Impact of CO2 induced warming on downhill skiing in the Laurentians". *Cahiers de Géographie du Québec* 31(82): 39–50.

McBoyle, Geoff and Wall, Geoffrey. 1992. "Great lakes skiing and climate change". In *Mountain Resort Development*, Gill, A. & Hartmann, R. (eds.). British Columbia: Centre for Tourism Policy and Research Burnaby, 70–81.

McBoyle, Geoff, Wall, Geoffrey, Harrison, Raymond and Quinlan, Christina. 1986. "Recreation and climate change: a Canadian case study". *Ontario Geography* 23: 51–68.

Moen, Jon and Fredman, Peter. 2007. "Effects of climate change on Alpine skiing in Sweden". *Journal of Sustainable Tourism* 15(4): 418–437.

Pickering, Catherine M. and Buckley, Ralf C. 2010. "Climate response by the ski industry: the shortcomings of snowmaking for Australian Resorts". *AMBIO* 39: 430–438.

Pons, Marc, Johnson, Peter A., Rosas-Casals, Marti, Sureda, Barbara and Jover, Eric. 2012. "Modeling climate change effects on winter ski tourism in Andorra". *Climate Research* 54(3): 197–207. DOI:10.3354/cr01117.

Pons, Marc, López-Moreno, Juan Ignacio, Rosas-Casals, Marti, and Jover, Eric E. 2015. The vulnerability of Pyrenean ski resorts to climate-induced changes in the snowpack, Climatic Change, 131, 591–605.

Research and Markets. 2018. *Global Ski Market: Size, Trends & Forecasts (2017 Edition)*. https://www.researchandmarkets.com/research/z564m4/global_ski_market?w=4.

Rixen, Christian, Teich, Michaela, Lardelli, Corina, Gallati, David, Pohl, Mandy, Pütz, Marco and Bebi, Peter. 2011. "Winter tourism and climate change in the Alps: an assessment of resource consumption, snow reliability and future snowmaking potential". *Mountain Research and Development* 31(3): 229–236.

Rutty, Michelle, Scott, Daniel, Johnson, Peter, Pons, Marc, Steiger, Robert and Vilella, Marc. 2017. "Using ski industry response to climatic variability to assess climate change risk: An analogue study in Eastern Canada". *Tourism Management* 58: 196–204. DOI:10.1016/j.tourman.2016.10.020.

Saarinen, Jarkko and Tervo, Kaarina. 2006. "Perceptions and adaptation strategies of the tourism industry to climate change: the case of Finnish nature-based tourism entrepreneurs". *International Journal of Innovation and Sustainable Development* 1(3): 214–228.

Sauter, Tobias, Weitzenkamp, Björn and Schneider, Christophe. 2009. "Spatio-temporal prediction of snow cover in the Black Forest mountain range using remote sensing and a recurrent neural network". *International Journal of Climatology*. DOI:10.1002/joc.2043.

Savills World Research. 2018. *Savills Ski Report – Climate Change, Resilient Resorts, Property Price Index*. https://pdf.euro.savills.co.uk/global-research/spotlight-savills-ski-report.pdf.

Schmidt, Philipp, Steiger, Robert and Matzarakis, Andreas. 2012. "Artificial snowmaking possibilities and climate change based on regional climate modeling in the Southern Black Forest". *Meteorologische Zeitschrift* 21(2): 167–172.

Scott, Daniel and Steiger, Robert. 2013. "Vulnerability of the ski industry". In: *Climate Vulnerability: Understanding and Addressing Threats to Essential Resources*, Pielke, R. A. (ed.). San Diego: Elsevier Inc. Academic Press, 305–313.

Scott, Daniel, Dawson, Jackie and Jones, Brenda. 2008. "Climate change vulnerability of the Northeast US winter tourism sector". *Mitigation and Adaptation Strategies to Global Change*, 13(5–6): 577–596.

Scott, Daniel, Gössling, Stefan and Hall, C. Michael. 2012. "International tourism and climate change". *Wiley Interdisciplinary Reviews – Climate Change* 3(3): 213–232.

Scott, Daniel, McBoyle, Geoff and Mills, Brian. 2003. "Climate change and the skiing industry in Southern Ontario (Canada): Exploring the importance of snowmaking as a technical adaptation". *Climate Research* 23: 171–181.

Scott, Daniel, McBoyle, Geoff and Minogue, Alanna. 2007. "The implications of climate change for the Québec ski industry". *Global Environmental Change* 17: 181–190.

Scott, Daniel, Steiger, Robert, Rutty, Michelle and Johnson, Petern Allan. 2014. "The future of the Olympic Winter Games in an era of climate change". *Current Issues in Tourism*. DOI: 10.1080/13683500.2014.887664.

Scott, Daniel, Steiger, Robert, Rutty, Michelle and Fang, Yan. 2018. "The changing geography of the Winter Olympic and Paralympic Games in a warmer world". *Current Issues in Tourism*. DOI:10.1080/13683500.2018.1436161.

Scott, Daniel, Steiger, Robert and Fang, Yan. 2018-submitted. "Regional ski tourism market vulnerability to climate change: An inter-comparison of Ontario (Canada), Quebec (Canada) and the Northeast USA". *Regional Environmental Change*.

Scott, Daniel, Steiger, Robert, Rutty, Michelle, Pons, Marc, Johnson, Peter A. and Vilella, Marc. 2017. "The differential futures of ski tourism in Ontario

(Canada) under climate change: the limits of snowmaking adaptation". *Current Issues in Tourism*. DOI:10.1080/13683500.2017.1401984.

Spandre, Pierre, Morin, Samuel, Lafaysse, Matthieu, Lejeune, Yves, François, Hugues and George-Marcelpoil, Emmanuelle. 2016. "Integration of snow management processes into a detailed snowpack model". *Cold Regions Science and Technology* 125: 48–64.

Steiger, Robert. 2010. "The impact of climate change on ski season length and snowmaking requirements". *Climate Research* 43(3): 251–262.

Steiger, Robert. 2012. "Scenarios for skiing tourism in Austria: integrating demographics with an analysis of climate change". *Journal of Sustainable Tourism* 20(6): 867–882.

Steiger, Robert and Abegg, Bruno. 2018. "Ski areas' competitiveness in the light of climate change: comparative analysis in the Eastern Alps". In *Tourism in Transition, Recovering from Decline and Managing Change*, Müller, Dieter K. and Więckowski, Marek (eds.). Springer, 187–196.

Steiger, Robert and Mayer, Marius. 2008. "Snowmaking and climate change. Future options for snow production in Tyrolean Ski Resorts". *Mountain Research and Development*, 28(3/4): 292–298. DOI:10.1659/mrd.0978.

Steiger, Robert, Scott, Daniel, Abegg, Bruno, Pons, Marc and Aall, Carlo. 2017. "A critical review of climate change risk for ski tourism". *Current Issues in Tourism*. DOI:10.1080/13683500.2017.1410110.

Steiger, Robert and Stötter, Johann. 2013. "Climate change impact assessment of ski tourism in Tyrol". *Tourism Geographies* 15(4): 577–600.

Töglhofer, Christoph, Eigner, Franz and Prettenthaler, Franz. 2011. "Climatic and economic impacts on tourism demand in Austrian ski areas". *Climate Research* 46(1): 1–14.

Tranos, Emmanouil and Davoudi, Simin. 2014. "The regional impact of climate change on winter tourism in Europe". *Tourism Planning & Development* 11(2): 163–178. DOI:10.1080/21568316.2013.864992.

Uhlmann, Bastienne, Goyette, Stephane and Beniston, Martin. 2008. "Sensitivity analysis of snow patterns in Swiss ski resorts to shifts in temperature, precipitation and humidity under conditions of climate change". *International Journal of Climatology* 29(8): 1048–1055. DOI:10.1002/joc.1786.

Vanat, Laurent. 2018. *2018 International Report on Snow & Mountain Tourism*. https://www.vanat.ch/international-report-on-snow-mountain-tourism.shtml.

Witmer, Urs. 1986. *Geographica Bernensia: G25. Erfassung, Bearbeitung und Kartierung von Schneedaten in der Schweiz.* Bern: Institute of Geography, University of Bern.

Wobus, Cameron, Small, Eric, Hosterman, Heather, Mills, Dave, Stein, Justin, Rissing, Matthew, Jones, Russel, Duckworth, Michael, Hall, Ronald, Kolian, Michael, Creason, Jared and Martinich, Jeremy. 2017. "Projected climate change impacts on skiing and snowmobiling. A case study of the United States". *Global Environmental Change Part A* 45: 1–14. DOI: 10.1016/j.gloenvcha.2017.04.006.

Wolfsegger, Christophe, Gössling, Stefan and Scott, Daniel. 2008. "Climate change risk appraisal in the Austrian Ski Industry". *Tourism Review International* 12(1): 13–23.

Zimmerman, Gregory, O'Brady, Caitlin and Hurlbutt, Bryan. 2006. *Regional Challenges of Future Climate Change: Endless Summer or Business As Usual?: The 2006 Colorado College State of the Rockies Report Card.* Colorado Springs: Colorado College.

III

Emerging Markets and Practices

3.1

Skiing, Climate Change, Regional Development and Terrorism: A GIS-Based Suitability Analysis for Future Ski Tourism Development in Turkey

CENK DEMIROGLU

Introduction

Ski tourism has conventionally been a core industry to many destinations in Alpine and Nordic Europe, North America and Japan. Most of these destinations today face the issues of stagnating markets, coupled with hypercompetition against urban and warm destinations and other leisure activities, such as home entertainment, on the demand side, and deteriorating snow reliability conditions due to the ongoing climate change phenomenon on the supply side. However, there are also some emerging destinations throughout the world such as China and Turkey, which have promising goals to contribute to the global ski tourism supply and demand.

In the case of Turkey, number of ski areas has been increasing exponentially in the last two decades while a growing public interest in snow sports is also observed. At the policy level, some *mega-projects* are on the agenda to grow the industry further with introduction and rejuvenation of ski destinations and, thereby, to contribute to the wider

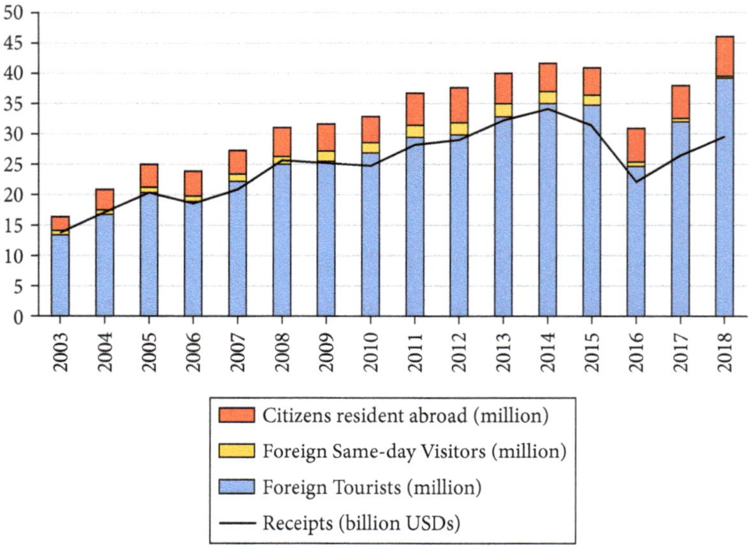

Fig. 31: Inbound visitors and receipts for Turkey 2003–2018

society and economy with intensified investments and job opportunities and eventually to overcome regional disparities especially in the least developed rural areas of the country. While this impetus could be justified in terms of certain human and physical factors, some critical challenges remain. This chapter looks into how the current ski tourism development schemes in Turkey could co-exist with these global and country-specific challenges. In doing so, it departs from recent expert debates and employs Geographical Information Systems (GIS) to offer a concrete model of the future of ski tourism in Turkey.

Background

Tourism in Turkey

Since the liberalization of economy in the 1980s, tourism sector in Turkey has pursued a rapid growth trend where annual inbound visitation figures have exceeded 40 million in the last decade (Fig. 31 based on Republic of Turkey Ministry of Culture and Tourism 2019a), ranking the country in the global top 10 in this respect (UNWTO 2019). Despite the various local and global shock periods, such as the earthquakes in 1999 and the avian influenza epidemic in 2006, the country has managed to recover immediately, even following a major series of internal and external security crises from 2015 to 2017 that have reduced the long-term stable (2009–2015) 6th place rank down to 10th and 8th in 2016 and 2017, respectively. The year 2018 has seen an all-time high figure of 46 million visits whereas the receipts have not recovered as fast since the average expenditure per capita has remained at an all-time low of 647 USDs in 2018 compared to the all-time high 850 USDs in 2003 for the span of 2003–2018, despite the recent increase in relatively higher spending Turkish citizens resident abroad from 2.4 million in 2003 to 6.6 million in 2018 (Republic of Turkey Ministry of Culture and Tourism 2019a).

According to the World Travel & Tourism Council (WTTC, 2018), total contribution of the travel and tourism sector, including domestic spendings, to Turkish Gross Domestic Product (GDP) was 12 % in 2017, directly and indirectly creating more than 2 million jobs. The total GDP contribution is forecasted to rise to 13 % by 2028, with the total number of jobs reaching 2.7 million. Such growth prospects, however, are much dependent on a spatiotemporal diversification of the tourism product by utilizing the various potentials the country has to offer, besides the traditional coastal, cultural, and more recently, medical attractions and activities. In this respect, efforts have intensified for promoting alternative tourism types, among which ski tourism has been one of the most trending (Göymen et al. 2017).

Ski tourism in Turkey

Turkic peoples have been highly acquainted with skiing in their fatherland, Central Asia, where horse-furred long skis, as well as horse sledges, have been in practice for millennia for commuting, battling and hunting purposes (Zhaojian and Bo, 2006; Allen, 2007). In the highly mountainous Anatolia, where modern Turkey lies, however, a connection to skiing cannot be traced back much, up until World War I, when Ottoman troops had become engaged more with skiing, especially around Erzurum and Kars provinces on the Eastern Front. Following a restoration of peace with the declaration of the Republic of Turkey in 1923, first touristic and recreational ski trips were observed for Mt. Uludağ in the Province of Bursa as well as the institutionalization of the sport with the foundation of the respective federation in the 1930s (Demiroglu, 2014; Kozak, 2018).

During 1950s to 70s, first basic mechanical lift installations were realized in areas such as Uludağ (Bursa), Erciyes (Kayseri), Palandöken (Erzurum), Sarıkamış (Kars), Elmadağ (Ankara), Kartalkaya (Bolu), Saklıkent (Antalya) and Çiftkaya (Bitlis). The *I. Tourism Development Period* of 1980s and 90s had laid the ground for designation of existing ski areas as *Tourism Centers* and the proposal of new ones by the central governments, yet physical developments were not as parallel as major efforts were concentrated on sun-sea-sand tourism. *II. Tourism Development Period* of the 21st century, however, has witnessed an exponential increase in the number of ski areas, evolving through both top-down and bottom-up initiatives, as well as major renovations and mega-events (Göymen et al., 2017). The 300 million USDs worth physical and institutional renovation of Erciyes Ski Resort in 2006 has been one of the major milestones in this respect, while Palandöken and its surrounding areas in Erzurum have received at least a similar amount of investments for hosting the *XXV Winter Universiade* in 2011. Reorganization and development of destinations through regional collaboration and differentiation via non-ski elements, as in the case of recently launched *Ski Silk Road* that overarches the three provinces of Erzincan, Erzurum and Kars in the northeast (Çetin et al., in review), are also among the new developments. The evolution of existing and proposed ski areas in Turkey is displayed on Fig. 32, based on an inventory maintained by Demiroglu (2008).

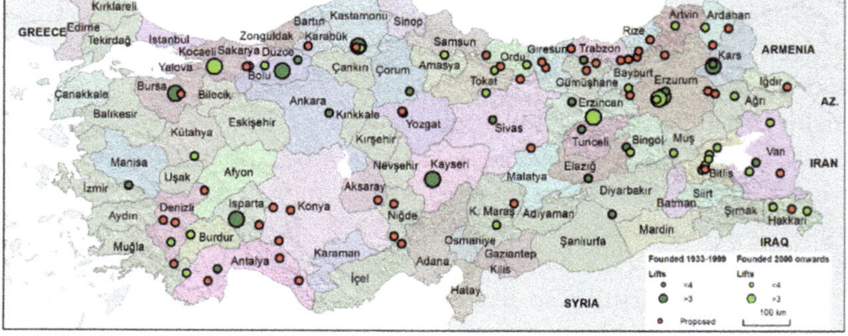

Fig. 32: Existing and proposed ski areas in Turkey

At the policy level, ski tourism has been a priority topic as exemplified within documents such as the *10th Development Plan* (2014–2018) of the central state government and the *Tourism Strategy of Turkey – 2023* approved by the Ministry of Culture and Tourism in 2007. The political will has remained consistent in this respect as ski tourism has once again been stressed as an important aspect during the discussions at the *3rd Tourism Council* meeting (Ministry of Culture and Tourism, 2017a), which is expected to set the stage for the *3rd Tourism Development Period.*

In line with the planning suggestions set forth by these policy documents of the last decade (Ministry of Culture and Tourism, 2007, 2017b), various authorities have been engaged in action especially towards suitability analyses for future ski tourism development. While the Ministry of Culture and Tourism (2019b) has proposed to increase ski tourism bed capacity from 11,459 to 80,175 at 29 destinations, 13 of which are still under either planning or construction, the Turkish Ski Federation, following their motto *ski-led development*, has undertaken a mandate to deliver an overall suitability assessment, in collaboration with leading French ski development firm Compagnie des Alpes Management. The ultimate aim of the task has been to propose 100 sites in 42 provinces as to foster a 48.5 billion USDs worth investment boom, including 5,000 hotels with 275,000 beds, within a 12 years span – a growth effort resembling the coastal tourism development tends in the 1980s and 90s. Besides, other bottom-up initiatives on ski resort development are also observed among municipalities and private entrepreneurs throughout the country. In this chapter, a comprehensive suitability analysis to aid and validate the existing efforts is aimed by accounting for the local and the global challenges discussed by stakeholders and experts in the recent years and the employment of Geographical Information Systems.

Methodology

Although site selection and suitability analyses are common practices in the professional ski industry, academic literature is still at an early stage of establishing guidelines in this respect. In the recent years, Chinese scholars have started to contribute to the literature, in line with their rapidly emerging ski tourism market. Li et al. (2016), informed by experts, have assessed ski resort development in Northern China through seven sub-indices on climate and snow, topography, ecology and natural disaster risks, location, construction and expansion, broken further down into 25 variables. More recently, a more comprehensive study by Deng et al. (in review) examines the whole Chinese territory in terms of natural and socioeconomic factors such as snow cover, temperature, topography, groundwater, vegetation, accessibility, and distance to tourist attractions and cities through a weighted overlay analysis.

In this study, we dwell on a preliminary proposal by Demiroglu et al. (2015), which has later on been developed through expert discussions on the field in Turkey and elsewhere through various completed and ongoing events and projects such as:

- *The Turkish Ski Tourism Forum* (Demiroglu, 2015)
- *SECTEUR: Sector Engagement for the Copernicus Climate Change Service: Translating European User Requirements* (ECMWF, 2015)
- *Uludağ Sustainable Tourism Master Plan* project (ATC Tourism Consultants, 2016)
- *Technical Assistance for the Development of a Winter Tourism Corridor in Erzurum, Erzincan and Kars* project in 2015–2017 (Republic of Turkey Ministry of Culture and Tourism, 2017c)
- *Sarıkamış Winter Tourism Workshop* (Kafkas University and SERKA, 2017)
- *Overall Assessment of the Turkish Mountains and Inventory of the Best Ski Sites* (Compagnie des Alpes, 2017)
- *Sustainable Tourism in Gümüşhane Workshop* (2018, outputs pending)
- *Copernicus Climate Change Services (C3S): European Tourism* (ECMWF, 2019)
- *Determination of Climate Change Adaptation Strategies for Winter Tourism in Erzurum with a Participatory Approach* (project recently approved by the Scientific and Technological Research Council of Turkey)

While the above experiences have confirmed the criticalities of the previously determined ski tourism suitability factors such as climate, topography, land cover, market accessibility and security (Demiroglu et al., 2015), further insights into the individual peculiarities as well as the mutual interventions of these criteria have been added. In this study, a suitability model is constructed in the light of these parameters and depending on the availability of data and tools. In essence, a GIS-based overlay analysis is performed. The most critical part on relevant snow and weather conditions is based on the *Ski Climate Index* (SCI), currently under development by Demiroglu et al. (in review), whereas other vital physical information on elevation, slope and aspect are derived from the *Shuttle Radar Topographic Mission (SRTM)* dataset provided by Consortium for Spatial Information (2018) at a 90 m horizontal resolution. Land cover and use data comes from the *Corine Land Cover* (CLC) program (European Environment Agency, 2012). On the human side, car and railroad network data are obtained from *OpenStreetMap* (Geofabrik GmbH, 2018) and a layer for existing and proposed airports is created according to the Republic of Turkey Ministry of Transport and Infrastructure (2019) database. Regarding regional *Socioeconomic Development Index* (SEDI) scores (Turkish Prime Ministry State Planning Organization, 2004), to be used as proxies for market catchments, relatively older district-level, rather than the more recent provincial scale, results were preferred for the sake of higher spatial resolution. Last but not least, as a special condition to Turkey, information on fatal terrorist attacks, which have mostly taken place in the mountainous regions with high potentials for ski tourism development, has been retrieved from a private database (University of Maryland, 2018) to be utilized for assessment of spatial concentrations.

The analysis is mostly undertaken in ArcGIS software, which provides users with various overlay techniques. In this study we aim for a *fuzzy overlay*, as with most criteria, trends and patterns are agreed upon by experts but discrete breakpoints are either unavailable or may be misleading. For this purpose, initially vector data were rasterized, wherever needed, and all raster inputs were either given a *fuzzy membership*, depending on their nature, or reclassified on a scale of 0 to 1. The overall overlay was based on a *fuzzy product*, in principle, in order to account for the most suitable locations as well the overriding effects of lack of snow and too flat or steep terrain to a nullifying degree. The *fuzzy sum* operator was also employed twice, when dealing with the synergic combinations of the *SCI* score and elevation, on the physical side, and *SEDI* score and accessibility, on the human side (for more technical information see Mitchell, 2012 and Esri, 2019).

Future suitability of Turkey for Ski tourism development

Climate (Change)

Climate, snow in particular, is the ultimate precondition to developing ski tourism. As noted by the industry stakeholder and experts, climate impacts on snow reliability has become more visible in the last decade, raising awareness and urging for adaptation. During 2013–2014 season, the second warmest winter of the 21st century in Turkey until then (Demiroglu, 2016a), many businesses have felt highly stressed for the first time due to snow shortage and season shrinkage, as was the case with the popular Uludağ ski resort, whose stakeholders refer to this period as the "worst season of their history" (Kozak, 2018). The consumers, on the other hand, have also coined the season as a "bad" one, rating it 2.1 on a scale of 5, and mentioned that they have substituted their usual ski trips temporally to later season or shortened their vacations (Demiroglu, 2016b). The issue has also been acknowledged at the highest policy level, as the Commission on Environment, Planning and Infrastructure have recommended consideration for climate change in preparation of the *Winter Tourism Strategy Document* (Ministry of Culture and Tourism, 2017d). The private sector representatives also note their awareness of the situation and mention that they are more in need of future projections-based information as they may be misled by past observations when determining locations for new and extension investments.

Regarding the climate layer in our suitability analysis, preliminary results from Demiroglu et al. (in review) who developed the *Ski Climate Index* (SCI) and applied it over the Turkish terrain were used. In their study, they computed *SCI* scores based on projections obtained by dynamical and double downscaling of the Earth system model MPI-ESM-MR (Max Planck Institüt für Meteorologie, 2012) by the regional climate model RegCM 4.4 (Giorgi et al., 2015) to an approximately 11 km resolution for the 1991–2000 and the 2021–2050 periods. While here we display both results (Fig. 33) to reflect the magnitude of future change, the *SCI* values for only the 2021–2050 period, which is based on the IPCC's business-as-usual RCP 8.5 trajectory (Riahi et al. 2011), are input for the overlay analysis to account for changing conditions under the most pessimistic scenario.

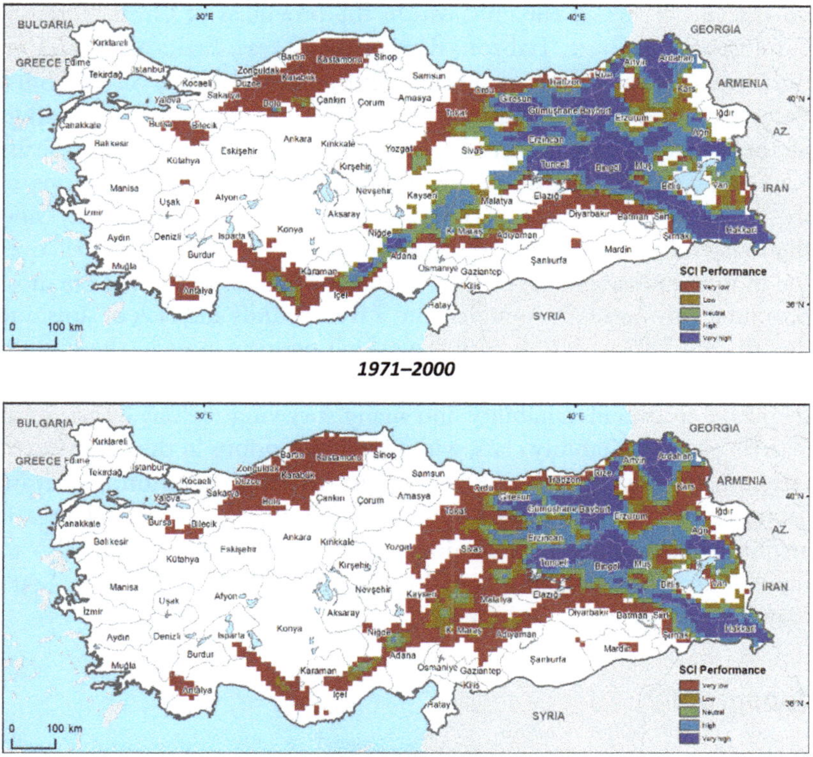

1971–2000

2021-2050 (RCP 8.5)

Fig. 33: Climatic suitability of Turkey for ski tourism

SCI is designed as a rather eliminative decision tree that rejects any location with a lack of any of the three snow reliability indicators – natural snow with a skiable depth, quality snowmaking potential, and ability to be skiable on New Year's Day. The *Snow-Reliability* facet score is further enhanced by a combination with the performance in the *Aesthetics and Comfort* facet, which is a fuzzy sum of thermal comfort, (lack of) adverse winds and sunshine duration. In summary, all snow and non-snow indicators contribute the performance of the index but a lack of any snow performance yields a null result.

According to the historical outputs on Fig. 33, a *Very High* (fuzzy membership between 0.8 to 1) climatic suitability for ski tourism development is observed partly along the İçel-Konya and the Adana-Niğde provincial borders and mainly within the provinces of Ordu, Giresun, Gümüşhane, Bayburt, Trabzon, Rize, Artvin, Sivas, Erzincan, Erzurum, Ağrı, Muş, Bingöl, Tunceli, Elazığ, Bitlis, Van, Hakkari, Şırnak, Siirt, Batman and Diyarbakır. High suitability is found along the surroundings of these regions and the intersection of Adana-Kahramanmaraş-Sivas-Kayseri provinces. Demiroglu et al. (in review) note that, in terms of the breakdown of these performances, natural snow reliability and thermal comfort have relatively homogeneous positive effects all over the mountainous regions, whereas high snowmaking capacities are most prominent around East Anatolia, and adverse winds and lack of sunshine due to cloud formation disturb the overall performances in the Eastern Black Sea region. The future projections on Fig. 33 highlight a major shrinkage in overall suitability and many surviving regions retreat to at least one lower suitability class, with partial exceptions in the provinces of Giresun, Bayburt, Trabzon, Rize, Artvin, Ardahan, Erzurum, Erzincan, Tunceli, Muş, Bingöl, Bitlis, Van, Şırnak and Hakkari. The change is attributed to a slightly weakening natural snow reliability but also to reductions in snowmaking capacities, especially before the New Year's Day (Demiroglu et al., in review).

Topography

Besides climate, topography is another integral and critical physical determinant for ski tourism suitability. Moreover, in this study, the relatively coarse resolution of the climate projections requires refinements via topographic inputs. Otherwise, the *SCI* results on Fig. 34 would be misleading due to vertical deviations within a general range of +/- 500 m, according to Demiroglu et al. (in review). For this reason, elevation values from the ultra-high resolution *SRTM* dataset (Consortium for Spatial Information, 2018) are combined with the *SCI* layer in order to distinguish suitable areas by almost the actual altitude bands (Fig. 34). Prior to the *fuzzy sum*, elevation values are fuzzified on a linear basis and within a range of 1300–3500 m. The bottom limit is based on the existing lowest altitude of commercial skiing in Turkey whereas 3500 m is set as an acclimatization limit, above which mountain sickness could prevail due to reduced oxygen amounts.

In further combination to elevation and climate, slope and aspect values need to be considered for a more comprehensive determination of physical suitability for skiing. According to Mieczkowski (1990),

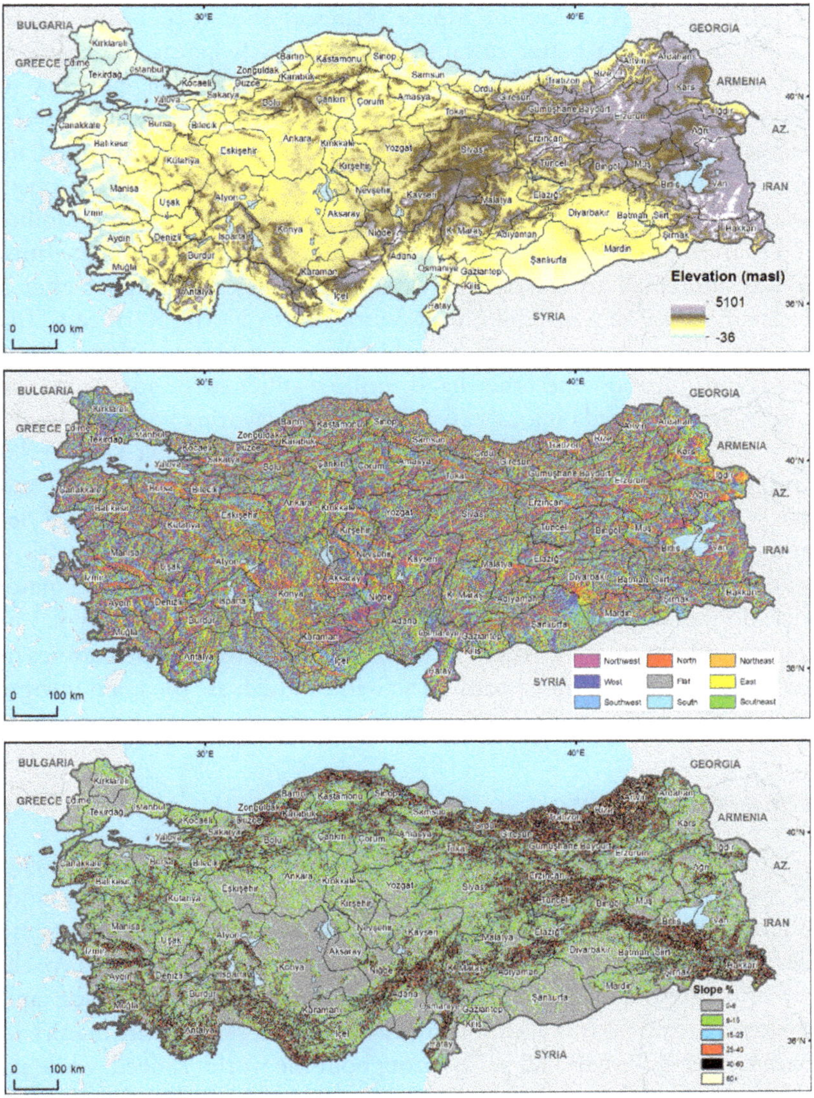

Fig. 34: Topographic features of Turkey relevant for ski tourism

an ideal North American or Alpine ski resort should be facing North to Northeast for a better sustained snow cover (whereas the opposite would be the case in Nordic Europe to favor sunshine over snow due to limited winter daylight) and bear various slopes for different skier levels, namely; maximum 25 % for beginners, 25–50 % for the intermediate and 50–70 % for the experts. The conventional North-Northeast aspect was also confirmed for Turkey by experts, especially since snow reliability is becoming more and more important under global warming. Thus, it was combined to the climate-elevation sum as a multiplier after a reclassification on a scale of South-Southwest to North-Northeast. The product has a decreasive effect to highlight the best locations. Slope reclassifications, on the other hand, utilized different standards for the case of Turkey, based on local expert knowledge, favoring potential novice runs of 8–15 % the most, followed by advanced beginner (15–25 %) and intermediate and advanced (25–40 %) runs. Expert (40–60 %) and extreme (60–180 %) runs were given the least values and flat (<8 %) and too steep (>180 %) areas were excluded from the analysis with 0 values assigned. The results were further multiplied into the climate-elevation-aspect trio to highlight suitability more and nullify any flat or dangerously steep areas. This approach was in line with the necessities of downhill skiing and snowboarding activities, which are the main income sources of mass winter tourism in Turkey.

Land cover and use

Tourism development competes with other forms of development as well as nature conservation efforts for the use of available land, often finding itself amidst several conflicts. Most times, ski resorts are developed on slightly vegetated terrain, to achieve the best of snowpack, yet also often close to or even within forest areas for aesthetical and sheltering reasons. It is also not surprising to find ski areas in vicinity of nature reserves or national parks throughout the world. Indeed, while ski tourism may claim itself to be *nature-based*, it often ends up urbanizing the very rurality that once was its resource and aim (Stoddart, 2012; Lasanta, 2013) by attracting masses of urban tourists and causing the building of extensive structures such as roads, lifts, hotels etc. In Turkey as well, ski tourism has been claimed to cause deforestation within protected areas (Kozak, 2018) and experts note that recently proposed mega ski destinations in ecologically sensitive and unique areas, such as

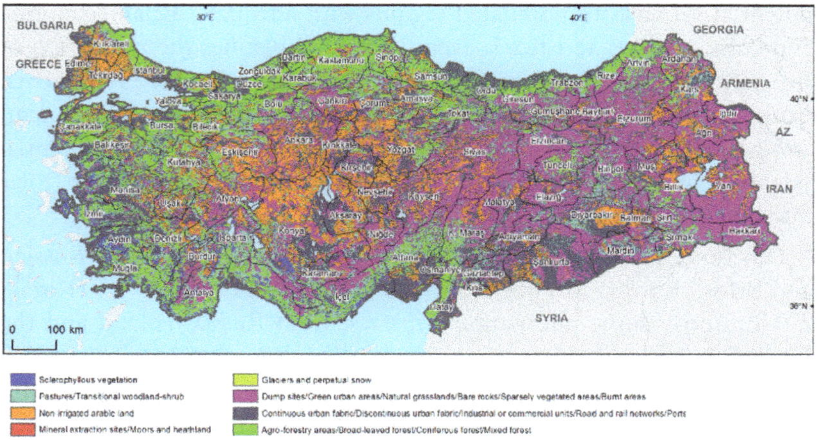

Fig. 35: Land cover and use in Turkey (2012)

the Kaçkar Mountains (Rize-Artvin), are prone to new conflicts in the near future.

In this study, several categories of the Corine Land Cover data (European Environment Agency, 2012) have been reclassified under groups of similar suitability, according to expert views favoring sustainability (Fig. 35). Areas with glaciers and perpetual snow have been assigned with the top values (although they often lie at ecologically sensitive, high-altitude areas), followed by moors-heathland and mines, the latter of which often becomes ideal ski venues once extraction is terminated, and less sensitive natural areas and alternative urban areas such as grasslands, sparsely vegetated areas, bare rocks, burnt areas, green urban areas and dump sites. Pastures and transitional woodland-shrub are also given high priority whereas the stance against non-irrigated arable land is neutral and sclerophyllous vegetation and forest areas are the least preferred. Built urban areas and infrastructure have been excluded from the analysis.

Market accessibility

While physical factors do play their vital roles in determination of locations most suitable for ski tourism development, the relatively more implicit and complex human factors also have their parts. Accessibility of potential ski areas to populated regions with adequate social, economic and cultural characteristics is eventually what justifies the realization of ski resort projects that usually require high fixed capital investments. Coupled with the fact, as noted by several experts, that ski tourism, compared to coastal tourism, is highly domestic in character, potential ski areas first would need to ensure local market bases within their immediate ranges to be viable.

In Turkey, some international interest to ski resorts from Russian and British tourists had been recorded in the 1980s and the 90s (Kozak, 2018). Today some international interest from Russia, Poland and the Netherlands is claimed to be on the rise. Tourists from Arabic countries and Iran, who frequented the areas especially during the shoulder and the summer seasons until the end of the 20th century (interrupted by the 1999 Earthquake) have also recovered in numbers – this time even participating in snow sports during the winter seasons. However, domestic market is still considered to be the main vein of ski business in Turkey, which also explains why many popular resorts are found within or close to major cities – such as Uludağ in Bursa, Kartalkaya in Bolu, Erciyes in Kayseri, Kartepe in Izmit and Palanadöken in Erzurum (Evren and Kozak, 2018) although physically equivalent or superior and vast areas are found especially in the eastern parts of the country. Number of nationwide skiers is mentioned to be on a recent rise (Kozak, 2018) and targeted to hit 4 million within the next decade (Hudson and Hudson, 2015; Göymen et al., 2017).

In order to reflect the domestic market potential, results of the *Socioeconomic Development Index* (SEDI) study by Turkish Prime Ministry State Planning Organization (2004) were utilized. Although there exists a more recent study by Republic of Turkey Ministry of Development (2013) for the year 2011, the version in use bears a higher resolution in terms of districts, rather than provinces, and does not account for accessibility and criminality indicators, as provided by the latter version, and avoids any duplications in our study as we introduce different and up-to-date measures when accounting for security and accessibility issues. The 32 variables in *SEDI 2004* represent demographics, employment, industrialization, health, education, agriculture and other welfare benefits, and categorize the districts under six groups (Fig. 36, 1 – most developed to 6 – least developed). In the final overlay analysis, individual *SEDI* scores of the districts ranging from -2.01 to 5.08 and a score of 10

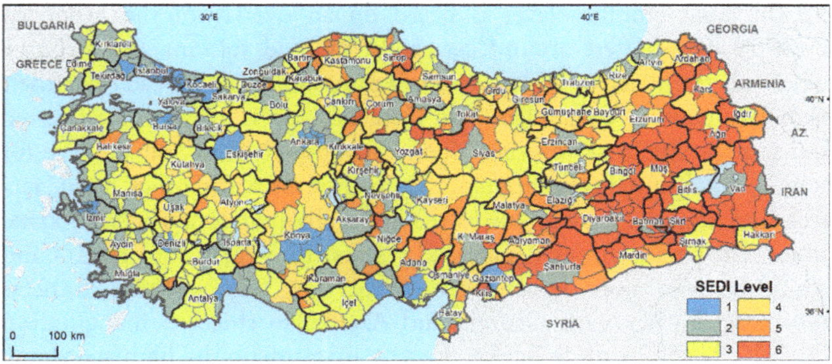

Fig. 36: Socioeconomic Development Index (SEDI) levels of districts in Turkey (2004)

for the metropolitan areas, excluded from the original analysis, are used as fuzzification inputs.

Besides proximity of the ski areas to market catchments, mobility of the potential markets to the ski areas constitutes a complementary factor to determine overall accessibility. In this respect, consideration for the coverage of major transport networks could provide a generalized indicator. Fig. 37 depicts such coverage by taking account of existing and proposed airports, excluding those serving for the Turkish Air Force, railroads and major car roads, including motorways, primary roads and trunks.

Fig. 37: Major car roads, railroads and airports in Turkey

The final mobility layer is based on an overlay of the rasterized multiple ring buffers around car roads, railroad stations and airports, where all three networks have been weighted equally as each has their own competitive advantage. For instance, car is the most preferred transport mode in ski tourism due to its door-to-door convenience for travel with heavy ski equipment. Train travel, on the other hand, is a trending mode in Turkey due to the culturally popularizing *East Express* route that passes through many ski resorts from Ankara to Kars – and would be even more demanded once the *Iron Silk Road* that connects Northeast Turkey with Georgia and Azerbaijan since its inauguration in 2017 becomes more active. Recent investments in high-speed rail networks throughout the country as well as the global climate change awareness that favors railroads over other transport modes for the sake of mitigation are the two other factors that give the railroads a competitive edge. However, the present and the proposed investments for air travel in Turkey cannot be neglected in the competition for travel mode choice, as the ultimate aim of establishing an airport at every 100 km distance is about to be reached. These investments would definitely bring western market catchments closer to the mountainous East, as already evident for destinations such as Palandöken (Erzurum) and Sarıkamış (Kars), but then the question would be the conventional hub and spoke structure in Turkey that leads most flights to reach their destinations via only the two hubs; Istanbul and Ankara.

Perceived security

Security concerns over Turkey, due to the unrest commonly observed within its wider region and its own internal conflicts, as well as natural disasters such as earthquakes severely hitting the country on certain occasions, have had their negative impacts on socioeconomic development for many decades. Regarding the consequences of terrorism to international tourism to Turkey, Yaya (2009) has assessed a cost of 700 million USDs for the year 2006 alone, and Alfonso-Rodríguez (2017) finds out that reduced contribution of inbound tourism to GDP could be as high as 10 %. However, both studies note that the effects of terrorism events last within a year. The fast recovery performance was evident even in the recent years (Fig. 38), when several consecutive attacks took place, including those directly targeting foreign tourists.

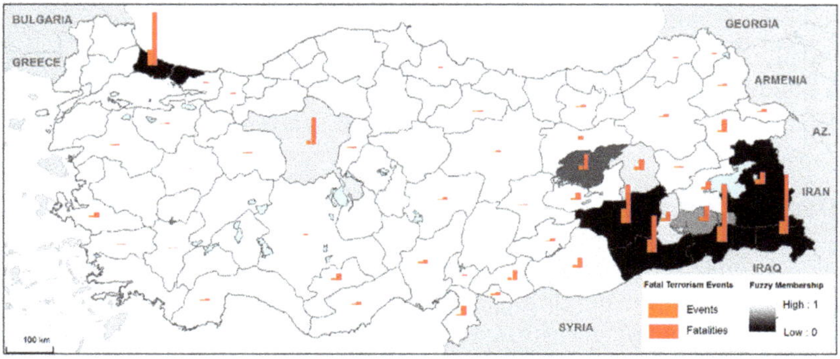

Fig. 38: Frequencies of Fatal Terrorist Attacks in Turkey (1983–2017)

Fig. 38 summarizes the frequencies of terrorism events and related fatalities per province in Turkey for the 1983–2017 period. Major concentration areas are the two largest cities in the West and several provinces in the East and the Southeast, where number of fatal events range between 160 and 193 and death tolls between 449 and 716 in the top five problematic provinces.

It should be noted that while major cities of the West suffer from transnational and domestic terrorism that drives away inbound tourism demand, even if for short whiles, the major cause of violence in the East results from separatist movements that destroy the security image of the region in both international, and more importantly, domestic markets. The latter concentration area also highly correlates with the mountainous regions with strong physical potential for ski tourism development. Indeed, some minor attacks have even targeted the ski area properties in the recent years, while those ski areas and many other potential ones could have acted as catalysts in transforming these least developed regions into higher levels, thereby alleviating major socioeconomic issues as well as some causes of the separatist violence. The power of such potential can be understood from relatively better-off districts which have been developed as tourism destinations along Turkey's southwestern coasts (Turkish Prime Ministry State Planning Organization, 2004) and through worldwide champions such as Tirol in Austria. However, in order for this transformation to take place, initially extra-regional domestic visitors need to be attracted, which in return would also improve the socioeconomic conditions in the local market base and drive the region to be even more resilient. However, as noted by experts, citizens associate mostly the *province names* with low security, depending on the actual frequency and more importantly the media coverage of the events, and tend to keep a (mental) distance away from those regions.

Overall suitability

The final overlay analysis to reflect the overall suitability of Turkey for future ski tourism development starts with a depiction of the physical suitability, based on climatic and topographic factors, and then moves on with *fuzzy product* based overlays of the land cover, market accessibility, and security layers in order to highlight their decreasive effects and to point out the best locations (Fig. 39). When physical suitability is combined with land cover and use, a certain suitability reduction is most apparent in forest areas and non-irrigated arable lands. Insertion of the market accessibility layer to the overlay, on the other hand, brings the final output one step closer to the current distribution of existing and proposed ski areas (Fig. 1). This is best exemplified by the stronger prominence of Uludağ (Bursa) where local demand and accessibility are strong. Note that the maximum fuzzy membership value for suitability retreats to 0.64, indicating strong decreasive influence of human factors to ski tourism development. The security layer further displays the vicious cycle inherent between development and terrorism, where, for instance, unrestored peace would pose a major threat to the development of Hakkari, which otherwise possesses a high suitability for ski tourism.

Conclusion

This study has preliminarily shown that the combined physical and human potential of Turkey for ski tourism development is promising for the future, where a mutually beneficial scenario can come into effect by utilization of the physically most suitable resources in the socioeconomically least developed regions. Although negative impacts of climate change are to be observed in the future in Turkey, as well as the rest of the world, the climatic diversity of the country will help it to focus its development efforts on the most resilient regions that would be likely to benefit from spatial substitution trends among the ski tourists. However, another question besides "*where* should ski tourism be developed"? is "*who* should develop ski tourism?". In this respect, establishment or

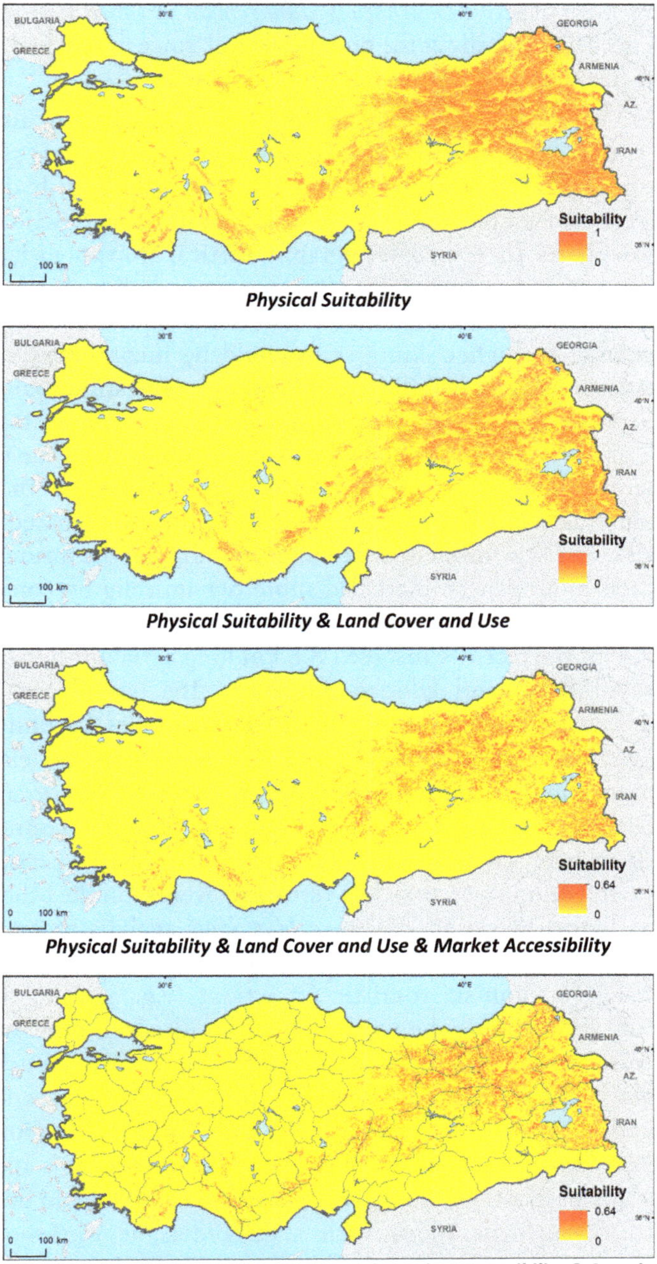

Fig. 39: Overall suitability of Turkey for ski tourism

encouragement of active *Destination Management Organizations* (DMOs) is much needed especially at the provincial level, since most issues, mainly from the human aspect, arise from within this scale.

From a methodological point of view, further suggestions are on the agenda to beat the limitations of this study, once relevant data and tools become available. In terms of climate layer, evolving non-hydrostatic models can be employed to assess *SCI* performance with higher resolution inputs. In terms of topography, vertical drop potentials, slope variety and flat area availability (for accommodation development) could be taken into account. When considering slope variety, evolution of the market to higher skiing skills could be incorporated and the percentage thresholds be adjusted accordingly. Regarding land cover and use, while deforestation could remain as an opposed factor for the sake of sustainability, the climate change led retreat of the tree line to higher altitudes could be taken into account. In terms of market accessibility assessments; firstly, population projections based on age pyramids could be utilized to get an understanding of the future market size. Accessibility can be modelled more dynamically by considering additional infrastructural investments justified by viable ski resorts, as in the case of the recently installed 8.8 km long aerial lift line between the City of Bursa and Uludağ ski resort. The overall accessibility coverage can be built via more accurate techniques beyond simple buffers such as the *Path Distance* tool, which accounts for elevation, or the more complex multimodal transport *Network Datasets* available in ArcGIS. The coverage can also include international connections, especially where cross-border collaboration potential exists. The security perception layer, based on terrorism events, on the other hand, could be modelled via an *Emerging Hot Spots* analysis that accounts for not only the spatial concentrations but also the temporal trends. As a final layer, non-ski touristic attractions (e.g. culture, thermal) that would differentiate the product from hundreds of other ski areas in the world could be included. Currently, the recently launched *Ski Silk Road* destination in Erzincan-Erzurum-Kars provinces and the collaboration between Erciyes Ski Resort and Cappadocia region under the *Cultural Skiing* theme are two concrete examples of this approach. Last but not least, all this study design discussed above could be further validated through iterations with stakeholders at workshops and events alike.

Acknowledgements

This study was supported by the Swedish Research Council for Sustainable Development, FORMAS, as part of the project *Mobilising the Rural: Post-productivism and the New Economy.* The author remains thankful to all actors and stakeholders engaged with this study's input projects and events listed in the Methodology section. "»

References

Alfonso-Rodríguez, Julio A. 2017. "Evaluating the dynamics and impact of terrorist attacks on tourism and economic growth for Turkey". *Journal of Policy Research in Tourism, Leisure and Events* 9: 56–81.

Allen, E. John B. 2007. *The Culture and Sport of Skiing: From Antiquity to World War ll.* University of Massachusetts Press, Amherst.

ATC Tourism Consultants GmbH. 2016. *New project awarded: Sustainable tourism plan for Uludağ, Turkey.* Available online at http://atc-consultants.com/news/item/176-new-project awarded-sustainable-tourism-plan-for-uludag-turkey (last accessed 27 March 2019).

Cetin, Gürel, Demiroglu, O. Cenk and Okumus, Fevzi. in review. "Multi-scalar destination strategies for the neonate: case of Erzurum-Erzincan-Kars Winter Tourism Corridor". *Journal of Destination Marketing & Management.*

Compagnie des Alpes. 2017. *The Turkish Ski Federation and Compagnie des Alpes sign a contract for the assessment of Turkey's "ski" potential and the identification of the promising sites.* Available online at https://www.compagniedesalpes.com/sites/default/files/actualites/cda_cp_tkf_-_en.pdf. (last accessed 23 March 2019).

Consortium for Spatial Information. 2018. Available online at http://srtm.csi.cgiar.org/srtmdata/ (last accessed 27 March 2019).

Demiroglu, O. Cenk. 2008. *Turkish Ski Areas.* Available online at http://www.skiturkey.info (last accessed 23 March 2019).

Demiroglu, O. Cenk. 2014. *Kış Turizmi.* Ankara: Detay.

Demiroglu, O. Cenk. 2016a. *Climate Change Vulnerability of Ski Tourism in Germany and Turkey.* Istanbul Policy Center, Istanbul.

Demiroglu, O. Cenk. 2016b. "Türkiye'deki kış sporları turistlerinin iklim değişikillği algı ve uyumları üzerine bir ampirik bir çalışma". In *I. Ulusal Alternatif Turizm Kongresi Bildiri Kitabı*, Güncü, Arzu, Güneş, Erkan,

Alagöz, Gürkan, Fırat, Mehmet Çağlar, Canbaba, İbrahim Erhan and Başar, Burak (eds.). Erzincan: Erzincan Üniversitesi Turizm ve Otelcilik Meslek Yüksekokulu Yayınları, 140–147.

Demiroglu, O. Cenk, Akbas, Abdullah, Ozturk, Tugba, Turp, M. Tufan, An, Nazan, Turkes, Murat and Kurnaz, M. Levent. 2015. "Site Selection for Ski Resorts and Areas: A GIS Application over the Turkish Mountain Terrain". *Association of American Geographers' Annual Meeting*, 21–25 April, Chicago.

Demiroglu, O. Cenk, Turp, M. Tufan, Kurnaz, M. Levent and Abegg, Bruno. (in review). "The Ski Climate Index (SCI): fuzzification and a regional climate modelling based application for Turkey". *International Journal of Biometeorology*.

Demiroğlu, O. Cenk. 2015. *Kayak Turizmi Forumundan Kayak Turizmi Politikasına Notlar*. Istanbul Policy Center.

Deng, Jie, Che, Tao, Xiao, Cunde, Wang, Shijin and Dai, Liyun. (in review). "Suitability analysis of ski areas in China: An integrated study based on natural and socioeconomic conditions". *The Cryosphere 13, 2149–2167.*

ECMWF. 2015. *SECTEUR: Sector Engagement for the Copernicus Climate Change Service: Translating European User Requirements*. Available online at http://climate.copernicus.eu/secteur#tourism (last accessed 23 March 2019).

ECMWF. 2019. *Copernicus Climate Change Services (C3S): European Tourism*. Available online at https://climate.copernicus.eu/european-tourism (last accessed 23 March 2019).

Esri. 2019. *Applying fuzzy logic to overlay rasters*. Available online at http://desktop.arcgis.com/en/arcmap/latest/tools/spatial-analyst-toolbox/applying-fuzzy-logic to-overlay-rasters.htm (last accessed 27 March 2019).

European Environment Agency. 2012. *Corine Land Cover 2012 raster data*. Available online at https://www.eea.europa.eu/data-and-maps/data/clc-2012-raster (last accessed 27 March 2019).

Evren, Savaş and Kozak, Nazmi. 2018. "Competitive positioning of winter tourism destinations: A comparative analysis of demand and supply sides perspectives–Cases from Turkey". *Journal of Destination Marketing & Management* 9: 247–257.

Geofabrik GmbH. 2018. *Download OpenStreetMap data for this region: Turkey*. Available online at https://www.eea.europa.eu/data-and-maps/data/clc-2012-raster (last accessed 27 March 2019).

Giorgi, Filippo, Elguindi, Nellie, Cozzini, Stefano, Solmon, Fabien and Giuliani, Graziano. 2015. *Regional Climate Model RegCM User's Guide Version 4.4*. Trieste: The Abdus Salam.

Göymen, Korel, Demiroglu, O. Cenk and Tulus, Cana. 2017. "Snow sports tourism development policies in Turkey". In *Lifestyle Sports and Public Policy*, Turner, Daniel and Carnicelli, Sandro (eds.). Routledge, London, 63–81.

Hudson, Simon and Hudson, Louise. 2015. *Winter Sport Tourism: Working in Winter Wonderlands*. Goodfellow Publishers, Oxford.

Kafkas University and SERKA. 2017. *Sarıkamış Kış Turizmi Çalıştayı*. Available online at https://www.serka.gov.tr/dokumanflipbook/sarikamis-kis-turizmi-calistayi/215 (last accessed 27 March 2019).

Kozak, Nazmi (ed.). 2018. *Türkiye Turizmi Sözlü Tarih Araştırması: Görüşme Metinleri Akçakoca, Erdek, Kapadokya, Uludağ* – v. 4. Istanbul: Yılmzlar Matbaası.

Lasanta, Teodoro, Beltran, Oriol and Vaccaro, Ismaël. 2013. "Socioeconomic and territorial impact of the ski industry in the Spanish Pyrenees: mountain development and leisure induced urbanization". *Pirineos: Journal of Mountain Ecology* 168: 103–128.

Li, Yu, Zhao, Minyan, Guo, Peng, Zheng, Ji, Li, Zehong, Li, Fujia, Shi, Yuanyuan and Dong, Suocheng. 2016. "Comprehensive evaluation of ski resort development conditions in northern China". *Chinese Geographical Science* 26: 401–409.

Max Planck Institüt für Meteorologie. 2012. New Earth system model of Max Planck Institute for Meteorology. Available online at https://www.mpimet.mpg.de/en/science/models/mpiesm.html (last accessed 23 March 2019).

Mieczkowski, Zbigniew. 1990. *World Trends in Tourism and Recreation*. New York: Peter Lang.

Mitchell, Andy. 2012. *The Esri Guide to GIS Analysis, Volume 3: Modeling Suitability, Movement, and Interaction*. Esri Press, Redlands.

Republic of Turkey Ministry of Culture and Tourism. 2007. *Tourism Strategy of Turkey –2023*. Available online at http://www.kultur.gov.tr/Eklenti/43537,turkeytourismstrategy2023pdf.pdf?0&_tag1=79668BB12A540BE0672E65E48D10C07D6DAE291 (last accessed 27 March 2019).

Republic of Turkey Ministry of Culture and Tourism. 2017a. 3. *Turizm Şurası*. Available online at http://turizmsurasi.kulturturizm.gov.tr (last accessed 27 March 2019).

Republic of Turkey Ministry of Culture and Tourism. 2017b. *Turizmde Ürün Çeşitliliği ve Sürdürülebilirlik Komisyonu Sonuç Raporu*. Available online at http://turizmsurasi.kulturturizm.gov.tr/Eklenti/57385,turizmdeuruncesitliligisurdurulebilirlikomisyonraporup-.pdf?0 (last accessed 27 March 2019).

Republic of Turkey Ministry of Culture and Tourism. 2017c. *Technical Assistance for the Development of a Winter Tourism Corridor in Erzurum, Erzincan and Kars*. Available online at http://www.skisilkroad.com/?lang=en (last accessed 27 March 2019).

Republic of Turkey Ministry of Culture and Tourism. 2017d. *Çevre-Planlama-Altyapı Komisyonu Sonuç Raporu*. Available online at http://turizmsurasi.kulturturizm.gov.tr/Eklenti/57380,cevreplanlamaaltyapikomisyonraporupf.pdf?0 (last accessed 27 March 2019).

Republic of Turkey Ministry of Culture and Tourism. 2019a. *2018 Yılı Genel Turizm İstatistikleri*. Available online at http://yigm.kulturturizm.gov.tr/Eklenti/62462,2018turizmgenelistatistiklerpdf.pdf?0 (last accessed 27 March 2019).

Republic of Turkey Ministry of Culture and Tourism. 2019b. *Kış Sporları Turizm Merkezlerine İlişkin Genel Bilgiler*. Available online at http://yigm.kulturturizm.gov.tr/TR10177/kis-sporlari-turizm-merkezlerine-iliskin-genel-bilgiler.html (last accessed 27 March 2019).

Republic of Turkey Ministry of Development. 2013. *10th Development Plan*. Available online http://www3.kalkinma.gov.tr/DocObjects/view/15089/Onuncu_Kalk%C4 %B1nma_Plan%C4 %B1.pdf (last accessed 27 March 2019).

Republic of Turkey Ministry of Development. 2013. *İllerin ve Bölgelerin Sosyo-ekonomik Gelişmişlik Sıralaması (SEGE-2011)*. Available online at http://www3.kalkinma.gov.tr/DocObjects/View/15310/SEGE-2011.pdf (last accessed 27 March 2019).

Republic of Turkey Ministry of Transport and Infrastructure. 2019. *Airports*. Available online at https://www.dhmi.gov.tr/sayfalar/havalimanlarimiz.aspx (last accessed 27 March 2019).

Riahi, Keywan, Rao, Shilpa, Volker, Krey, Cho, Cheolhung, Chirkov, Vadim, Fischer, Guenther, Kindermann, Georg, Nakicenovic, Nebosja and Rafaj, Peter. 2011. "RCP 8.5—a scenario of comparatively high greenhouse gas emissions". *Climatic Change* 109: 33.

Stoddart, Mark C. J. 2012. *Making Meaning out of Mountains: The Political Ecology of Skiing*. UBC Press, Vancouver.

Turkish Prime Ministry State Planning Organization. 2004. *İlçelerin Sosyo-ekonomik Gelişmişlik Sıralaması Araştırması*. Available online at http://www3.kalkinma.gov.tr/DocObjects/View/15310/SEGE-2011.pdf (last accessed 27 March 2019).

University of Maryland. 2018. *Global Terrorism Database*. Available online at https://www.start.umd.edu/gtd/search/Results.aspx?page=1&casualties_type=&casualties_mx=&country=209&charttype=line&chart=regions&ob=GTDID&od=desc&expanded=yes#reults-table (last accessed 27 March 2019).

UNWTO. 2019. *Yearbook of Tourism Statistics, Data 2013 – 2017*. Madrid: UNWTO.

WTTC. 2018. *Travel and Tourism Economic Impact 2018: Turkey*. Available online at https://www.wttc.org/-/media/files/reports/economic-impact-research/countries2018/turkey2018.pdf (last accessed 27 March 2019).

Yaya, Mehmet E. 2009. "Terrorism and tourism: the case of Turkey". *Defence and Peace Economics* 20: 477–497.

Zhaojian, Shan and Bo, Wang (eds.). 2006. *The Original Place of Skiing – Altay Prefecture of Xinjiang, China*. Beijing and Urumqi: People's Sports Publishing House and Xinjiang People's Publishing House.

3.2

Ski Tourism in China: Historical Development and Challenges in an Era of Climate Change

Yan Fang, Daniel Scott, and Robert Steiger

Introduction

In contrast to the mature ski tourism markets in many countries of Europe and North America, China's ski industry is relatively new. Ski tourism in China began in 1996 (Wu and Wei, 2018) and has experienced a massive boom in its first 20 years of development. This rapid development has been largely driven by increased personal wealth in China that has lead to massive growth in domestic and international tourism. China continues to lead global outbound travel after recording double-digit growth in tourism expenditure every year since 2007 (UNWTO, 2017). Being awarded the 2022 Winter Olympic Games and the evolution of the *principal contradiction*[1] prompted the Chinese government to develop policies to support the development of the ski industry. The Chinese central government has created ambitious goals to reach 800 ski resorts by 2022 and involve over 300 million Chinese people in winter sports by 2025 in order to develop a 1 trillion Yuan (about US$155.4 billion)

---

1 A report delivered by Xi Jinping at the 19th National Congress of the Communist Party of China suggested that China now face is the contradiction between unbalanced and inadequate development and the people's ever-growing needs for a better life.

ski industry (The State Council of the People's Republic of China, 2016). With the considerable growth over the last two decades, China is now estimated to rank among the largest ski tourism markets in the world, reaching 17.5 million skier visits and 703 ski resorts in 2017 (Wu & Wei, 2018), and it will soon surpass some of the global leaders like Canada (around 19 million skier visits per season) and Switzerland (21.2 million skier visits in 2016–17 season) (Vanat, 2017).

China's ski industry faces several challenges to continued growth, including limited facilities and services, less profitable ski resorts, and insufficient rate of new skier conversion to increase the domestic market. Climate change is also a potentially important risk to the long-term development of ski tourism in China, one that has not yet been assessed in the scientific literature. This chapter will summarize the stages of China's ski tourism development, including its current status and trends, and discuss the main challenges to future development in the early era of climate change.

The stages of China's ski tourism development

Although the first ski areas opened in the northeastern part of China during the 1980s, China's ski tourism began in 1996 when ski areas were first opened to the general public. Since the birth of ski tourism in China, three phases of ski tourism development can be identified in the growth of operating ski areas and skier visits in Fig. 40.

The emergence of the ski tourism market in China coincided with its decision to host the Third Asian Winter Games in Harbin (Heilongjiang Province) in 1996. After the conclusion of the games, some of the newly developed and upgraded ski areas were opened to the general public, initiating the emergence phase (1996–2000). Over this initial phase other ski areas were also gradually opened for public use and ski tourism was recognized as an economic development strategy in rural and mountain regions (Sun et al., 2016). At the end of the 20th century, China still had a very limited ski tourism industry with only 50 ski areas and 300,000 annual skier visits (Fig. 40).

The early development phase (2001 to 2010) was characterized by a five-fold increase in the number of operational ski areas, from 50 in 2000 to almost 270 in 2010 (Fig. 40). Skier visits also grew substantially from 300,000 at the end of the emergence phase to over 6 million in 2010

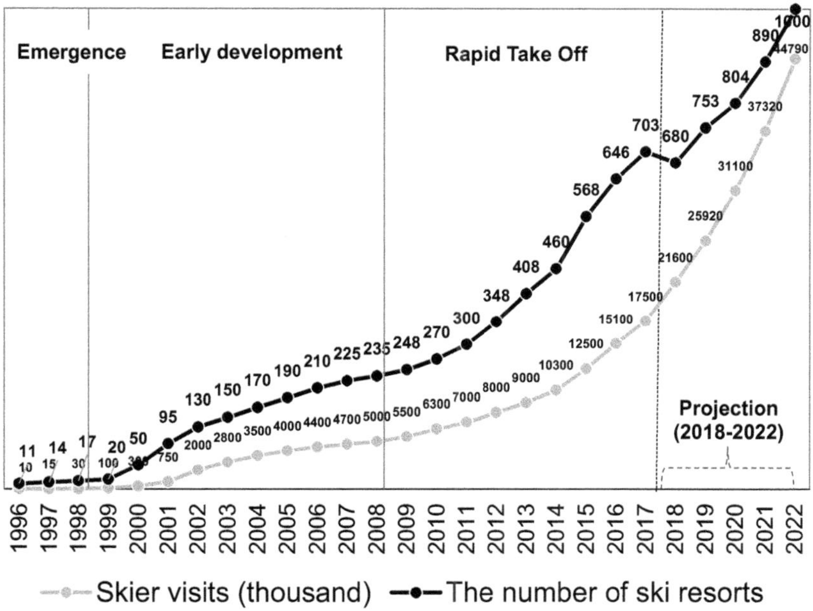

Fig. 40: Evolution of China's ski industry (1996–2022)

(Fig. 40). Recognition of the local development potential of ski tourism continued to grow after the Chinese government held the "International Ski Industry Forum" in 2001. Although several resorts were built in some south provinces (e.g. Yunnan, Xinjiang), Beijing and Hebei were still the main investment areas because of relative high ski demand from large urban markets and better snow conditions. For example, there were 13 ski resorts with tens of millions of Yuan investment just outside of Beijing, and an early ski industry cluster (e.g. Wanlong, Duolemeidi ski areas) created in Chongli county of Zhangjiakou (Hebei province). These early developments laid the foundation for the ski facilities that were part of China's successful bid to host the 2022 Winter Olympic Games. At the end of the early development phase, and only 15 years after opening as a tourism market, ski tourism in China was larger than traditional winter sports market of Norway (5.8 million skier visits) (Vanat, 2010).

The rapid take off or boom phase (2011 to present) has seen the number of ski areas double in only five years (2010 to 2015) and continues to increase toward the government of China's goal of over 800 ski areas by

2022 (Fig. 40). The geography of the ski industry in China also continues to evolve, with ski areas now operating in 24 provinces. Skier visits have already tripled in this phase, to over 17 million in 2017 (Fig. 40). Growth in winter sports participation is anticipated to continue with a series of supportive policies by the Chinese government, including for example the requirement of primary and secondary schools to open winter sports courses and/or cooperate with local ice and snow stadiums and clubs to encourage children to join more winter sports. This policy to encourage greater participation in winter sports will undoubtedly be bolstered by hosting the 2022 Winter Olympic Games. By 2022, it is estimated that as many as 1,000 ski areas will be operational, which would generate more than 44 million visits (Fig. 40).

Current status of ski tourism in China

Spatial distribution of China's ski areas and skier visits

China's ski industry has grown tremendously over the past 25 years. Figure 41 shows the location of China's over 600 outdoor operational ski areas in 2016 (Wu and Wei, 2018) and indicates that ski areas are unevenly distributed among its provinces. In 2017, Heilongjiang has the largest number of ski areas in China (124), with Shandong ranking as distant second (61), followed by Xin (59) and Hebei (58) with significant clusters of ski areas (Tab. 1). The majority of ski areas are located in the north and northeast of China, accounting for 30 % (provinces of Hebei, Beijing, Inner Mongolia, Shaanxi, Tianjin and Shanxi) and 29 % (provinces of Heilongjiang, Jilin, and Liaoning) respectively, while less than 1 % were built in the south China (provinces of Guangxi and Guangdong) (Fig. 42).

Skier visits show a similar geographic pattern, with ski areas in Heilongjiang and Hebei possessing the largest market share (11.2 % and 10.1 % respectively). The Beijing region is an important anomaly with 9.5 % of skier visits at only 24 (3.4 %) of the nation's ski areas, illustrating the importance of proximity and quality of facilities. Less than 1 % of skiers chose to ski in Southern China (i.e. Guangxi, Guangdong provinces). Strong growth in skier visits continues in north and northeast regions, with the provinces of Shaanxi, Hebei, and Jilin experiencing growth of 25 % or more from 2016 to 2017.

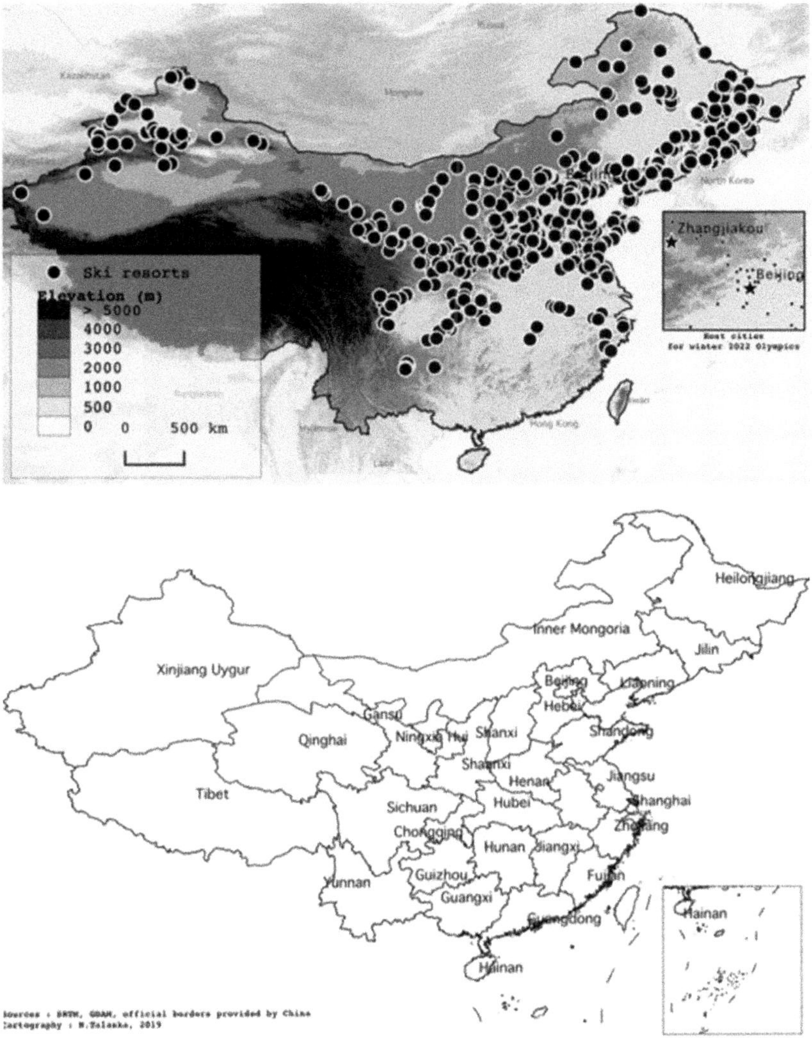

Fig. 41: The location of China's ski areas (in 2016)

Characteristics of skiers in China

Limited information is available on the socio-demographics of the Chinese ski market. The only available study was conducted by the Social Science Academy in its 2017 report on the development of ski industry

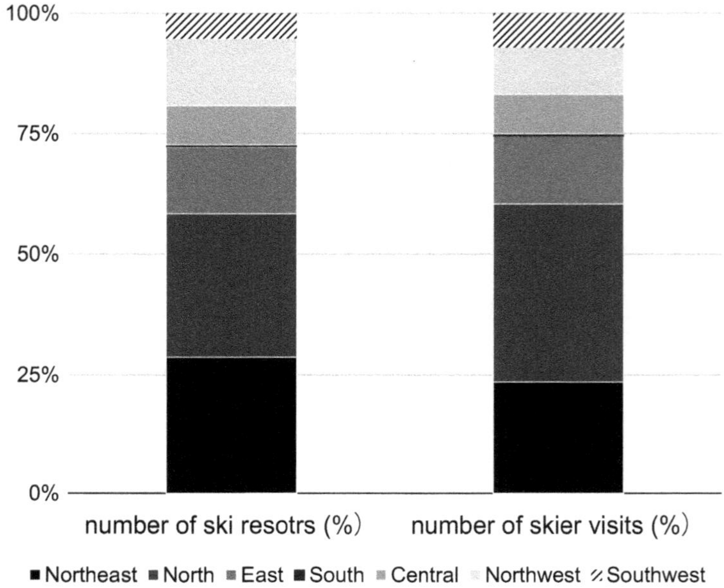

Fig. 42: Distribution of the number of ski resorts and skier visits in each region (in 2017)

in China (Sun et al., 2017). A small-scale survey was conducted with skiers (529 respondents in the Beijing region) to gain insights into trends and habits in winter sports. Because this survey was limited to the Beijing region and had a limited sample, it is unlikely to be representative of the national market. A much larger survey of the Chinese ski market should be a research priority. Table 2 provides the characteristics of Chines skiers participating in the survey.

Demographics

Beijing region skiers tended to be at the younger end of adult age spectrum, with 85 % between the ages of 18 and 39, and only 12 % over 40 and 3 % less than 18. While the baby-boomer generation has been a major cohort of the ski markets of Europe and North America for over 20 years, the post-World War 2 generation in China has almost no experience with skiing given the recency of rapid economic development and the emergence of winter sports in the mid-1990s. Since China's ski industry is relatively recent to develop, it is not surprising that almost all skiers (99 %)

Tab. 7: Distribution of ski areas and skier visits in each province (in 2017)

Region	Province	Number of skier visits		Number of ski resorts	
		2017	%	2017[1]	%
Northeast	Heilongjiang	1960000	11,2 %	124	17,6 %
	Jilin	1470000	8,4 %	41	5,8 %
	Liaoning	690000	3,9 %	37	5,3 %
North	Hebei	1760000	10,1 %	58	8,3 %
	Beijing	1670000	9,5 %	24	3,4 %
	Shanxi	1100000	6,3 %	45	6,4 %
	Inner Mongolia	840000	4,8 %	37	5,3 %
	Shaanxi	680000	3,9 %	31	4,4 %
	Tianjin	400000	2,3 %	13	1,9 %
Northwest	Xinjiang	860000	4,9 %	59	8,4 %
	Gansu	560000	3,2 %	20	2,8 %
	Ningxia	180000	1,0 %	12	1,7 %
	Qinghai	100000	0,6 %	7	1,0 %
East	Shandong	1040000	5,9 %	61	8,7 %
	Zhejiang	910000	5,2 %	18	2,6 %
	Jiangsu	390000	2,2 %	15	2,1 %
	Anhui	100000	0,6 %	3	0,4 %
	Fujian	30000	0,2 %	1	0,1 %
Central	Henan	900000	5,1 %	42	6,0 %
	Hunan	330000	1,9 %	8	1,1 %
	Hubei	210000	1,2 %	7	1,0 %
Southwest	Sichuan	580000	3,3 %	11	1,6 %
	Chongqing	330000	1,9 %	14	2,0 %
	Guizhou	300000	1,7 %	10	1,4 %
	Yunnan	40000	0,2 %	2	0,3 %
South	Guangxi	60000	0,3 %	2	0,3 %
	Guangdong	30000	0,2 %	1	0,1 %
	TOTAL	**17520000**	**100 %**	**703**	**100 %**

[1] ***Indoor and outdoor ski resorts are included.***

are domestic tourists according to the survey. In these respects, the Chinese market is very different than the mature markets overseas.

Skiers and snowboarders span the spectrum of income and education, but as in most international markets are concentrated at the highly education and affluent groups. However, nearly one-third of respondents fell into a relatively low-income group (under 5000 Yuan/month), possibly because of the over-representation of younger people in this market segment. Similarly, over 70 % of respondents had at least one university degree and a further 16 % had other form of post-secondary education.

Ski habits, motives and expenditures

The breakdown of annual skier participation was found to have similar patterns as some other international markets, with 82 % of the sample skiing less than five times per year and 18 % skiing five or more times per year.

From a list of motives to ski the top five included: leisure and entertainment (53 %), sports fitness (37 %), ski learning (24 %), friends and family social purpose (23 %), and taste improvement for leisure/recreation (15 %). In terms of expenditures, almost half of the respondents (47 %) spent less than 500 Yuan per person, per day for on-and off-mountain goods and services, while 25 % reported spending about 500 to 999 Yuan, 20 % between 1000–1999 Yuan, and 8 % more than 2000 Yuan per day.

Factors influencing choice of ski resort and international destination choice

A range of factors have been examined to understand ski destination choice behavior (Tab. 8), including the relatively tangible attributes of products, to the intangible benefits, needs, and personal values people seek to satisfy through their choice behavior (e.g. Klenosky, Gengler, and Mulvey, 1993; Won and Hwang, 2009). The survey found that in the most important factor influencing a ski/snowboard destination choice in China is snow condition (64 %). This is consistent with other international markets (Steiger et al., 2017) and not surprising considering the limited natural snowfall in most provinces where winter is the dry season. Other highly rated decision criteria include transportation convenience (48 %), services (47 %), price (39 %), vacation condition (26 %), and reputation (26 %). Transportation convenience appears to be more important in the Chinese market than in mature markets (Klenosky, Gengler and Mulvey, 1993; Konu, Laukkanen and Komppula, 2011; Richards, 1996), which is likely due to relatively higher rate of car ownership and better vehicle rental systems in other markets. The most popular international ski destinations for Chinese skiers/snowboarder were Japan (48 %), followed by Europe (23 %), and North America (19 %).

Tab. 8: Characteristics of skiers in the Beijing region survey (N=529).
Data Source: Compiled from Sun et al., 2017

Gender	%	**Occupation**	%	**Geographic distribution (top six)**	%
Men	52	Businessman	23	Beijing	43
Women	48	Technician	16	Heilongjiang	13
Age	%	Service industry	14	Hebei	10
Under 18	3	Education Industry	9	Liaoning	6
18–29 years	44	Student	8	Shandong	5
30–39 years	41	Job-waiting	2	Jilin	5
40+ years	12	Others	28	**Motives (top five)**	%
Monthly Income / yuan	%	**Education background**	%	Leisure and entertainment	53
Under 5000	30	Junior high and below	13	Sports fitness	37
5000–9999	43	High school	16	Ski learning	24
10000–14999	20	Bachelor's	50	Friends and family social purpose	23
15000 or more	7	Master's	19	Taste improvement	15
Expenditure/yuan	%	PhD	2	**Abroad ski destination choice**	%
Under 500	47	**Factors influencing choice of ski resorts**	%	Japan	48
500–999	25	Snow condition	64	Europe	23
1000–1999	20	Transportation convenience	48	North America	19
2000 or more	8	Services	47	New Zealand	4
Times of visiting for one ski season	%	Price	39	Korean	5
Once	33	Vacation condition	26	Others	1
Two-four times	49	Reputation	26		
Five-nine times	7	Others	6		
Ten times and more	11				

Economic actors involved in the ski resorts

Companies in China for machines involved in ski resorts

The whole ski resorts in China cover an area of 3080 hectares, and snowmaking machines are required for almost all these places (99 %) as well as groomer. Although China's companies have made tremendous progresses in recent years due to the rapid development of China's ski industry (Tab. 9), foreign brands still dominate the whole market for products involved in the ski resorts such as lifts, groomer, and snow guns. For example, the technology of detachable lift has been successfully developed by Beijing Materials Handling Research Institute in recent years, making the number of domestic detachable lifts used in China's ski resort increase quickly (Fig. 43). However, the figure for imported detachable lifts is still twice that of imported ones.

Ski equipment manufacturer

Ski equipment manufacturers in China are mainly focused on ski clothing, and they prefer to cooperate with international partners for seizing the huge demand market. Taking the 361°as an example, it formed a joint venture with One Way, a well-known supplier of ski poles and accessories founded in Finland.

Tab. 9: Companies in China for machines involved in ski resorts

The name of companies	Products
XueBawang	Snowmaking machine
Carving Ski	Snowmaking machine, groomer
EnternalStar Ice & Snow	Snowmaking machine, magic carpet
Nortec	Snowmaking machine, magic carpet
XueCheng	Snowmaking machine， snowmobile
HanNiu	Groomer
NiWeite	Snowmobile
Yahao	Magic carpet
Beijing Materials Handling Research Institute	Detachable lift
GuoYou	Drag-lift, magic carpet
HongJi	Aerial lift, surface lift

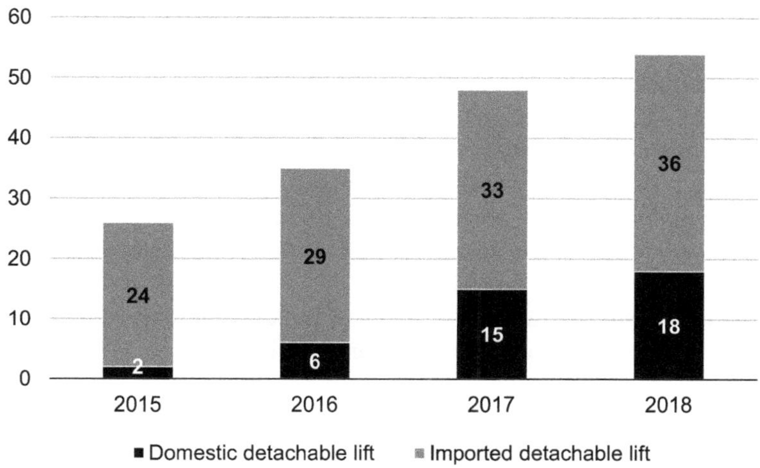

Fig. 43: The number of detachable lifts used in China's ski resort

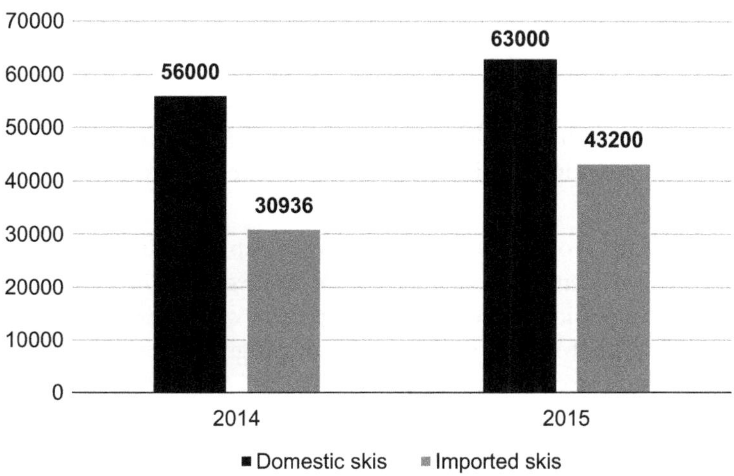

Fig. 44: The number of skis for renting in China's ski resorts

In terms of the production of skis/snowboard, most are original equipment manufacturers for foreign companies. With the technological accumulation, some firms began to produce skis/snowboard, which occupied more advantageous positions in rental market (Fig. 44).

The role of government in China's ski industry

Government involvement has greatly influenced China's tourism industry. The Chinese government has played several roles in the development of the tourism sector through policy documents, including operator, regulator, investment stimulator, coordinator and educator (Zhang and al., 1999). Regarding the development of ski tourism in China, policy also is considered as an important factor (Kan and Wang, 2016). After its successful bid for the 2022 Winter Olympic Games, the Chinese government released a series of policy documents focusing on the ski industry to support its development (Tab. 10).

National policy documents mentioned overall goals, which contain both qualitative and quantified aims of China's ski industry which should be achieved over a specific time period. According to the national targets, provincial and municipal authorities or bodies set corresponding sub-goals and developed action plans to achieve these objectives. Only the national goals are discussed in this study (see Fig. 45). The Chinese central government created ambitious goals to reach 800 ski resorts by 2022 and involve 300 million Chinese people in winter sports, hitting 1 trillion yuan (about $155.4 billion) by 2025 in terms of the scale of this industry. The goals also include the ski manufacturing industry and ski sport competition. Although the goals for recreational skiing and ski tourism are not explicitly proposed, they are important parts in task sections.

In order to involve 300 million Chinese people in winter sports, some measures have been adopted to get people hooked on skiing. For example, the plans for 'millions of teenagers take part in winter sports on a regular basis' and 'school winter sport education program' were mentioned to get children hooked on skiing at a young age. Winter Olympic model schools targeting primary and secondary ones were named in some provinces (e.g. Beijing, Hebei, Jilin), and this plan tries to incorporate education on winter sports into their curriculum.

Seven policy documents were released by more than one body, which means they are joint policies. Table 5 gives all the issuing and/or leading bodies of these documents. In terms of national level, an administration or a ministry is the main body issuing the documents while bureaus and commissions are the main ones at a subnational level. Although several policies officially were issued by the government and Party Committee, the leading body for implementation also was a ministry/bureau in these

Tab. 10: Overview of policy documents included in the content analysis

	Title	Publication Time	Level
1	Planning for Ice-Snow Sports Development 2016–2025	2016.11	National
2	Planning for Facilities Construction of Ice-Snow Sport Resorts 2016–2022	2016.11	National
3	Programme for Winter Sports Promotion and Popularization 2016–2020	2016.11	National
4	Opinions of the Beijing Government on Promoting Ice-Snow Sports Development 2016–2022	2016.3	Direct-controlled municipality
5	Planning for Ice-Snow Sports Development in Hebei Province 2016–2022	2016.11	Province
6	Opinions of the Hebei Provincial Government on Promoting Ice-Snow Sports and Industry Development	2017.3	Province
7	Planning for Ice-Snow Sports Development in Heilongjiang Province 2016–2022	2016.12	Province
8	Planning for Ice-Snow Sports Goods Industry Development in Heilongjiang Province 2017–2022	2017.8	Province
9	Five Actions for Ice-Snow Tourism in Heilongjiang Province	2018.2	Province
10	Opinions of the Jilin Provincial Government on Promoting Ice-Snow Industry Development	2016.9	Province
11	Planning for Ice-Ski Sports Development in Tianjin Province 2016–2025	2017.4	Direct-controlled municipality

cases. In addition, some authorities or bodies were involved in some cases but the form of collaboration was not developed in two-thirds of the documents. Moreover, some key policy-makers among these issuing bodies play different roles with regard to their impacts on the policy (Tab. 11). The Tourism of the People's Republic of China as well as Sport are the central government agencies directly responsible to the State Council for policy-making, development and administration while National Development and Reform Commission as a supra-ministerial cabinet

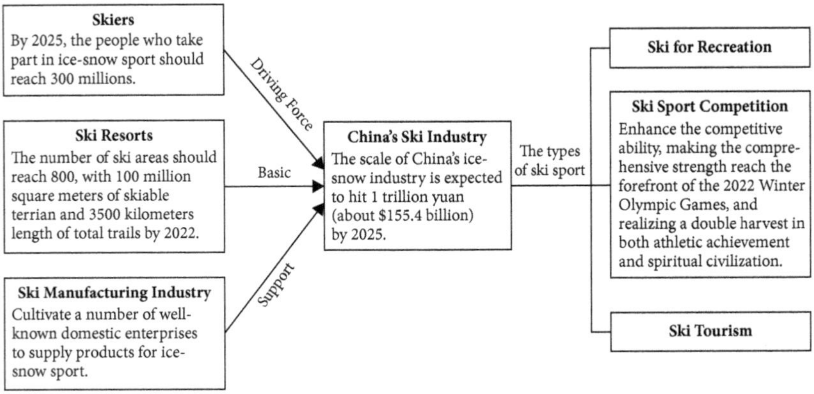

Fig. 45: Development goals of ski industry in China

agency in charge of the macro-management of economic and social development, considers ski industry issue from a strategic perspective. Apart from administrative orders, local government has a degree of flexibility in central decisions and indeed is encouraged to formulate their own strategies. Take Beijing as an example, specific tax and land policies (e.g. 15 % reduction in corporate income tax) are given to support the development of ski resorts in Beijing.

There is no doubt that policy plays a positive role that could be confirmed by leap development in recent years. Due to these clear support policy documents, capital moves more quickly to this industry, which has been one of the hottest investment sectors since 2015 (Rao, 2017). Ski demand, a driving force in the growth of China's ski industry, thus is met due to supply-side structural reform which offers expected ski areas guided by these policies.

Current challenges facing China's ski industry

The scarcity of high-quality ski resorts

China currently has over 700 operating ski areas nationwide (Wu an Wei, 2018), but less than 10 % could be considered high-quality by international standard. Most are only poorly equipped in terms of lift capacity with one or a few magic carpets, that are only suited to those first learning to ski. Jing (2018) contends that China's ski resorts are much less competitive than overseas destinations for serious skiers/snowboarders, and that only 25 % of ski areas in China meet international standards of lift capacity, variety of terrain (easy-intermediate-difficult), and other services that are common in the mature markets of Europe, North America and Japan. As a result, more than 600,000 Chinese skiers and snowboarders travelled abroad in 2017, with Japan currently the most favored destination, ahead of Switzerland, Italy and Canada (Toropov, 2018). Without further development of high-quality ski areas in China, there is the risk that more Chinese skiers will travel outside the domestic market and that a portion of beginner skiers, particularly young skiers, will lose interest as they outgrow the majority of ski areas designed for those first learning to ski, slowing the development of the industry.

Low rate of ski participation and beginner conversion

Although more and more Chinese have been taking to the slopes in recent years, skiing and snowboarding holidays are still a niche market in China. China ranks fourth worldwide with regard to skier visits in 2017 (Vanat, 2017) but its ratio of skiers to population remains much lower that other markets where snow sports have been a long cultural tradition (Tab. 6). According to Lan (2017), the rate of beginner conversion in China is low (less than one percent), which is far below the level in Western countries. In other words, only one percent of Chinese that try skiing turn into active participants in the ski tourism marketplace (Tab. 12). The current low conversion rate is partly due to the lack of a ski and winter sports culture, but may also be related to an initial bad skiing experience at poorly equipped and crowded ski areas that lack sufficient instruction to teach new skiers. The 2017 China Ski Industry White Book showed that most people (56 %) were not inclined to hire instructors because of high cost (Sun and al., 2017) and there is not enough qualified ski instructors in China to offer services to all beginners. Strategies to convert more first-time skiers to become active in the ski marketplace is a key challenge for the growing Chinese ski market. Positively, China's demographics, including the massive untapped market of those who have not tried skiing and younger population, provide much greater opportunities than the ageing populations of many European countries and Japan.

Tab. 11: Issuing and/or leading body of include policy documents

Body	Number documents issued	%	Body	Number documents issued	%
National level (n = 3)					
Administration of:			**Ministry of:**		
Sport of the People's Republic of China (PRC)	3	100	Housing and Urban-Rural Development of the PRC	2	66.67
Tourism of the PRC	3	100	Education of the PRC	2	66.67
Customs of the PRC	1	33.33	Finance of the PRC	2	66.67
Taxation of the PRC	1	33.33	Land and Resources of the PRC	2	66.67
Industry and Commerce of the PRC	1	33.33	Industry and Information Technology of the PRC	2	66.67
Forestry of the PRC	1	33.33	Human Resources and Social Security of the PRC	1	33.33
China Federation of	1	33.33	Water Resources of the PRC	1	33.33
Trade Unions	1	33.33	Agriculture of the PRC	1	33.33
Disabled Persons	1	33.33	Culture of the PRC	1	33.33
Women	1	33.33	Civil Affairs of the PRC	1	33.33
China Insurance Regulatory Commission of the PRC	1	33.33	National Development and Reform Commission of the PRC	3	100
The People's Bank of the PRC	1	33.33	The Communist Youth League of the PRC	1	33.33
Provincial/ Municipal level (n = 9)					
Bureau of:			**Commission of:**		
Sports	4	44.44	Development and Reform	3	33.33
Tourism	2	22.22	Industry and Information Technology	1	11.11

Tab. 11: Continued

Body	Number documents issued	%	Body	Number documents issued	%
Finance	1	11.11	Urban-Rural Development	1	11.11
Land Resources and Housing Administrative	1	11.11	Provincial/ Municipal Government	3	33.33
Education	2	22.22	Party Committee	1	11.11

Tab. 12: Global comparison of skier participation rate (skiers to population). ***Data source: Vanat, 2014***

Top 10 countries by skier visits	Skier visits	Number of skiers	Participation rate (% population)
USA	57 092 127	13 616 748	4.3 %
France	56 226 000	8 573 709	13 %
Austria	53 155 600	2 959 793	36 %
Japan	34 432 389	11 452 777	9 %
Italy	28 100 000	4 918 584	8 %
Switzerland	26 538 264	2 958 530	37 %
Canada	18 700 400	4 307 199	12.5 %
Germany	14 922 000	14 606 508	18 %
Czech Republic	8 700 000	2 032 584	20 %
Sweden	8 070 800	1 823 885	20 %
Emerging markets	**Skier visits**	**Number of skiers**	**Participation rate (% population)**
China	6 880 000	5 128 426	0.4 %
Turkey	1 200 000	806 945	1.0 %
Russia	3 653 870	3 562 512	2.5 %
Azerbaijan	50 000	47 951	0.5 %
Georgia	100 000	91 118	2 %

Lack of domestic ski culture

Altay of Xinjiang, China is considered as one of the world's birthplaces of skiing, where the first written description of skiing dates back to the Western Han Dynasty, 206 BC to 24 AD ("Origin Story", 2017). However, a modern winter sports and ski culture has not developed in China in the same way that it has in other birthplaces of skiing like Norway and Austria.

Tab. 13: Comparison of the annual snowfall in major cities and world-class ski areas. ***Data source: "42 Ski", 2015; World Weather Online*** **(https://www.worldweatheronline.com/)**

Well-known ski resorts	Country	Annual snowfall in meters
Grand Targhee	USA	11.7
Whistler Blackcomb	Canada	11.6
Squaw Valley	USA	11.4
Niseko	Japan	11.0
Warth-Schröcken	Austria	10.5
Big Sky	USA	10.1
Lake Louise/Banff	Canada	10.1
Chamonix	France	9.6
Braunwald	Switzerland	9.0
Damüls	Austria	9.0
Major ski towns/cities	**Country**	**Annual snowfall in meters**
Chamonix-Mont-Blanc	France	4.8
Niseko	Japan	3.1
Québec	Canada	2.1
Yabuli	China	1.0

A primary reason for the limited development of snow sports in China relative to other winter sports nations is the limited natural snow resources in most areas of China. As there is no information on snowfall in China's ski resorts, we compare the five-year average annual snowfall in major ski towns/cities worldwide. The annual snowfall is considerably lower in China than that in other countries (Tab. 13). Clearly, famous ski towns or cities in traditional winter sports markets (e.g., Canada, Austria, France, Japan) have average annual snowfall of 2–5m while it is only around 1m in Yabuli town, an origin region for ski tourism that has relatively good snow condition in China (Tab. 13). The high cost of skiing relative to many other sports (Sun and al., 2016) was also a major deterrent to skiing in China until the economic revolution began in the 1980s.

Low profitability of ski resorts

The majority of ski resorts in China are struggling to make profits because of limited facilities and services and a reliance on a single type of tourism product. Most ski areas in China have not begun to diversify their revenue streams into ski tourism related services (e.g., retail, hospitality, spas) or four season tourism (e.g., green season activities like mountain hiking and biking, golf) or real estate related investments like many rebranded mountain resorts have in Europe and North America. For example, Wanlong Ski Resort is one of the oldest ski areas in China, and it only begun making a profit in 2014 (less than 10 % of its revenue) (Chen, 2017). Around 70 % of its revenues come directly from entrance tickets and equipment rentals, with the remaining 30 % from hotels and catering. With rising costs of staff, energy, environmental regulations, and water putting pressure on profits (Li and Glenn, 2018), many smaller ski resorts are likely to be under financial stress and many may close in the near future due to growing competition and the lack of capital to diversify. This would follow similar trends of diversification and consolidation in North America (Scott, Dawson and Jones, 2008; "Ski resorts", 2018) and Europe (Haanpää and al., 2015; Nordin and Svensson, 2007). Larger ski area clusters with well developed transportation access and climate advantages will need to follow the economic diversification strategy toward the four season mountain resort model that leading ski destinations in mature international markets have pursued over the last 25 years.

Environment impacts

The environmental impact of the ski industry has been criticized by many scholars and industry observers (Duglio and Beltramo, 2016; Beaudry, 2018; Comai, 2014). Environmental management systems to improve sustainable operations is considered an important market opportunity for leading ski areas (Duglio and Beltramo 2016; Pröbstl-Haider and al., 2018). The environment impact of ski areas has far been largely ignored in China. Although several studies show that the construction of some ski resorts in China exerted localized impact on the environment (Liang and al., 2005; Hu and Yang, 2011; Zou and al., 2005), considerably larger adverse effects might arise if stakeholders do not pay enough attention to sustainability as China's ski resorts continue to expand their terrain and water use for snowmaking. In order to mitigate the potential negative effects of China's ski resorts, an environmental monitoring and reporting system will need to be developed, including effective regulations to guide ski resort construction and operations.

The vulnerability of China's ski industry to climate change

Climate change is recognized as a major threat to the ski industry (Scott, Gössling and Hall, 2012; Steiger and al., 2017). With limited natural snow resources, impacts of projected climate change are potentially very significant in China. Fang et al. (2019) provide the first analysis of the potential impact of climate change on China's ski industry. The study examined 115 ski areas that represented the top 25 % of ski areas by lift capacity (i.e., very small ski areas with only magic carpets for lift capacity were not included), and consistent with studies in other countries that found average ski seasons became more variable and shorter on average by the 2050s (Steiger and al., 2017). Ski areas located in more northerly latitudes (i.e. many ski areas in Heilongjiang, Inner Mongolia and Xinjiang) show longer season lengths under all climate change scenarios for all future time periods than those resorts located in southern regions (i.e. many ski areas in Shandong and Henan) (Fig. 46). With advanced snowmaking, between 78 % and 73 % of the ski areas studied were able to maintain an average ski season of at least 100-day season length under the RCP4.5 and RCP8.5 emission scenarios (representing a successful Paris Climate Agreement outcome and business as usual emissions pathways respectively). These figures drop to between 72 % and 66 % in the 2080s, which is higher than several international markets studied previously (see Steiger and al., 2017).

The average amount of machine-made snow (in cm) required to maintain a 100-day season is higher in most areas of China because of the

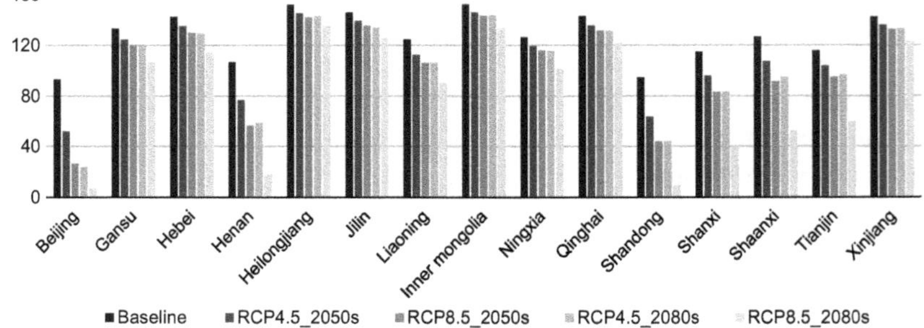

Fig. 46: Projected ski season length in China under different GHG emissions scenarios

limited amount of natural snowfall compared to many other high profile ski destinations world wide (see Tab. 7). With the current climatology, ski areas located in Beijing, Henan, Shandong, and Shaanxi depended more heavily on snowmaking, requiring more than 150 cm per season on average, while in Heilongjiang, Jilin, Inner Mongolia, Qinghai and Xinjiang less than 60 cm of snowmaking is required. By mid-century required snow production will increase 13–36 % in the RCP 4.5 emission scenario and 25–58 % under high emission scenario (RCP 8.5) by mid-century. Ski areas in one-third of Provinces will require more than 200 cm of snowmaking per year under the RCP 8.5 scenario. The increased costs of snowmaking and continued access to water supplies will create uncertainties for the sustainability of some ski areas, particularly in warmer southern regions and lower elevation ski areas.

Conclusion

China is a relatively new and emerging ski tourism market on the international stage, but one that has the potential to become the largest domestic market (measured by skier visits) in the next two decades. Its rapid growth is characterized by several challenges that are different from those commonly faced in the mature markets of Europe, North America and Japan. Further research is required to understand these near-term challenges and assess the longer-term destination level development potential under a changing climate will assist the government and tourism industry to develop strategies and regulatory that supports the sustainable development of ski tourism in China.

References

42 ski resorts that get the most snow in the world. 2015, October 7. *Unofficial Networks*. Retrieved from https://unofficialnetworks.com/2015/10/07/the-42-ski-resorts-that-get-the-most-snow-in-the-world/ (Last access: 14 february 2020).

Beaudry, Frederic. 2018. "Ski Resorts and their impact on the environment". *Thought Co*. Retrieved from https://www.thoughtco.com/ski-resorts-and-the-environment-1203969 (Last access: 14 february 2020).

Chen, Meiling. 2017. "China is 'biggest market for entry-level skiers'". *China Daily*. Retrieved from http://www.chinadaily.com.cn/a/201712/25/WS5a404e0ba31008cf16da3344.html (Last access: 14 february 2020).

Comai, Ben. 2014. “Assessing the environmental impact of ski resorts”. *Greening the Ski Industry*. Retrieved from http://blog.uvm.edu/bcomai/2014/04/15/assessing-the-environmental-impact-of-ski-resorts/ (Last access: 14 february 2020).

Duglio, Stefano and Beltramo, Riccardo. 2016. “Environmental management and sustainable labels in the ski industry: A critical review”. *Sustainability* 8(9): 851.

Fang, Yan, Scott, Daniel and Steiger, Robert. 2019. “The impact of climate change on ski resorts in China”, *International Journal of Biometeorology* 63(11): 1–13.

Haanpää, Simo, Juhola, Sirkku and Landauer, Mia. 2015. “Adapting to climate change: Perceptions of vulnerability of down-hill ski area operators in Southern and Middle Finland”. *Current Issues in Tourism* 18(10): 966–978.

Hu, Jiping and Yang, Junlin. 2011. “Impacts of ski area construction in northeast China: a case study of Changbai Mountain resort”. *Protection Forest Science and Technology* 103(4): 89–92.

Jing, Travel. 2018. “China's low-quality ski resorts are an opportunity for Japan”. *Jing Travel*. Retrieved from https://jingtravel.com/chinas-low-quality-ski-resorts-are-an-opportunity-for-japan/ (Last access: 14 february 2020).

Kan, Jun-chang and Wang, Fei. 2016. “Driving factors and innovation paths of skiing industry upgrading under the strategy of Winter Olympic Games in China”. *China Sport Science* 36(6), 11–20.

Klenosky, David B., Gengler, Charles E. and Mulvey, Michael S. 1993. “Understanding the factors influencing ski destination choice: A means-end analytic approach”. *Journal of leisure research* 25(4): 362–379.

Konu, Henna, Laukkanen, Tommi and Komppula, Raija. 2011. “Using ski destination choice criteria to segment Finnish ski resort customers”. *Tourism Management* 32(5): 1096–1105.

Lan, Xiaoxiong. 2017. “2017 Snow Season Report III: How do skiers turn into enthusiasts, training and apps?”. *Lan Xiong Sport*. Retrieved from http://www.lanxiongsports.com/?c=posts&a=view&id=5550 (Last access: 14 february 2020).

Li, Pei and Glenn, Elias. 2018 February 5. “China's winter resorts face tougher terrain despite Olympics excitement”. *Reuters*. Retrieved from https://www.reuters.com/article/us-china-economy-skiing-insight/

chinas-winter-resorts-face-tougher-terrain-despite-olympics-excitement-idUSKBN1FQ07K (Last access: 14 february 2020).

Liang, Xiongbing, Cheng, Shenggao, Song, LiJun and Zhang, Zhongwang. 2005. "Study on the assessment of ecological impact based on RS & GIS: a case of construction of ski field in NiuChangping of Shennongjia". *Journal of Xiangfan University* 26(2): 75–79.

Nordin, Sara and Svensson, Bo. 2007. "Innovative destination governance: The Swedish ski resort of Åre". *The International Journal of Entrepreneurship and Innovation* 8(1): 53–66.

Origin Story: Where did skiing begin. 2017 June 12. *International Skiing History Association.* Retrieved from https://www.skiinghistory.org/news/origin-story-where-did-skiing-begin (Last access: 14 february 2020).

Pröbstl-Haider, Ulrike, Brom, Monika, Dorsch, Claudia and Jiricka-Pürrer, Alexandra. 2018. *Environmental Management in Ski Areas: Procedure—Requirements—Exemplary Solutions.* Springer, New York.

Rao, Xianjun. 2017. "Ski industry has emerged as hottest investment sector in China". *EEO.* Retrieved from http://www.eeo.com.cn/2017/1117/316984.shtml (Last access: 14 february 2020).

Richards, Greg. 1996. "Skilled consumption and UK ski holidays". *Tourism management* 17(1): 25–34.

Scott, Daniel, Dawson, Jackie and Jones, Brenda. 2008. "Climate change vulnerability of the US Northeast winter recreation–tourism sector". *Mitigation and adaptation strategies for global change* 13(5–6): 577–596.

Scott, Daniel, Gössling, Stefan and Hall, Michael C. 2012. "International tourism and climate change". *Wiley Interdisciplinary Reviews: Climate Change* 3(3): 213–232.

Ski resorts: Years they were founded. 2018, February 14. *International Skiing History Association.* Retrieved from https://www.skiinghistory.org/history/ski-resorts-years-they-were-founded (Last access: 14 february 2020).

Steiger, Robert, Scott, Daniel, Abegg, Bruno, Pons, Marc and Aall, Carlo. 2017. "A critical review of climate change risk for ski tourism". *Current Issues in Tourism 22*(5): 1–37.

Sun, Chenghua, Wu, Bin, Wei, Qinghua and Zhang, Hongjun. 2017. *Annual report on development of ski industry in China (2017).* China: Social Science Academic Press.

Sun, Chenghua, Wu, Bin, Wei, Qinghua, Zhang, Hongjun, Yin, Zhenhua and Yu, Yang. 2016. *Annual report on development of ski industry in China (2016)*. China: Social Science Academic Press.

The State Council of the People's Republic of China. 2016. *Planning for Ice-Snow Sports Development 2016-2025*. Retrieved 10 June 2018, from http://www.gov.cn/xinwen/2016-11/25/content_5137611.htm (Last access: 14 february 2020).

Toropov, Pavel. 2018. "Why most of the 1.2 million skiers in China avoid local ski resorts and head for Japan instead". *South China Morning Post*. Retrieved from http://www.scmp.com/lifestyle/health-wellness/article/2142872/why-most-12-million-skiers-china-avoid-local-ski-resorts (Last access: 14 february 2020).

UNWTO. Asia Tourism Trends (2017 Edition). Retrieved 16 June 2018, from https://www.e-unwto.org/doi/pdf/10.18111/9789284419142 (Last access: 14 february 2020).

Vanat, Laurent. 2010. *2010 International report on mountain tourism*. Retrieved 10 June 2018, from http://www.vanat.ch/ (Last access: 14 february 2020).

Vanat, Laurent. 2014. *2014 International Report on Snow & Mountain Tourism – Overview of the key industry figures for ski resorts*. Retrieved from http://www.ropeways.net/rnn/konfig/uploads/pdf/59.pdf (Last access: 14 february 2020).

Vanat, Laurent. 2017. *2017 International Report on Snow & Mountain Tourism– Overview of the key industry figures for ski resorts*. Retrieved from https://www.vanat.ch/RM-world-report-2017-vanat.pdf (Last access: 14 february 2020).

Won, Doyeon and Hwang, Sunhwan. 2009. "Factors influencing the college skiers and snowboarders' choice of a ski destination in Korea: A conjoint study". *Managing Leisure* 14(1): 17–27.

Wu, Bin and Wei, Qinghua. 2018. *China Ski Industry White Book (2017 Annual Report)*. Retrieved 5 June 2018, from http://www.vanat.ch/China%20Ski%20Industry%20White%20Book%202017.pdf.

Zhang, Hanqiu, Chong, King and Ap, John. 1999. "An analysis of tourism policy development in modern China.". *Tourism Management* 20(4): 471–485.

Zou, Ning, Cheng, Shenggao, Zhu, Gang, Huang, LiHua and Guo, Feng. 2005. "Influence of ski area construction on environment and potential countermeasures". *Journal of Xiangtan Normal University (Natural Science Edition)* 27(2): 35–39.

3.3

The Future of Ski Resorts in the Light of Climate Change: A Consideration from the Tourist's Point of View

Coralie Achin

Introduction

Following a few initiatives in the 19th century, the 20th century saw the development of many tourist resorts in mountain areas on different continents. At the time, these resorts centred on winter sports and ski areas – factors that appealed to tourists eager for the outdoors and a feeling of modernity – and had economic benefits for many local actors and communities. At first, this development mainly concerned the Alps: In 1970, 60 % of the 20,000 ski lifts in operation in the world were in this region (Barbier, 1978, quoted in Pearce, 2002). The development subsequently spread to become standardised around the world: North America, Northern Europe, Japan and Australia gradually emerged as new destinations for those seeking to do winter sports, which diminished the importance of the European Alps. Today, the Alps region "only" has 39 % of the world's ski lifts (Vanat, 2017). In addition, on all continents, ski resorts have been key drivers of local economic development, which boosts the supporting territories' economies (Lasanta, 2010; Lasanta et al., 2007; Veyret-Verner, 1959). Although most of these resorts have entered a management phase, it is worth emphasising that new tourist destinations are also coming on the scene: China and Croatia, in particular, are hoping to boost their economies through tourism and have committed themselves to having the facilities required.

After an initial development phase driven by plans to develop this activity, the first challenges came in the 1980s, which had the first slump in sales in the real estate sector and the first winters without snow (Lorit, 1991). Although the lack of snow had a significant impact on winter resorts, it was ascribed to a mix of weather conditions: Back then, bad winters were little more than an intermittent phenomenon and were not viewed as harbingers of a lack of snowfall in winters to come. However, since the mid-2000s, climate change has become a real source of concern. Several factors have played a role in increasing awareness: First, the Intergovernmental Panel on Climate Change's Fourth Assessment Report (IPCC, 2007) predicted that climate change would lead to a decrease in snowfall in mountain areas. In the future, this decrease will have a strong, adverse impact on activities that depend on snow, foremost among which are ski resorts. In addition to these irrefutable scientific conclusions, the release of the film *An Inconvenient Truth* in 2006, as well as the many debates surrounding the entry into force of the Kyoto Protocol (2005), also contributed to raising the general public's awareness of climate change and its impact (Pickering et al., 2010). After that, a number of studies have subsequently focused on an issue that had not previously been discussed at great length: the impact of climate change on tourism. Work targeted at the "supply" side of the equation has made it possible to explore, with increasing delicacy and precision, the consequences of a scarcity of snow regarding ski areas' viability. To this end, research teams applied the climate change scenarios presented by the IPCC to the snow conditions at resorts, including in different territorial areas. Thus, whether in North America (Dawson and Scott, 2007; Dawson and Scott, 2010; Dawson et al., 2009; Scott et al., 2007), Western Europe (François et al., 2014; Koenig and Abegg, 1997; Moen and Fredman, 2007; Pons-Pons et al., 2012), Eastern Europe (Demiroglu et al., 2015; Demiroglu et al., 2016; Ogrin et al., 2011), Asia (Fukushima et al., 2003) or Australia (Hendrikx et al., 2013; Hopkins et al., 2013; Koenig, 1998), the extensive research that was undertaken to identify the impact of climate change on ski resorts' viability has stated the expected decline in the number of ski areas that will remain viable in the future. However, these studies, which provide valuable information on the viability of ski areas, have to be considered with some circumspection within a comparative framework: The information does not make it possible to address with much precision the intraregional specificities linked to differences in orientation and altitude or to microclimates.

In turn, the analysis of the consequences of climate change with respect to the "demand" side (that is, tourists) has received very little attention and still has major gaps. Only a few works have focused on tourists' views of climate change or the adaptation strategies that can be used in this context. In this article, on an international scale and on the basis of a literature review, we will examine the appeal of ski resorts in the context of climate change by focusing on tourists' perceptions and behaviours.

Climate change, between representations and projected impacts

In mountain regions, climate change is threatening the winter activities of ski resorts. However, little work has looked into the effects as perceived/anticipated by the various stakeholders in tourist activities. Several of these studies have looked at the perceptions of ski lift operators, while only a few have tackled the issue from the tourists' perspective.

Ski lift managers' minimal reaction to climate change

The work carried out on the supply side has examined the perceptions of climate change on the part of the actors in the sector. A comparative analysis of them makes clear that they have a limited level of awareness regarding climate change as a major risk factor. In her 2014 article, Trawöger (2014) uses semi-direct interviews to look at ski lift operators' (n=24) perception of climate change. The collected data enabled her to present a four-part classification of the station managers according to their sensitivity to climate change: "convinced planners", "annoyed deniers", "ambivalent optimists" and "convinced wait-and-seers", of which the latter group was the largest (close to 46 %). Although these "convinced wait-and-seers" do not question climate change, they do not report seeing any particular effect on their resort and, therefore, are not putting in place any response mechanisms. In this respect, Trawöger's work is in line with the conclusions that Wolfsegger et al. (2008) drew a few years earlier: They had questioned 36 ski lift operators at the most vulnerable Austrian resorts (among the Top 71), that is, those that have more than 10 km of ski trails and whose ski area is situated at least 50 % below 1,500 m. The authors reached the conclusion that climate change was not perceived as a major threat, and bad winters were, for the most part, equated with cyclical meteorological behaviour – part of the natural order of things and without a specific connection to climate change. These findings have to be nuanced a bit, as the work that Saarinen and Tervo carried out in 2006 and 2010 (Saarinen and Tervo, 2006; Tervo and Saarinen, 2010) shows a heightened awareness of climate change after a snowless winter.

This awareness, which is partial but growing, illustrates the challenge ski lift managers have to face to extricate themselves from a short-term (specifically, economic) reality. The main goal is today for them to preserve the direct economic benefits and the jobs generated by the activity: Therefore, the structuring of the local economy heavily depends on the ski resorts being operational. Some of the leading markers of climate change are the increasing costs that have to be incurred to protect against bad snowfall conditions and the effect of bad winters on tourists' behaviour (Luthe, 2009).

The tourists' growing awareness of climate change

Tourists are the key players at tourism-oriented destinations. Nevertheless, very little work has been carried out to explore their views on climate change. As a result, this issue has to be approached by settling for elements of limited scope resulting from work done on a small territorial scale among a sample of tourists whose representative dimension has to be treated with caution. The work that Behringer et al. (2000) carried out at five resorts in Central Switzerland at the beginning of the 2000s showed that tourists had a strong sense of the impact of climate change on ski resorts. Out of the 1,000 individuals surveyed, 83 % reacted in the affirmative when asked whether climate change has an impact on ski resorts. In response to the question of the time scale, 41 % of those surveyed estimated that the effects of climate change would be felt between 2000 and 2030. Ten years later, in a different territorial context, Pickering et al. (2010) found similar results in Australia: In response to the question, "Do you believe that global warming could threaten skiing in Australia? Please circle: No, Yes. If Yes, when? Before 2030, Before 2060, Between 2060 and 2100", 60 %

of those surveyed responded 'Yes' and chose the option "Before 2030"; in 1998 in the same geographical context, this figure had been 52 % of respondents (Konig, 1998, quoted in Pickering, 2010). At the same time, the number of people questioned who did not consider climate change to be a threat fell from 22 % to 12 %. While these responses might suggest that tourists have a strong sensitivity to climate change, research carried out at other tourist destinations has yielded some mixed results. At the Acadia National Park (Maine, USA), De Urioste-Stone et al. (2016) asked tourists visiting Mount Desert Island how they perceived the impact of climate change on tourism in the park. Among those who responded, 13 % said they were sceptical that climate change was a threat, while 25 % (the believers) had an above-average belief that climate change was a threat. The remaining 62 % were neutral, saying climate change would not have any particular effect on tourism in the park.

It is unquestionably a good thing that these elements relating to the tourists' perception or representation of climate change exist. However, this approach is subjective, to say the least, since the criteria for risk representation include not only perception (the five senses) but also beliefs and the socio-economic situation (Peretti-Watel, 2000), which the work by De Urioste et al. (2016) confirms: The tourists most sceptical of the effects of climate change are, for the most part, older men, as well as low-income individuals. These results make it important to publicise the work that has been done on climate change to the widest possible audience. It also enables us to ask questions about the methodologies used. Finally, does asking tourists about their behaviour not make it possible to rid ourselves of the abstraction of the concepts and to provide, indirectly, useful information on representations and perceptions?

(How) do tourists adapt to climate change?

The issue of whom tourist behaviour evolves with respect to climate change is covered in a second part of the work examining ski resorts visitors. More detailed than the first, this set of data is very heterogeneous, which makes a comparison of the results more complex. This heterogeneity concerns both the territorial contexts targeted by the research and the methodologies and approaches used.

An emphasis on substitution strategies for tourists in the event of a lack of snow

Tourists' projected behaviours in the event of a lack of snow can be summarised by making use of the substitution strategies identified by Iso-Ahola (1986), which have since been reworked and refined: spatial substitution, temporal substitution and activity substitution. The first two can be understood as "within" substitutions (Iso-Ahola, 1986): There is no change in the activity as it is traditionally practised – only the location (spatial substitution) and the timing of the practice (temporal substitution) are changed. The last kind, activity substitution, refers to a "between" substitution. Its impact strikes at the very root of the practice since the traditional activity (winter sports) is abandoned in favour of another, whatever it may be. In terms of how they are applied to winter tourism at ski resorts, these forms of substitution have different consequences for ski resorts and their supporting territories. Spatial substitution, which entails a change in location, will involve a preference for higher altitudes, where there will be a better guarantee of snowfall. As a result, ski resorts located at higher altitudes will be favoured. Temporal substitution suggests modifying practice periods in order to take advantage of the optimal climate and weather conditions for downhill ski. What is modified here is not the resort where the activity is carried out but only the time period when it is done, as the tourist seeks out the most favourable conditions to practise winter sports. Lastly, activity substitution leads to a shift away from skiing (and, more broadly, from activities at ski areas) and towards other activities that, on the one hand, do not rely solely on snow and, on the other hand, are not systematically practised on the grounds of the ski resort.

All of these changes and adaptation strategies that have been highlighted will have an impact on ski resorts. However, this impact will vary in intensity depending on the particular substitution strategies and the relevant territorial contexts. In this regard, Steiger et al. (2017) note that taking an analysis of the consequences of tourists using a substitution strategy and turning it from the micro-level (resort) to the macro-level (territory) reveals a stable number of skiers and, more broadly, a stability in the share of tourism in the regional GDP. This will only be confirmed in a territorial context where the density of existing resorts allows visitors to move between different resorts inside a given territorial perimeter. In addition, this movement of the tourists, which allows, at least at the outset, the associated economy to remain intact, is very likely to change because of congestion phenomena that will probably occur if a small number of resorts profit from the movement of the individuals (Rutty et al., 2015; Steiger et al., 2017), whether it is of the spatial or the temporal variety. The third type of substitution, namely activity substitution, will have a strong impact on the economic structuring of the territories affected by the activity. The abandonment of ski resorts is in line with the many challenges facing the economic model of tourist diversification and, in particular, territories' capacity to take advantage of enough economic benefits to ensure the sustainability of local jobs and services. In order to examine the tourists' behaviour and, in so doing, "predict" the impact of climate change on ski resorts and their territories, several studies have adopted an approach using questionnaires distributed to tourists.

Questionnaires and tourists' preferences for adaptation

This methodology based on questionnaires distributed to tourists has been used in different territorial contexts since the end of the 1990s. Most of the studies have focused on the main winter sports markets (United States and Austria), although more marginal destinations (like Australia) have not been excluded (see the average annual number of skier-days for these countries; Vanat, 2017). However, it should be noted that, as of yet, some of the market leaders (France and Italy) have not been studied in depth, even as the number of tourist skier-days and nights at resorts in these countries suggest there will be serious consequences for local economies in the future. This geographical breakdown also raises the question of so-called emerging markets: For example, since 2006, China has seen its skier-days increase by about 250 %, and equipment manufacturers expect a steady rise in the number of skiers. That being said, how these new converts to winter sports will react in the event of a lack of (sufficient) snowfall is still largely unknown. This geographical diversity, which is essential to understanding resorts' viability (and to ensuring a holistic approach to the reliability of resorts by comparing information on the viability of ski areas and the expected development of their visitors), makes a universal analysis more complex, as contextual differences (position of winter sports in national cultures, density of resorts, costs of practice etc.) influence the tourists' expected behaviour. In a similar vein, the number of years (19) between the small number of surveys carried out (eight in 2018) also makes any comparison rather tenuous: Between 1998 (Konig, 1998) and 2015 (Rutty et al., 2015), new challenges emerged: sustainable development and, along with it, a growing awareness of the environment, but especially climate change, which has benefitted from detailed knowledge elements and increasing media coverage. In addition, even though environmental concerns and the rise in the price of oil barrels tend to mitigate the phenomenon, what should be noted is the major advances made in the world of transport, which has made it easier for tourists to go on holiday and, as a result, facilitated research into the viability of tourist destinations abroad. Lastly, the heterogeneity of the methodologies used is also a factor that has to be taken into account before comparing the results that have been obtained. In particular, two methodologies are primarily used to survey the tourists. The goal of the first one is to ask tourists what they did in previous seasons when there was little snowfall (Dawson et al., 2013). The second, used more broadly, questions the tourists about what their reactions might be in winters in the future if there is very little snowfall: many winters with little snowfall, three in five winters or five consecutive winters without snowfall (Behringer et al., 2000; Bürki, 2000; Dawson and Scott, 2010; Pickering et al., 2010; Rutty et al., 2015; Unbehaun et al., 2008). Here, the analysis leads tourists to imagine what their behaviour might be in unknown, unfamiliar situations, in which case the frequency with which the situation occurs appears to be a discriminating factor in the different protocols. In fact, because the tourists do not have to answer the same questions, they necessarily give different answers, including any of the substitution strategies.

Despite the reservations laid out above, the results obtained in the different studies provide some important initial insights into this question of how tourists adapt to a lack of snowfall. Firstly, activity substitution, which, as a reminder, is likely to hit the relevant ski resorts the hardest, is planned primarily in territorial contexts where skiing is responsible for only a modest number of national skier-days. By way of illustration, this substitution strategy seems more mobilized in Australia than in Switzerland or in the United States (Behringer et al., 2000; Dawson et al., 2013; Pickering et al., 2010). Similarly, tourists' commitment to a spatial substitution strategy – i.e. continuing to ski but opting for more viable resorts than the ones they usually go to – seems to correlate strongly with the density of nearby ski resorts that have heterogeneous characteristics. The reason is the increase in travel costs, which can limit their participation in this type of strategy. Hopkins et al. (2013) used a quantitative analysis to examine Australian tourists' trips to New Zealand. They were able to show that New Zealand, which has a more positive perception with regard to snowfall than Australia does, would be more likely to attract skiers from its neighbour than the other way around. The interviews also highlight the (prohibitive) cost of a possible stay for Australian skiers in the northern hemisphere – a cost that is much lower in the context of an Australia–New Zealand flight, although it certainly depends on the willingness to pay of the different segments of tourists. Outside of this particular territorial context, spatial substitution appears to have a stronger position as the preferred adaptive solution: 58 % of the excursionist-respondents are prepared to travel three hours to go skiing (Rutty et al., 2015), and 68 % of skiers are in favour of this strategy when faced with a binary choice between spatial substitution and no longer skiing (Unbehaun et al., 2008). The last kind of substitution strategy, namely temporal substitution, is examined less often in the various surveys that have been carried out. These surveys have targeted the possibility of reducing the amount of skiing (skiing less; cf. Rutty et al., 2015, for example) rather than moving/changing the ski period, although this phenomenon certainly seems worth examining: Vivian (2011) showed that, in cases where the season starts later than usual, only 10 % of skiers plan to visit the site more often, while 80 % do not expect to change their plans at all, which yields a lower number of ski days overall.

Are ski resorts at risk?

With tourists' growing awareness of climate change and its effects, as well as these same tourists considering various substitution strategies, ski resorts are seeing their future being called into question. Does this spell the end for these resorts? The different studies that have been carried out confirm that major changes should be expected, even though it seems unlikely that all of the resorts will have to face the same challenges with the same intensity.

Modelling to predict the change in tourists' behaviour

In parallel with the distribution of questionnaires to tourists in order to understand how their behaviour is changing in response to climate change, other studies have focused on identifying correlations between climate conditions (temperatures, meteorology, snow depth) and tourist numbers. To this end, different variables have been used, from snow depth (Hamilton et al., 2007; Shih et al., 2009) to temperatures (Hamilton et al., 2007) for the climate aspect, and the number of ski passes sold (Dawson et al., 2009; Pickering, 2011; Steiger, 2011; Rutty et al., 2017) and tourist nights (Steiger, 2011) for the ski resorts' tourist numbers. The analysis of previous winters has revealed a significant relationship between snowfall and the number of day visitors (Pickering, 2011). The days with the least amount of snowfall are also the days when the ski resorts are visited by the fewest people, and the same is true on the scale of a tourist season: The winters with the least amount of snowfall are also the winters that see the fewest visitors. Dawson et al. (2009) show a 10 % decrease in attendance during the 2001–2002 season, while Steiger (2011) shows a 25 % decrease in visitors in more than 40 % of the smallest Tyrolean resorts included in the study during the 2006–2007 winter. The size of the resort is a differentiating factor (the smallest resorts are the ones most affected by climate change) and so, too, is the altitude at which the winter sports resort is located (Steiger, 2011; Rutty et al., 2017). Pickering (2011) notes that the lowest-lying resorts saw a decrease in visitors of close to 70 % during the 2011–12 winter, while the highest-lying resorts were able to take advantage of the tourists who had decided to go elsewhere and saw their numbers rise by 10 %. However, the resorts' snow conditions alone are not enough to explain the variations in attendance: Hamilton et al. showed that the snowiest days in Boston were also the busiest days at the nearby ski resorts (Hamilton et al., 2007).

In addition to these analyses of periods of time in the past, various research studies have been conducted to identify the effects of climate change on resorts' visitor numbers. Thanks to climate modelling using climate scenarios provided by the IPCC, the winters that produced a lack of snowfall, namely 2001–2002, 2006–2007 and 2011–12, were defined as representative of average observable conditions within the next 50 years (Rutty et al., 2017). As a result, the expected decrease in attendance (around 10 %, although this figure may well be much higher for the smallest and the lowest-lying ski resorts) appears to be lower than the results obtained through the questionnaires distributed to the tourists. However, while this method has the definite advantage of focusing on the real practices of tourists responding to low snowfall (unlike the previously cited questionnaires, whose aim, for the most part, was to collect expected behaviours), it does not allow for an assessment of the possible consequences of a repeated lack of snow (Steiger, 2011). To date, bad winters have been an intermittent phenomenon and have been followed by winters with better snow conditions. For this reason, the resorts' visitors have not had to change their practices on a regular basis. However, if such bad winters become the norm over the next 50 years, what will the tourists do? Isn't there a good chance that a number of them will give up skiing when the snow conditions deteriorate?

(How) will ski resorts still be appealing in the future?

As we have seen, climate change will cause mountain territories to lose their natural volume of snow. Of course, ski resorts will be impacted, but to what extent will tourist numbers be affected? Several studies have looked into the factors that make a ski resort appealing to visitors. Since the mid-1990s, snowfall has been considered the leading factor making winter resorts appealing (Carmichael, 1996; Won et al., 2008). Among a range of factors that have been examined (diversity of ski tracks, snow conditions, value for money, ski lifts, reception and accessibility), snowfall is viewed as the primary factor making a resort appealing to skiers staying for either short or longer periods of time. However, a distinction has to be made according to the technical level of the skiers: While the skiers who have the best level of skiing will also be the most sensitive to the snow quality (and to the diversity of the ski area), the less experienced ones will focus first of all on the costs associated with their time at the resort (Godfrey, 1999; Richards, 1996). Beyond that, the tourist accommodation, the amenities of the territory and the distance between their lodgings and the resort are all important factors affecting the ski resorts' appeal.

If the leading factor of ski resorts' appeal remains snowfall, resorts will want to make their ski areas more reliable by refining their artificial snow networks. Combined with other techniques to make ski areas sustainable (preparing and reshaping the slopes etc.), artificial snow makes it possible to improve the viability of the ski areas (Scott et al., 2006; Steiger and Stötter, 2013). For visitors, opinions are more mixed (Hopkins, 2014; Pütz et al., 2011). The role that artificial snow plays in keeping ski areas open and the quality of the snow cover on the slopes are both favourable arguments for tourists. By contrast, this does not hide the negative perception of artificial snow with regard to the environment (consumption of the resource) and finances. The development of these new technologies leads to an increase in the cost of equipment for resorts, possibly shouldered, at least in part, by the consumer. Moreover, as we have seen, since the financial factors are critical for skiers who are beginners, one challenge for ski areas that depend on the reliability of their ski area is to renew their clienteles. In addition, this perception of artificial snow seems to be linked to the particular type of practice: downhill skiers (who have to ski on this manmade snow) will accept it more readily than those who are mostly off-piste skiers (Hopkins, 2014).

Conclusion

While enthusiasm for skiing tends to be somewhat less than for other activities or other relaxing or nature-oriented experiences (Mauri and Turci, 2018), skiing (in combination with snowboarding) continues to be the main activity (Steiger, 2012) in ski resorts. And yet, given this observation, there are many uncertainties. While the resorts' offering to tourists tends to benefit from increasingly precise knowledge elements, a full understanding of the demand is still on the side-lines of these studies that look into how ski resorts are adapting. However, knowledge of their behaviour and perceptions seems to be crucial to ensuring the viability of ski areas equipped with artificial snow or other tools and techniques with a similar purpose. In this regard, we have started work on the sensitivity of tourists and confirmed/refuted a number of hypotheses on the tourists and their future activities. The diversity of these actors, along with the specific characteristics of each of the territorial contexts, requires territorially limited approaches that are crucial to understanding the soundness of the adaptation strategies that ski resorts have undertaken. This is certainly germane to resorts organised according to the community model (Flagestad and Hope, 2001), especially in France, where the public stakeholder and, therefore, the citizens are called on to participate in funding the activity. We have to be sure that the ski areas should stay viable to ensure, at a larger scale, the ski resort sustainability.

References

Behringer, Jeannette, Buerki, Rolf and Fuhrer, Jürg. 2000. "Participatory integrated assessment of adaptation to climate change in Alpine tourism and mountain agriculture". *Integrated assessment* 1: 331–338.

Bürki, Rolf. 2000. "Klimaänderung und Wintertourismus in Obertoggenburg". *Natur-Forschung in der Region St.Gallen, Berichte der St.Gallischen Naturwissenschaftlichen Gesellschaft Band 89*. https://www.researchgate.net/publication/315756054_Klimaanderung_und_Wintertourismus_im_Obertoggenburg (Last access 14 february 2020).

Carmichael, Barbara A. 1996. "Conjoint analysis of downhill skiers used to improve data collection for market segmentation". *Journal of Travel & Tourism Marketing* 5: 187–206.

Dawson, Jackie and Scott, Daniel. 2007. "Climate change vulnerability of the Vermont Ski Tourism Industry (USA)". *Annals of Leisure Research*, — volume 10: 550–572.

Dawson, Jackie and Scott, Daniel. 2010. "Systems analysis of climate change vulnerability for the US Northeast ski sector". *Tourism and Hospitality Planning & Development*, 7: 219–235.

Dawson, Jackie, Scott, Daniel and Havitz, Mark. 2013. "Skier demand and behavioural adaptation to climate change in the US Northeast". *Leisure/Loisir* 37:127–143.

Dawson, Jackie, Scott, Daniel and McBoyle, Geoff. 2009. "Climate change analogue analysis of ski tourism in the northeastern USA". *Climate Research* 39:1–9.

De Urioste-kStone, Sandra, Le, Lena, Scaccia, Matt D. and Wilkins, Emily. 2016. "Nature-based tourism and climate change risk: Visitors' perceptions in Mount Desert Island, Maine". *Journal of Outdoor Recreation and Tourism* 13: 57–65.

Demiroglu, O. Cenk, Kucerová, Jana and Ozcelebi, Oguzhan. 2015. "Snow reliability and climate elasticity: case of a Slovak ski resort". *Tourism Review* 70: 1–12.

Demiroglu, Osman C., Turp, Mustafa T., Ozturk, Tugba and Kurnaz, Mehmet L. 2016. "Impact of climate change on natural snow reliability, snowmaking capacities, and wind conditions of ski resorts in Northeast Turkey: A dynamical downscaling approach". *Atmosphere* 7(52): 1–12.

Flagestad, Arvid and Hope, Christine A. 2001. "Strategic success in winter sports destinations: a sustainable value creation perspective". *Tourism Management* 22: 445–461.

François, Hugues, Morin, Samuel, Lafaysse, Matthieu and George-Marcelpoil, Emmanuelle. 2014. "Crossing numerical simulations of snow conditions with a spatially-resolved socio-economic database of ski resorts: A proof of concept in the French Alps". *Cold regions science and technology* 108: 98–112.

Fukishima, Takehiko, Kureha, Masaaki, Ozaki, Noriatsu, Fujimori, Y. and Harasawa, Hideo. 2003. "Influences of air temperature change on leisure industries–case study on ski activities". *Mitigation and Adaptation Strategies for Global Change* 7: 173–189.

Godfrey, Kerry B. 1999. "Attributes of destination choice: British skiing in Canada". *Journal of Vacation Marketing* 5: 18–30.

Hamilton, Lawrence C,. Brown, Cliff and Keim, Barry D. 2007. "Ski areas, weather and climate: time series models for New England case studies". *International Journal of Climatology* 27: 2113–2124.

Hendrikx, Jordy, Zammit, Christian, Hreinsson, Einar Ö. and Becken, Susanne. 2013. "A comparative assessment of the potential impact of climate change on the ski industry in New Zealand and Australia". *Climatic Change* 119: 965–978.

Hopkins, Debbie. 2014. "The sustainability of climate change adaptation strategies in New Zealand's ski industry: a range of stakeholder perceptions". *Journal of Sustainable Tourism* 22: 107–126.

Hopkins, Debbie, Higham, James E. and Becken, Susanne. 2013. "Climate change in a regional context: relative vulnerability in the Australasian skier market". *Regional Environmental Change* 13: 449–458.

IPCC 2007. *Climate change 2007. Synthesis report. Contribution of Working Groups I, II and III to the fourth assessment report*, Geneva, Switzerland: Intergovernmental Panel on Climate Change.

Iso-Ahola, Seppo E. 1986. "A theory of substitutability of leisure behavior". *Leisure sciences* 8: 367–389.

König, Urs. 1998. Climate change and the Australian ski industry. In Green Ken (ed.) *Snow: A Natural History; An Uncertain Future*, Australian Alps Liaison Committee, Chipping Norton: 207–223.

König, Urs. 1998. *Tourism in a Warmer World: Implications of Climate Change Due to Enhanced Greenhouse Effect for the Ski Industry in the Australian Alps*. Wirtscaftsgeographie und Raumplanung. Vol. 28. University of Zurich, Zurich.

König, Urs and Abegg, Bruno. 1997. "Impacts of climate change on winter tourism in the Swiss Alps". *Journal of Sustainable Tourism* 5(1): 46–58.

Lasanta, Teodoro. 2010. "El turismo de nieve como estrategia de desarrollo en el Pirineo aragonés". *Cuadernos de investigación geográfica* 36: 145–163.

Lasanta, Teodoro, Laguna, María and Vicente-Serrano, Sergio M. 2007. "Do tourism-based ski resorts contribute to the homogeneous development of the Mediterranean mountains? A case study in the Central Spanish Pyrenees". *Tourism Management* 28: 1326–1339.

Lorit, Jean-François. 1991. *Les difficultés financières des stations de sports d'hiver.* Rapport pour le Ministère de l'Intérieur et l'Inspection Générale de l'Administration.

Luthe, Tobias. 2009. *Vulnerability to Global Change and Sustainable Adaptation of Ski Tourism: SkiSustain*, Series "Outdoor Sports and Environmental Science". Institute of Outdoor Sports and Environmental Science. German Sports University Cologne, Cologne.

Mauri, Chiara and Turci, Lorenzo. 2018. "From ski to snow: rethinking package holidays in a winter mountain destination". *Worldwide Hospitality and Tourism Themes* 10: 201–210.

Moen, Jon and Fredman, Peter. 2007. "Effects of climate change on Alpine skiing in Sweden". *Journal of Sustainable Tourism* 15: 418–437.

Ogrin, Matej, Ogrin, Darko, Rodman, Nastja, Mocnik, Matic, Vengar, Rok, Smolej, Andraž and Buncic, Gregor. 2011. "Climate change and the future of winter tourism in Slovenia". *Croatian Geographical Bulletin* 73(1): 215-228.

Pearce, Douglas G. 2002. *Tourist development*, Harlow, Longman Scientific & Technical.

Peretti-Watel, Patrick. 2000. *Sociologie du risque*, Armand Colin, Paris.

Pickering, Catherine. 2011. "Changes in demand for tourism with climate change: a case study of visitation patterns to six ski resorts in Australia". *Journal of Sustainable Tourism* 19: 767–781.

Pickering, Catherine, Castley, Guy and Burtt, Michelle. 2010. "Skiing less often in a warmer world: Attitudes of tourists to climate change in an Australian ski resort". *Geographical Research* 48: 137–147.

Pons-Pons, Marc, Johnson, Peter A, Rosas-Casals, Martí, Sureda, Bàrbara and Jover, Èric. 2012. "Modeling climate change effects on winter ski tourism in Andorra". *Climate research* 54: 197–207.

Pütz, Marco, Gallati, David, Kytzia, Susanne, Elsasser, Hans, Lardelli, Corina, Teich, Michaela, Waltert, F and Rixen, Christian. 2011. "Winter tourism, climate change, and snowmaking in the Swiss Alps: tourists' attitudes and regional economic impacts". *Mountain Research and Development* 31: 357–362.

Richards, Greg. 1996. "Skilled consumption and UK ski holidays". *Tourism management* 17: 25–34.

Rutty, Michelle, Scott, Daniel, Johnson, Peter A., Jover, Eric, Pons, Marc, Steiger, Robert. 2015. "The geography of skier adaptation to adverse conditions in the Ontario ski market". *The Canadian Geographer/Le Géographe canadien* 59: 1–403.

Rutty, Michelle, Scott, Daniel, Johnson, Peter A., Pons, Marc, Steiger, Robert and Vilella, Marc. 2017. "Using ski industry response to climatic variability to assess climate change risk: An analogue study in Eastern Canada". *Tourism Management* 58: 196–204.

Saarinen, Jarkko and Tervo-Kankare, Kaarina. 2006. "Perceptions and adaptation strategies of the tourism industry to climate change: The case of Finnish nature-based tourism entrepreneurs". *International Journal of Innovation and Sustainable Development* 1: 214–228.

Saarinen, Jarkko, & Tervo, Kaarina. 2010. "Sustainability and Emerging Awareness of a Changing Climate: The Tourism Industry's Knowledge and Perceptions of the Future of Nature-based Winter Tourism in Finland". In *Tourism and Change in Polar Region. Routledge, Abingdon-on-Thames*, 165–182.

Scott, Daniel, McBoyle, Geoff and Minogue, Alanna. 2007. "Climate change and Quebec's ski industry". *Global Environmental Change* 17: 181–190.

Scott, Daniel, McBoyle, Geoff, Minogue, Alanna and Mills, Brian. 2006. "Climate change and the sustainability of ski-based tourism in eastern North America: A reassessment". *Journal of sustainable tourism* 14: 376–398.

Shih, Charles, Nicholls, Sarah and Holecek, Donald F. 2009. "Impact of weather on downhill ski lift ticket sales". *Journal of Travel Research* 47: 359–372.

Steiger, Robert. 2011. "The impact of snow scarcity on ski tourism: an analysis of the record warm season 2006/2007 in Tyrol (Austria)". *Tourism Review* 66: 4–13.

Steiger, Robert. 2012. "Scenarios for skiing tourism in Austria: integrating demographics with an analysis of climate change". *Journal of Sustainable Tourism* 20: 867–882.

Steiger, Robert, Scott, Daniel, Abegg, Bruno, Pons, Marc and Aall, Carlo. 2017. "A critical review of climate change risk for ski tourism". *Current Issues in Tourism*: 1–37.

Steiger, Robert and Stötter, Johann. 2013. "Climate change impact assessment of ski tourism in Tyrol". *Tourism Geographies* 15: 577–600.

Steiger, Robert, Scott, Daniel, Abegg, Bruno, Pons, Marc and Aall, Carlo. 2017. "A critical review of climate change risk for ski tourism". *Current Issues in Tourism* 22(11): 1343–1379.

Tervo, Kaarina and Saarinen, Jarkko. 2010. "Sustainability and emerging awareness of a changing climate: the tourism industry's knowledge and perceptions of the future of nature-based winter tourism in Finland". *Tourism and Change in Polar Regions*. Routledge.

Trawöger, Lisa. 2014. "Convinced, ambivalent or annoyed: Tyrolean ski tourism stakeholders and their perceptions of climate change". *Tourism Management* 40: 338–351.

Unbehaun, Wiebke, Pröbstl, Ulrike and Haider, Wolfgang. 2008. "Trends in winter sport tourism: challenges for the future". *Tourism Review* 63: 36–47.

Vanat, Laurent. 2017. *International report on snow & mountain tourism – Overview of the key industry figures for ski resorts.*

Veyret-Verner, Germaine. 1959. "La deuxième révolution économique et démographique des Alpes du Nord: Les "sports d'hiver. Réflexions et suggestions"". *Revue de géographie alpine* 47: 293–305.

Vivian, K.-A. 2011. *Behavioural Adaptation of Skiers and Snowboarders in the US Northeast to Climate Variability and Change.* (Unpublished Master's thesis). University of Waterloo, Waterloo, ON.

Wolfsegger, Christophe, Gössling, Stefan and Scott, Daniel. 2008. "Climate change risk appraisal in the Austrian ski industry". *Tourism Review International* 12: 13–23.

Won, Doyeon, Bang, Hyejin and Shonk, David J. 2008. "Relative importance of factors involved in choosing a regional ski destination: Influence of consumption situation and recreation specialization". *Journal of Sport & Tourism* 13: 249–271.

3.4

Environmental Management in Ski Resorts

Ulrike Pröbstl

Introduction

Slogans such as "skiing kills the European alps" have had a substantial impact on the perception of winter tourism in alpine areas. The "Alpine Report" published in 1998 presented research findings describing a significant and increasing feeling of guilt among the group of active skiers (CIPRA, 1998). The promotion and marketing of environmental management helps decrease this gap between the perceived experience of fun and the feeling of contributing to something that harms the alpine environment. In this chapter, we will focus on the situation in the European Alps, mainly the situation in Germany, Austria, Switzerland and South Tyrol (fig. 47).

Especially in countries in which winter tourism significantly contributes to economic wealth, political parties and NGOs also promote environmental management (Umweltbundesamt, 2016; WWF Austria, 2018).

Figure 48 illustrates the interactions between various players and shows how political representatives, alongside demand and supply, influence environmental management in ski resorts. In the last decade, the influence exerted by political recommendations as well as by respective guidelines has markedly increased. The discussion on possible effects of

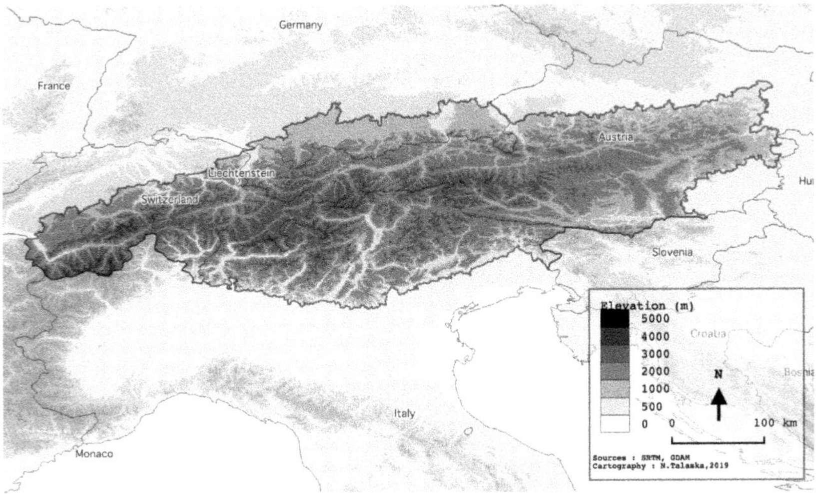

Fig. 47: Studied area in European Alps (Germany, Austria, Switerland, South Tyrol)

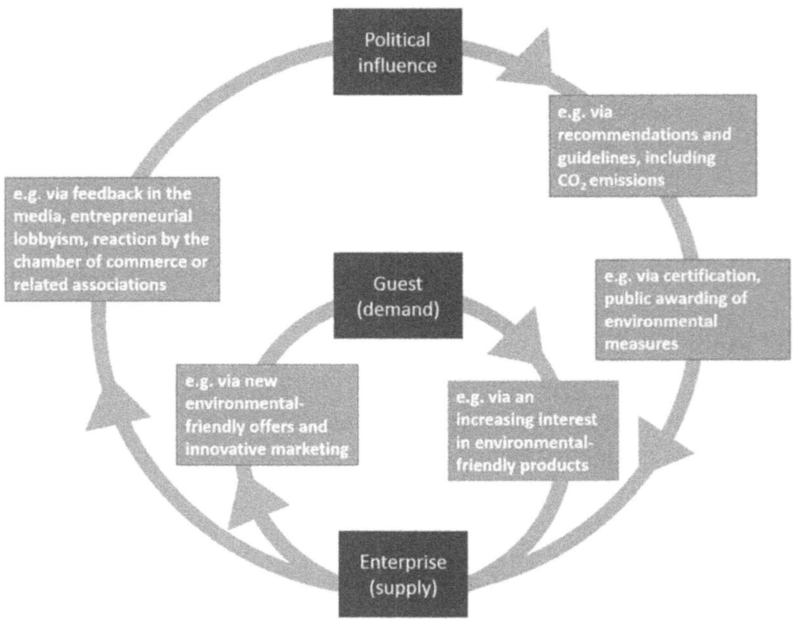

Fig. 48: Factors influencing certified environmental management

climate change, the increasing amount of electricity required for snow making and the related CO_2 emissions have also been critically discussed in the media, especially in alpine countries. Since climate change and climate change adaptation are fairly recent additions to various guidelines such as the Environmental Impact Assessment (EIA) or the Strategic Environmental Assessment (SEA), the consideration of environmental issues has also been supported by political and administrative bodies. Meanwhile, entrepreneurs who have implemented remarkable adaptation strategies and environmental management are continuously awarded and promoted in public (see fig. 49, Tourismuspresse 2017).

Environmental engagement is also influenced by the demand side. Since the 1990s, many environmental organisations highlighted the relevance of environmental management in alpine ski resorts. In this context, different organisations, such as STOKE in Partnership with Mountain Rider's Alliance (STOKE Certified, 2008), the foundation "pro natura pro ski" (Pröbstl and al., 2003), and also in the context of

Fig. 49: In Schönbrunn palace, Vienna, the minister of environment, agriculture, forestry and water management honoured the ski resort "Schmittenhöhebahn" and its representatives for their eco-certified environmental management (Mein Bezirk 2015)

the alpine convention (Pröbstl and al. 2010), highlighted and/or awarded environmental certification and related management.

However, one should also mention the critical feedback loop by the enterprises and the chamber of commerce stating that certified environmental engagement should not be mandatory but still part of a free decision by the respective ski resort.

Since many ski resorts today advertise with seals of approval, titles and awards we want to explain in this chapter the difference between these various awards and an eco-certified ski resort. Looking at general awards it is often unclear which institutions and criteria stand behind these appraisals. Both skiers and ski areas criticise this practise of "green-washing", since a cautious use of ecologically and scenically sensitive mountain areas is more important now than ever. An environmental management system that builds on international or European standards ensures "real" improvements and external evaluation. It relies on the personal responsibility of the providers rather than on regulations imposed by authorities. In the following we will explain eco-certification with respect to ski resorts and what it implies.

Environmental certification: main components and processes

When speaking of environmental management and its certification, we refer to measures prescribed by two sets of rules and standards, which have the same goals but differ in their historical development, their areas of application and (to a slight degree) their requirements:

- The international standard for environmental management, ISO 14001, lays down the criteria for an environmental management system and is part of a wider network of standards. It was first developed in 1996 as 14001:1996. The current version of ISO 14001 came into effect in September 2015 as ISO 14001:2015.
- EMAS stands for "Eco-Management and Audit Scheme". It is a voluntary system in which businesses, as well as other organisations and institutions within the EU member states can choose to participate. The regulation has been in effect since April 1995 and has already been revised twice. EMAS III has been operative since 11 January 2010 (Regulation (EC) No 1221/2009). With this revision, it became possible for organisations outside of the

EU to participate. The general aim of this regulation (just as ISO 14001) is to promote continuous improvement of organisations' environmental performance. EMAS places greater emphasis on informing the public.

In the following the article focusses on EMAS only. However both regulations do not dictate concrete details as to the form environmental policies, programmes or management should take, nor does it define any material parameters such as thresholds or environmental standards, to which businesses would be bound beyond the legal requirements. Such stipulations are hardly feasible, especially in light of the diversity of businesses participating in the EMAS system. Thus, it is up to the individual business to analyse its environmental impact responsibly and to take steps and measures in accordance with its self-set goals. This is significant, because it allows businesses to set goals for themselves that are timed to match economic requirements and aligned with their existing environmental protection measures. At the same time, the business can satisfy the regulation's call for continued improvements by steadily increasing its own requirements for optimisation of environmental performance with each cycle (cf. fig. 50)

The main stages of this cycle are explained in Pröbstl-Haider and al. (2018): Environmental policy, the roof that encompasses all environmental activities, usually marks the first step of the procedure. Basic environmental guidelines are hereby formulated and integrated into corporate policy. This ensures that environmental and economic corporate goals are aligned and pursued in equal measure.

The centrepiece of the environmental management system is the environmental review. It constitutes a first appraisal of an organisation's environmental performance and impact. Aside from compiling all of the organisation's environmentally relevant data, the source data is also analysed and evaluated to determine weak points. The environmental review further includes an assessment of the organisations' compliance with environmental regulatory requirements. The environmental programme is based on the results of this (first) environmental review. The environmental programme contains all targets and measures aimed at overcoming the identified weaknesses, as well as specific deadlines. It also specifies the individuals responsible for the implementation of these measures and the available resources. In the next step, measures listed in the environmental programme begin to be implemented.

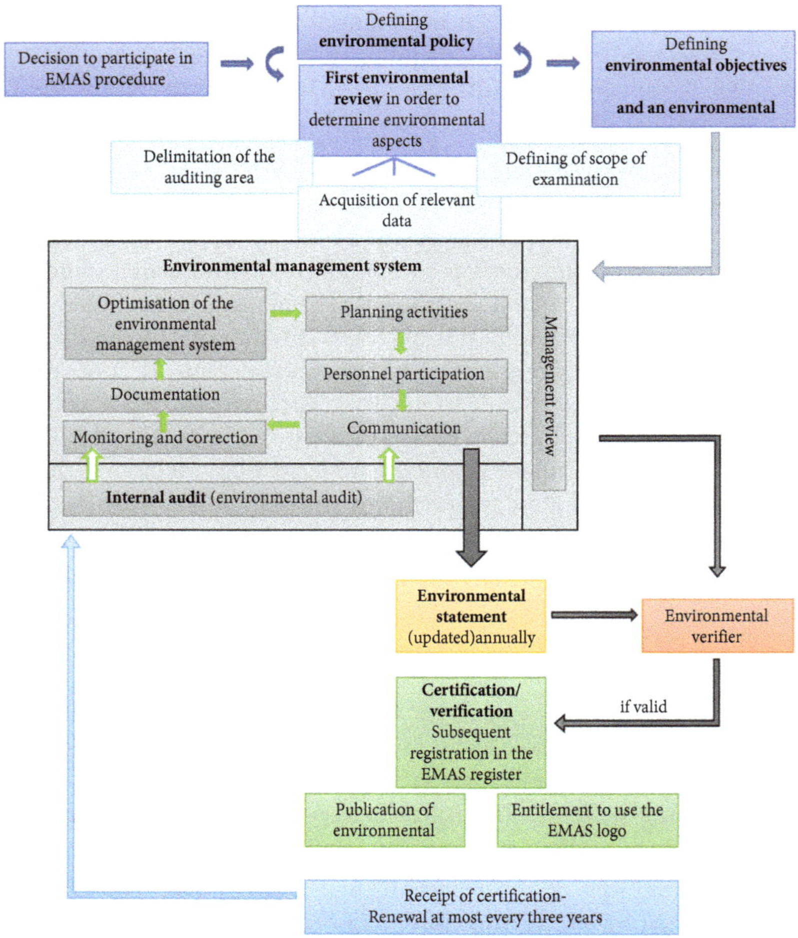

Fig. 50: Procedure according to the EMAS regulation (Pröbstl-Haider et al., 2018)

All structures for the implementation of the organisation's internal environmental protection policy are defined in the environmental management system as part of the overall management system. Thus, the environmental management system contains personnel-related organisation structures and responsibilities, as well as process control. It also ensures documentation of all environmental activities.

The organisation's entire environmental activities are to be presented to the public in regular environmental statements. The environmental statement, thereby, functions as a gateway to the extra-organisational level and ensures effective communication with different stakeholder groups. It should therefore be kept simple and easy to understand and should summarise the environmental performance in brief.

Verification of the organisation is carried out based on the environmental statement and the environmental management system. The process involves an independent, external environmental verifier who assesses whether an organisation's environmental policy, (first) environmental review, environmental programme, environmental management and environmental statement comply with the requirements of the EMAS regulation. If all requirements are met, the organisation can take part in the EMAS system and be added to the EMAS register. Following its registration through the national competent body, the organisation is entitled to use the EMAS logo.

In order to maintain its eligibility in the long term, the organisation is required to repeat the procedure in regular intervals, at least every three years. SME with low environmental impact are allowed to extend the audit cycle to four years.

Expected advantages for the enterprise

In order to establish a suitable environmental management system in ski areas, a distinct set of methods and a specially designed review framework need to be applied. EMAS and ISO 14001, with their wide range of possible applications, allow considerable freedom to develop such individual approaches. Yet, this flexibility also means that each business (or each business type) has to find its own specific way to integrate operative aspects with ecological and landscape-related aspects in a responsible manner. This requires an independent methodological design.

The methodological elements described in the following can also be used for ecological improvements irrespective of environmental management procedures (ISO 14001 or EMAS) (Pröbstl-Haider and al., 2018). Specific challenges that suggest the need for a separate approach for ski areas can be summarised as follows:

- a combination of high standards regarding passenger transportation as well as system safety;
- the influence of weather conditions on operation and usage of installations
- the need to take safety precautions in the mountains and during activities
- the high resource input required for operations
- the increased sensitivity of the affected natural habitats due to the high altitude (e.g. vegetation, water, soil, wild animals)

In addition, the issue of waste generation and disposal, as well as the supply of drinking water need to be considered, both of which are particularly complicated in mountainous areas.

From an organisation's point of view, there are a number of arguments in favour of introducing an environmental management system (cf. Tab. 14):

A significant aspect in favour of introducing an environmental management system is an increase of business competitiveness, largely thanks to the image boost triggered by an environment-oriented

Tab. 14: Potential benefits of participating in the EMAS system for ski enterprises (Pröbstl-Haider et al., 2018)

Potential benefits for the ski enterprise			
Competitiveness	**Cost reduction**	**Risk reduction**	**Improvement of organisation**
Appeal to an environmentally sensitive circle of customers Environmental commitment as a key qualification Improved mutual trust with conservation authorities Easier candidacy for major winter sports events	Reduced costs for preparing approval documents Lower insurance rates Cheaper loans from banks Prevention of cost-intensive restoration in case of large-scale damages Application of cost-effective innovation measures	Knowledge about the potential for large-scale damages Reduced risk of damages due to erosion Clarity regarding liability in case of damages	Improved knowledge of personnel Detailed knowledge about impact on nature and the landscape Reduced bureaucratic effort Knowledge about the relative contributions of winter and summer tourism

corporate policy. Resorts which regularly apply to host major international events use their environmental certification and related reports to ensure that the event will be environment-friendly and to prove their experience in eco-friendly management. After intense criticism of events in Albertville (France), the Trentino Region (Italy), Bansko (Bulgaria) and several other destinations, the International Ski Federation (FIS) has become more cautious concerning potential environmental threats and negative media coverage (FIS 2009, 2016). Furthermore, it has been observed that the resorts have also started to invest in "green promotion" of their ski destination and careful, cooperative planning processes (for an example see Ignatzi, 2017; Åre 2019 offices, 2017).

Even if EMAS involves an initial financial burden for the business, the procedure often reveals savings potential regarding energy, water or basic materials which can lead to future cost reduction.

The (first) environmental review uncovers weak points that are then addressed by the business's environmental policy and environmental programme. The compulsory provision of incident – and emergency preparedness also contributes to risk reduction. The establishment of an information system containing all environmentally relevant corporate activities assists efficient internal controlling. This involves breaking down internal processes into manageable components, namely planning, implementation, monitoring and correction. This information system and the aforementioned risk management have a positive effect on legal security and insurance obligations.

Figure 51 depicts an overall evaluation of all major management issues and resources used. Based on this evaluation, the enterprise can identify areas which are in need of significant intervention. In addition, documentation of environmental issues and analyses in the form of GIS mapping allow businesses to respond more effectively in case of conflicts, emergencies and allegations lodged by third parties.

The GIS mapping in Fig. 52 visualises the proximity of vegetation units on ski slopes to their natural state. Red colours denote degradation and a need for restauration. Green vegetation units are characterised by typical alpine vegetation. Areas very close to nature are highlighted in dark green colours. Such databases are crucial tools for the documentation of various environmental issues. They are also deployed to fulfil legal requirements and obligations such as monitoring or compensation duties.

Environmental impact / Aspect	Soil		Water		Flora		Fauna		Air		Noise		Waste		Use of resources: Packaging		Water		Energy		Operating materials		Need for action
	1	2	1	2	1	2	1	2	1	2	1	2	1	2	1	2	1	2	1	2	1	2	
Snowmaking	C	B	B	A	B	B	B	B	C	C	A	A	C	C	C	C	A	A	A	A	C	B	III
Winter piste maintenance	B	B	B	B	B	A	C	B	C	C	A	B	C	C	C	C	C	C	B	B	B	A	III
Skiing off-piste	C	C	C	C	B	A	C	B	C	C	C	B	B	B	C	C	C	C	C	C	C	C	III
Summer piste maintenance	B	A	A	A	A	A	B	B	C	C	C	C	C	B	C	C	C	C	C	C	C	C	I
Cableway operation	C	C	C	C	C	C	C	C	C	C	C	B	C	C	C	C	C	C	A	A	B	B	III
Gastronomy	C	C	B	B	C	C	C	C	C	B	C	C	B	B	B	B	B	B	C	C	C	B	III
Shop/sales	C	C	C	C	C	C	C	C	C	C	C	C	C	B	B	B	C	C	C	C	C	C	IV
Skirental	C	C	C	C	C	C	C	C	C	C	C	C	C	C	C	C	C	C	C	C	C	C	IV
Adminstration	C	C	C	C	C	C	C	C	C	C	C	C	C	C	C	C	C	C	C	C	C	C	IV
Waste water treatment	C	B	A	A	B	A	C	B	C	B	C	C	C	C	C	C	C	B	C	C	C	C	III
Drinking water supply	C	B	A	A	C	C	C	C	C	B	C	C	C	C	C	C	C	B	C	C	C	C	III
Organisation of tournaments	C	C	B	B	C	C	C	C	C	C	C	C	B	B	B	B	B	B	B	B	B	B	III

Environmental (risk) for normal operative situation (1) and abnormal operative situation or emergency (2)
A= high relevance
B= medium relevance
C= low relevance

Need for action: I= short-term, II= mid-term, III= long-term, IV= none

Fig. 51: Evaluation of all management issues and resources needed
(1: under normal management condition; 2: In case of an accident or emergency; A: high relevance, B: somewhat relevant, C: low relevance; I = requires immediate action, II = requires action soon, III = requires action in the long run, IV = action currently not required) (Pröbstl-Haider et al., 2018)

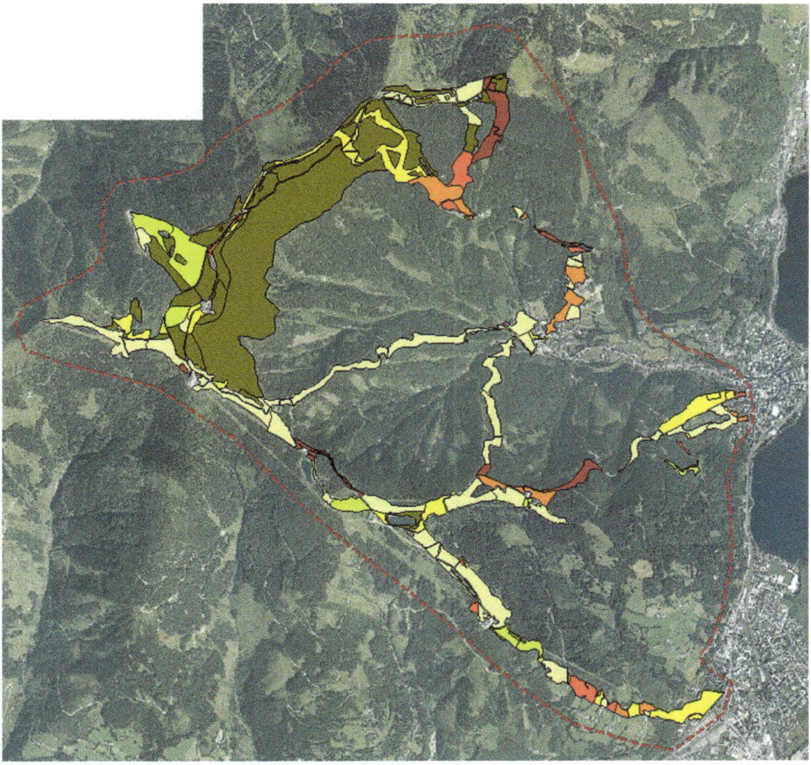

Fig. 52: Vegetation on ski slopes, inventory in GIS with evaluation of its naturalness
(red: low quality, not natural, dark green, high value, very close to nature) (Pröbstl-Haider et al., 2018)

Establishing different process components at all levels of the organisation improves communication with and among personnel. Integration of environmentally oriented principles and their practical implementation boost employees' motivation and significantly influence their identification with the business.

The assessment of a business's entire environmental impacts in the first environmental review, as well as the way these impacts are dealt with in its environmental policy and environmental programme, create the basis for sustainable economic activity. Furthermore, increased transparency regarding environmental impacts and their publication in the environmental statement can help improve a business's public image.

The environmental statement represents a crucial document providing an overview of external and internal communication. It summarises the entire process and all of its fundamental elements. The following Tab. 15 illustrates the typical structure of environmental statements released by ski resorts and explains their basic elements in a few key words.

A properly structured environmental statement goes a long way towards fulfilling requirements regarding appropriate documentation. A further component of appropriate documentation consists of the preparation of documents, charts, tables and data compilations for the first environmental statement in order to be able to continuously update them in digital form. The ski area information system (in GIS) and the documents mentioned above are required for further planning, as well as the execution and monitoring of the processes that comprise environmental management.

Furthermore, suitable media need to be selected for the dissemination of internal and external documentation. Usually, the environmental statement (see Fig. 53) is publicly available in printed form and as a download.

The publication of an environmental statement contributes to an open dialogue with the public, specifically interested parties such as local communities and customers. An open dialogue regarding possible environmental impacts, products and services helps identify the public's and other interested parties' concerns. The environmental report therefore contributes to openness and transparency in the sector and reaffirms the commitment to regularly provide information on environmental impacts. These factors also play an important role in the way organisations build confidence and trust among interested parties.

Potential catalysts to increase the number of certified ski enterprises in the future

Increasing the public's interest

The choice of ski areas and winter holiday destinations has increased steadily over the past years, especially in Eastern Europe and in Turkey. North American ski areas are also increasingly courting European skiers, enticing them with powder skiing, "heli-skiing" and a "pristine back country".

Tab. 15: Structure of an environmental statement (example after Pröbstl-Haider et al., 2018)

Structure of the environmental statement	Content in key words
Foreword	- Objectives - Naming of the type of certification - Relevance for internal and external communication
Characterisation of the business in figures and facts	- Characterisation of the business or organisation - Portrayal of the special characteristics of the natural environment, the pistes, recreation facilities, etc. - Size and structure of the space used for skiing, and demarcation of the area affected by the environmental management using map details
Systematic environmental management	- Responsibilities for environmental management - Involvement of external contributors and committees - Organisational diagram
Description of the environmental management system	- Indication of the underlying basis (e.g. EMAS regulation or ISO 14001:2015) - Tasks and composition of the prepared documents - Reference to environmental policy, programme, objectives and measures as basic structure - Reference to trainings, awareness raising and financial resources - Explanation of the environmental database and described risks - Environmental database for potential direct impacts - Reference to potential indirect environmental impacts
Environmental policy	- Basic principle - Scope of efforts - Delimitation - Environmental management system - Monitoring - Information
Direct environmental effects	- Snowmaking - Piste maintenance in winter - Safety - Piste maintenance in summer - Lift and cableway operation - Drinking water supply and waste water treatment - Waste management - Use of consumables and natural resources (electricity, water, petroleum products, emissions)

Continued on next page

Tab. 15: Continued

Structure of the environmental statement	Content in key words
Indirect environmental effects	- Customer traffic on access roads and parking areas - Supply chain and partner businesses
Environmental programme	- Illustration in table form of targets/objectives, measures, responsibilities and realisation timeframes for relevant environmental issues, e.g. piste maintenance and environmental protection, safety, personnel training
Summary of available data on environmental performance	- Input-output analysis - Core indicators
Reference to compliance with legal requirements, reference to environmental regulations	- Information on measurements and thresholds - Reference to significant legal regulations - Description of the legal database
Declaration of validity	- Name and licence number of the environmental verifier and date of validation
Information and feedback	- Contact persons within the business, additional information

It will become more important in future, therefore, for ski areas to distinguish themselves from other regions and to develop a unique profile. Since the sports possibilities and snow reliability are comparable in many destinations, other aspects need to be included in destinations' outward presentation.

An increasingly important argument in this respect is the environmental compatibility of the business and its so-called green profile. Therefore, the issue of environmental management in ski areas is also gaining increased attention in the media.

Of course, the question remains how many customers can be attracted by a green profile and environmental involvement. According to our results, such attributes appeal foremost to the quality-oriented tourist, who wishes to combine sports, recreation and an experience of a natural winter landscape. These customers are particularly important with regards to regional value added. Roth and al. (2018) showed that the proportion of environmentally sensitive customers is increasing, particularly in the Alpine region. But a change in thinking is not just taking place in Europe. In North America, as well, large ski areas

Fig. 53: Example of an environmental statement addressed to the public (Schmittenhöhebahn AG, 2018)

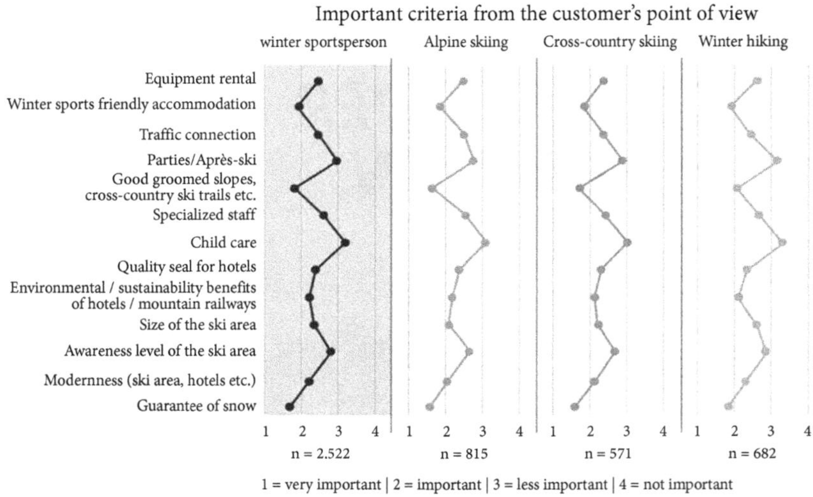

Fig. 54: Importance of environmental management relating to various achieved goals, according to a representative national sample of German winter tourists (after Roth et al., 2018)

like Whistler in Canada are placing more and more importance on environmental aspects[1].

Figure 54 illustrates that, no matter which activity is preferred, environmental and sustainability issues are ranked among the five most important issues (after Roth and al., 2018). Sustainable management in a winter sports business must be a matter of course. The way a forward-thinking management conveys its achievements in the protection of nature and environment needs to become more proactive, provocative and exciting.

A convincing aspect from a business's point of view continues to be that user groups are very different from provisions imposed by an authority, as they suggest voluntary, market economy-oriented measures focused on environmental precaution. The extent of chosen measures, as well as their focus and orientation are largely within the responsibility

---

[1] Further details can be found under the header "environment" on the homepage of Whistler Blackcomb: https://www.whistlerblackcomb.com/explore-the-resort/about-the-resort/environment.aspx (accessed: 06.06.2018).

of the business. The extensive range of possible applications creates an incentive for both severely and less severely impacted areas.

The external certification and evaluation provide a level of credibility that cannot be reached by the many self-proclaimed awards and seals of approval. The EMAS evaluation or the ISO-certificate are therefore easy to distinguish from "green-washing" via seals of approval.

Increasing relevance of digitalisation and cost saving management

In Europe, environmental management systems (EMAS and ISO 14001) have established themselves as important instruments for sustainable development. Decisive arguments for participating in these systems are resource savings, improvement of legal security including environmental liability, and motivation of personnel.

After their success and wide acceptance in production, these instruments are now also gradually becoming more widespread in tourism.

Environmental management is also part of the digital management strategy of ski resorts. Summer slope management, for example, may consider a readjustment of mowing frequencies, a reduction of fertilisers and the use of biomass for energy production.

The use of layered digital data from various sources such as vegetation mapping, intensity of use based on skier-data, snow depth and artificial snowmaking offers new opportunities to enhance effective and efficient management and to save money.

The experiences of several Austrian ski resorts show that GPS based snow measuring saves about 10 % of the average production of artificial snow. A resort of about 70–100 km of slopes can, therefore, recoup its initial investment in digital data in less than a year.

This example shows that digitalisation may be a key driver of the increase in certified ski resorts and the development of a more sophisticated management.

Increasing the legal relevance of environmental concerns

A wide range of European environmental directives are relevant and applicable in European ski resorts, such as the Water Directive, the

Habitats Directive and the Birds Directive. These directives and their application need to be respected. In addition, the term "environmental damage" was coined due to the requirement under Community law to implement the Environmental Liability Directive, which is valid in all European Community member states. The term pertains to potential impacts on protected natural assets such as water, soil, animal and plant species, and natural habitats (Voets, 2009; European Union, 2013; Pröbstl-Haider and al., 2018).

As such, environmental damage, e.g. a damage to protected species and natural habitats under European law, may also be a concern in ski resorts. Here, it describes any damage that has significant adverse effects on reaching or maintaining the desired conservation status of habitats or species. The significance of such effects is to be assessed with reference to the baseline condition, taking into account the criteria set out in Annex I of the Environmental Liability Directive. In the case of an animal or plant species, these criteria include their number or density and the area covered, the species' capacity for propagation, the habitat's capacity for natural regeneration, and the significance of the damage (at a local, regional and higher level, etc.).

Furthermore, the term "environmental damage" also pertains to water damages, i.e. any damage that significantly adversely affects the ecological, chemical or quantitative status or the ecological potential of the waters concerned. Operators are liable for the reparation of damages if these are found to be significant and if a causal connection between the damages and the activities of the business or operator can be determined (fig. 55).

Regarding environmental liability, another new aspect is the so-called presumption of causation. Operations that intentionally or negligently cause biodiversity damages must remedy those damages in addition to any other environmental liability. If biodiversity damage cannot be fully remedied, those liable must take compensatory measures. If a facility is deemed inherently suited to cause a given damage, then the facility is presumed to have actually caused the damage in question. The burden is on the facility operator to disprove damage causation. Strict standards apply for this procedure. The significance of environmental management systems has increased because they facilitate such disproval procedures.

Thanks to the Environmental Liability Directive, non-governmental organisations (NGOs) and citizens are now also allowed to inform relevant

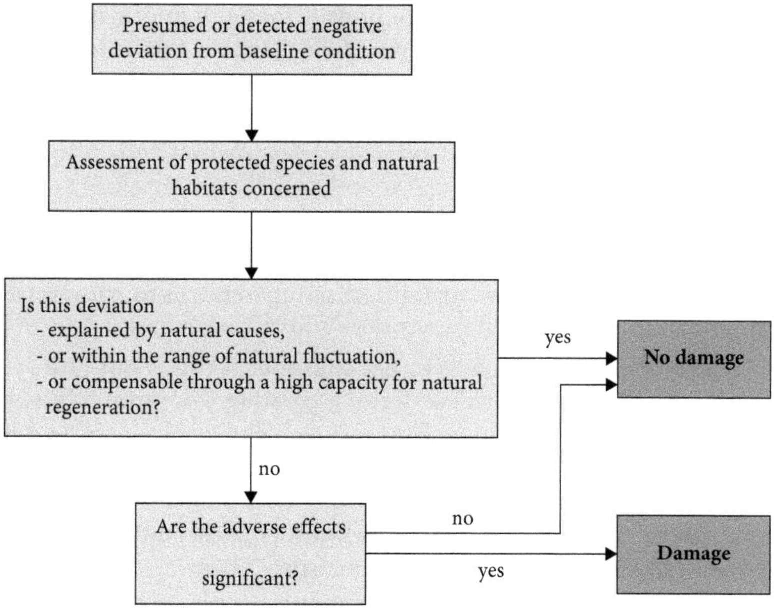

Fig. 55: Steps for assessment of environmental damages (following Voets, 2009)

authorities of any environmental damages (or the imminent danger of environmental damage). They can further take legal action against measures or the omission of measures by such authorities, to ensure in the public interest that those authorities fulfil their responsibilities to prevent environmental damages with all suitable methods and take all necessary safety measures (European Union, 2013:6).

This example shows that environmental management and certification might be necessary to avoid unjustified accusations or to help explain possible impact.

Discussion

An analysis of various international websites shows that, even though environmental issues are of increasing relevance, only few enterprises invest in comprehensive, publicly available information campaigns (Busch, 2014).

Among the few resorts which do provide in-depth information about environmental issues, several illustrate their measures and action plans – mainly based on certification programmes – in detail, while others primarily present first achievements and future visions. For this second group of resorts Whistler, Canada, and Laax, Switzerland, offer good examples.

Many ski resorts such as Whistler promote environmental visions and highflying goals. However, their environmental "top 10 list" is not as transparent as a certification, mainly indicating trends of recent years and new development goals. On its website, Whistler promotes:

"Zero waste, zero carbon, zero net emissions – this is our goal – for Whistler Blackcomb to achieve Zero Operating Footprint. This will only be possible with the support of our employees, our guests and our community. For all of us, Every Step Counts. This collective mission inspires guests to visit us who are factoring environmental impacts into their travel choices" (https://www.whistlerblackcomb.com/explore-the-resort/about-the-resort/environment.aspx). Laax, "Weisse Arena Gruppe", also promotes itself with impressive goals of becoming "the world's first self-sufficient ski resort". However, according to recent press releases (Laax 23.3.2018) there is still a long way to go: "Two years ago, LAAX revealed its vision to become the world's first self-sufficient ski resort. An ambitious feat that is already bearing fruit today – over the last seven years, the winter destination in Graubünden has reduced its energy consumption by 15 per cent". Despite its relatively small environmental successes, the enterprise has invested a lot in their visualisation (see Fig. 56 and https://www.laax.com/infos/greenstyle).

The emphasis these leading resorts place on the presentation of their efforts to pursue environment-friendly management demonstrates that environmental issues will play a key role in the way resorts will market and position themselves in the future. The initiative of Laax to promote its – still fairly limited – achievements in the form of appealing videos may offer an example of how environmental issues should be promoted. However, this promotion should supplement a sound environmental certification, rather than replace it.

Finally, a number of studies in ski areas (Pröbstl, 2001; Dietmann and Kohler, 2005) have shown that the project area must be adequately defined. Exclusion of alternative off-piste or lesser frequented routes, for example, is not justifiable from a technical point of view, because

Fig. 56: Sequence from a promotion video for environment-friendly resort management in Laax. Below other videos e.g. on food and waste (see https://www.laax.com/infos/greenstyle)

valuable habitat for wild animals could be disturbed as a consequence of the operation. Broad delimitation is also important when it comes to management, especially where neighbouring forms of use (hunting, forestry, nature conservation) could be impacted.

Such broad delimitation is also strongly recommended for the establishment of an environmental management system. Many ecological questions, for instance regarding water balance and fauna, but also recreation, can only be properly answered if considered on a sufficiently large scale (Pröbstl-Haider and al., 2018).

Conclusion

The paper introduces the certification of environmental management in ski resorts as a crucial step towards enhancing their competitiveness, cost reduction, risk management and organisational benefits. The increasing relevance and availability of digital data offer new opportunities for cost-efficient management and provide new tools. Eco-friendly environmental management seems to be of increasing relevance to skiers and should, therefore, be addressed by customised measures and supported by engaging marketing such as the promotional films released by Laax. Marketing films alone, however, will not satisfy the interest of the increasingly well-travelled and sophisticated skiers of the future.

References

Åre 2019. 2017. "How sustainable can FIS Alpine World Ski Championships really be?" <http://are2019.com/sustainable-can-fis-alpine-world-ski-championships-really/> accessed: 06 June 2018.

Busch, Mathias. 2014. *Umweltkommunikation in Skigebieten*. Master thesis, University of Natural Resources and Life Sciences, Vienna.

CIPRA Internationale Alpenschutzkommission. 1998. *1. Alpenreport*. Berne, Stuttgart, Vienna: Paul Haupt.

Dietmann, Thomas, Kohler, Ulrich and Lutz, Gernot. 2005. "Die Skigebiete in den bayerischen Alpen. Ökologischer Zustand, Konfliktbereiche, Lösungsmöglichkeiten – eine Schlussauswertung der Skipistenuntersuchung Bayern". *Jahrbuch des Vereins zum Schutz der Bergwelt* 70: 45–60.

European Union. 2013. *Umwelthaftungsrichtlinie – Schutz der natürlichen Ressourcen Europas*. Luxemburg: Amt für Veröffentlichungen der Europäischen Union.

Federation International de Ski (FIS). 2009. *Environmental Guidelines for Sustainable FIS World Championships*. <http://www.fis-ski.com/mm/Document/documentlibrary/MajorEvents/03/66/27/wsc-candidates-2019_environmental-guidelines_Neutral.pdf> accessed: 06 June 2018.

Federation International de Ski (FIS). 2016. *FIS environmental rules*. <http://www.fis-ski.com/mm/Document/documentlibrary/Administrative/04/22/77/10fisrulesforconductandenvironment_02.08.2016_English.pdf> accessed: 06 June 2018.

Ignatzi, Christian. 2017. "Zwei alte Bäume standen der Ski-WM im Weg". <https://www.welt.de/sport/wintersport/ski-alpin/article161833527/Zwei-alte-Baeume-standen-der-Ski-WM-im-Weg.html> accessed: 06 June 2018.

Nothdurfter, Christa. 2015. "EMAS-Zertifizierung für die Schmittenhöhebahn AG". <https://www.meinbezirk.at/pinzgau/wirtschaft/emas-

zertifizierung-fuer-die-schmittenhoehebahn-ag-d1380668.html> accessed: 06 June 2018.

Pröbstl, Ulrike. 1998. "Ist umweltgerechter Skisport möglich?". In *1. Alpenreport: Daten –Fakten – Probleme – Lösungsansätze*, Commission Internationale pour la Protection des Alpes (CIPRA) (eds.). Berne, Stuttgart, Vienna: Paul Haupt, 265–271.

Pröbstl, Ulrike. 2001. *Skigebiete in den bayerischen Alpen, Ergebnisse einer ökologischen Studie*. Weilheim: Stöppel Verlag.

Pröbstl, Ulrike, Jiricka, Alexandra and Pichler, Irina. 2010. *Stiftung Pro natura - pro ski, Zweijahresbericht*. Vienna: University of Natural Resources and Life Sciences.

Pröbstl, Ulrike, Roth, Ralf, Schlegel, Heiner and Staub, Rudolf. 2003. *Auditing in Skigebieten – Leitfaden zur ökologischen Aufwertung*. Liechtenstein: Stiftung pro natura – pro ski.

Pröbstl-Haider, Ulrike, Brom, Monika, Dorsch, Claudia and Jiricka-Pürrer, Alexandra. 2018. *Environmental Management in Ski Areas: Procedure – Requirements – Exemplary Solutions*. Berlin: Springer International Publishing.

Pröbstl-Haider, Ulrike, Brom, Monika, Dorsch, Claudia and Jiricka-Pürrer, Alexandra. 2018. *Umweltmanagement in Skigebieten*. Switzerland: Springer Spektrum.

Roth, Ralf, Krämer, Alexander and Severiens, Julia. 2018. *Zweite Nationale Grundlagenstudie Wintersport Deutschland 2018*. Schriftenreihe SIS.

Schmittenhöhebahn AG. 2018. "Ökobericht & Umwelterklärung der Schmittenhöhenbahn AG. Eins mit der Schmitten. 2017/18". <https://www.schmitten.at/01_service/downloads/pdf/downloads/Schmitten_Okobericht_A4_FINAL.pdf> accessed: 06 June 2018.

STOKE Certified. 2018. *Riders. Operators.* <https://www.stokecertified.com/certification/snow/> accessed: 06 June 2018.

Tourismuspresse. 2017. "Nachhaltig erfolgreich – Die Schmitten in Zell am See". <https://www.tourismuspresse.at/presseaussendung/TPT_20170622_TPT0006/nachhaltig-erfolgreich-die-schmitten-in-zell-am-see-bild> accessed: 06 June 2018.

Umweltbundesamt. 2016. "Umweltschutz auf Reisen". <https://www.umweltbundesamt.de/themen/umweltschutz-auf-reisen-0> accessed: 06 June 2018.

Vail Resorts Management Company. 2018. "Our mission environment". <https://www.whistlerblackcomb.com/explore-the-resort/about-the-resort/environment.aspx> accessed: 25 May 2018.

Voets, Cornelia. 2009. "Welche Auswirkungen hat das neue Umweltschadensgesetz auf Planungs- und Umweltprüfinstrumente wie z.B. FFH-VP und UVP und welchen Beitrag können diese Instrumente leisten, Umweltschäden zu vermeiden?". UVP Report 23(5): 240–244.

Weisse Arena Gruppe. 2015. "Umwelt und Nachhaltigkeit. Greenstyle". <https://www.laax.com/infos/greenstyle> accessed: 25 May 2018.

WWF Österreich. 2018. "WWF-Urlaubstipps. Wie Sie Ihren Urlaub ökologisch verträglich gestalten". <https://www.wwf.at/de/menu673/subartikel2516/> accessed: 06 June 2018.

Conclusion

A look back at the chapters in the book

The three parts of this book examine resorts' economic and managerial models, the diversity of the forms of adaptation and markets and emerging issues. They cover the global changes affecting winter sports resorts, as well as their responses to these changes. By reviewing the English-language work[1] that has been done over the past three decades, this work seeks to contribute to a broad-based understanding of international research on resorts. In these few concluding lines, the authors also wanted to position the research as it relates to the work that has been carried out in France, in particular by Labex ITTEM. What are the main lessons that can be drawn, and what are the most important questions that arise from the different contributions to this book? Three sets have been identified: the influence of resorts' historical trajectories and governance models; the complex relationships between globalisation and territorialisation, which lead to various forms of innovation and adaptation; and finally, the blind spots that call for new research and perhaps an even more international approach to work and collaboration.

In the first part, Michael Zemla showed that formerly communist countries' models are not limited to the corporate and the community-based varieties that are generally used to think about and objectify both the structuring and the governance models of resorts. In the next chapter, Sara Nordin and Bo Svensson insisted that local actors be involved in governance models in order to prevent the economic aspects from riding roughshod over the social and environmental issues of sustainable development. Next, Doyeon Won described the involvement of public and private actors in the development of skiing in South Korea. Skiing, which started in the country in the 1970s, was initially the domain of private investors, who were attracted by the largely profitable market.

---

1 Only Achin's article, which deals directly with diversification, reports on French research, and it very specifically deals with the theme of tourism diversification and its support by public policies.

Following the economic crisis that hit the country between 1997 and 2001, the government took charge – in particular, to host the Olympic Games and to breathe new economic and demographic life into some areas in the country. Today, however, because of the South Koreans' waning appetite for winter sports, the construction of new resorts has slowed down significantly. Peter Williams, Alison Gill and Chris Sheppard presented the evolutionary trajectory of resorts in British Columbia. Made up of local, national and supranational stakeholders from both the private and the public sector, these resorts can be viewed as hybrid systems in historical contexts that have a differentiated impact on the evolution of each of them. In the last chapter, Ismaël Vaccaro and Beltran Oriol showed that it can be difficult to break off a development trajectory, even if it has its flaws. Indeed, despite skiing having little effect in the Catalan Pyrenees and the context having many difficulties, the Spanish government provides unconditional support because it does not foresee any other form of development for the region. Generally speaking, this first part looked at the factors shaping resorts' trajectories and governance models while exposing the development models' relative inertia.

In the second part of the book, the authors reviewed the changes affecting the ski market and the possibilities it has to adapt. In the first three chapters, Andrea Macchiavelli, Michel Archambault, Romano Wyss and Tobias Luthe showed that the evolution of tourists' expectations have as much of an impact on the future of resorts as climate change does. In order to respond to these cultural and social changes, the authors recommended that the resorts propose environmental, commercial and organisational innovations – especially the most vulnerable resorts, which are located at lower altitudes and do not have the necessary investments to make snow and/or diversify their activity offer *in situ.* Daniel Scott and Robert Steiger looked at models developed to predict the impact of climate change on resorts' viability. Because the models have many weaknesses (including inaccurate data, omission of parameters like snowmaking etc.), it is still complicated to use them to get an overall view and make relevant recommendations regarding the reality of the management styles at the resorts. This second part explored the relationship between globalisation and the territorialisation of the forms of innovation and adaptation that the resorts have adopted: Are the resorts characterised by any global models? Do these models spread uniformly around the world, or do they change according to local circumstances?

The last part of the book focused on markets and issues that would benefit from further research. The first two chapters examined emerging markets, which have thus far received much less attention than mature markets (United States, Europe, Japan and Australia). While Cenk Demiroglu proposed a way to identify the opportunities and the obstacles related to the development of winter sports resorts in Turkey, Yan Fang, Daniel Scott and Robert Steiger provided an overview of the development of winter sports in China over the past 20 years. Even though the growth of this market has been remarkable, like its potential, China has to face numerous challenges that could stall its growth – in particular, the difficulty of increasing the activity when skiing has no historical roots in the country. The subsequent chapters looked at emerging practices. Coralie Achin showed that tourists' adaptation and practices, just like their perceptions of climate change, have been poorly studied, even though these developments are now considered an essential part of the strategies that the resorts are implementing and are often encouraged by public policies. Lastly, Ulrike Pröbstl described a kind of environmental management – EMAS certification – whose value lies in external accreditation. Although this type of certification can be advantageous for resorts (by reducing costs, offering a competitive advantage etc.), too few resorts are involved in these voluntary processes of environmental certification. However, the younger generations, who can ensure the future of the market, are sensitive to the environmental impact of their sports practices (Achin, Macchiavelli, Pröbstl, Won). It is becoming essential to understand what is behind the spread of these environmental practices, as well as the bottlenecks linked to the resorts' own model, which may have difficulty influencing their trajectory. This final part revealed the blind spots affecting the study of winter sports resorts and called for greater collaboration between researchers on an international level in order to establish comparative studies that can lead to greater generalisation.

Which concepts can be used to think about the way the resorts are organised?

Globalisation, hybridisation and the emergence of economic and managerial models

Michael Zemla pointed out that the two main economic and managerial models governing resort construction – namely, integrated resorts/catalytic resorts (corporate model/community based-model – essentially correspond to a specific era (the 20th century) and a specific region (the West). As such, they are not necessarily applicable to all other regions of the world, particularly those with a communist past or those, like China, that offer an unusual hybrid between communism and capitalism. Yan, Scott and Steiger showed that although Chinese authorities supervise and encourage the development of skiing, private actors are involved in this undertaking – not the local populations. More data will have to be collected regarding the involvement of public and private actors, but at first blush neither the corporate nor the community-based model (Flagestad et al., 2006) fully covers the situation of skiing in China. Moreover, this binary vision can oversimplify and does not consider the different nuances of the situation if there is no consideration of the way in which these models overlap with each other or have been embedded in realities flowing from the establishment of resorts since the 19th century (Tissot, 2003; Boyer, 2000; Gigase, Humair and Tissot, 2014).

Despite possible inadequacies, it is on the basis of these conceptual references that the resorts are designed and considered part of a mature or an emerging destination, which is often characterised by political or economic operations that are very different from those in the West. This way of looking at the situation reflects a globalised way of thinking that goes hand in hand with the circulation of goods, capital and people. It also raises the question of the effects that the internationalisation of knowledge production systems can have, as well as the responsibility of researchers in the field – in particular, what kind of analytical results the dissemination of such a reading grid might turn up and what it might potentially conceal. In the end, the implementation of this corporate/community grid helps to highlight the diversity of the resorts' modes of organisation. It also underscores the diversity of the stakeholders involved in the light of their motivations and development approach. The analyses in this book tend to confirm the observation that is often made in the West regarding the primary interest of community/corporate categories: These models help to gain a better understanding of the diversity of resorts, both in their creative history and in their current developments, which sometimes leads them to shift from one model to another. For example, the resort of Les Deux Alpes (Isère, France) recently changed from a community model, which is unusual for large resorts, to the more classic corporate model, under the aegis of the Compagnie des Alpes.

Although they do not fully reflect the reality of the situations, these two models nevertheless suggest the degree to which local populations are involved in the resorts' development, and, therefore, also the number of actors involved. In a somewhat caricatured way, the corporate model is more in line with large resorts, which have significant investment resources, are located at high altitudes and are sometimes considered to be less rooted in the territory they are situated in. These resorts can hope to continue living off their economic model by making adjustments on the margins and without having to undertake a major bifurcation. However, such a favourable view of these mountain resorts because of their lower vulnerability to snowfall depletion is debatable. Are these resorts really as resilient as they seem, especially given the environmental issues that go far beyond climate change alone? In practice, the articles focusing on adaptation involve cases that are similar to the community-based model because they are small resorts, particularly vulnerable to global changes and limited in their investments. For them, organisational innovation might be even more necessary than for large resorts, especially since more players are involved (Macchiavelli, Wyss and Luthe, Vaccaro and Oriol). Also, from a research perspective, a major challenge would be to study in greater depth how these large resorts innovate and adapt, for example through an international panel of "laboratory" resorts to be monitored over the long term.

In the examples presented in this book, we can also see a new model emerging: resorts anchored in a territorial project that form a network with other sectors of activity (whether tourism-oriented or not) through which they differentiate themselves while also adding value to the collective. Wyss and Luthe's notion of circular economy thinking perfectly illustrates this trend. This model is certainly closer to the community-based model in that it involves the local actors and is built thanks to them, but the comparison is not perfect because as its central unit, the tourism project no longer has the ski resort but rather a territory with several types of attractive amenities (agricultural products, landscape attractions, cultural and heritage offers etc.). By diversifying, the challenge for the territories is to maintain their dependence on skiing alone, and what used to be a disadvantage for smaller resorts may turn into sure-fire appeal. In the words of Macchiavelli, this model, which remains largely undefined, is based on two pillars: cooperation (between actors) and specialisation (of the territory). It also refers to the evolution of the resorts' drivers of development.

In this new approach to tourism, the territory, its inhabitants and its activities are once again taking centre stage. In many cases, deterritorialised tourism models, like leisure parks, have proved to be unsustainable (Macchiavelli). In response, the empowerment of local actors is a necessary component to ensure the sustainability of tourism activities – and especially of local benefits – within the framework of anchoring development. Moreover, innovations offering varied responses to global trends are also being created at the local level. Thus, we see how useful studies relating to the governance of resorts – and, more broadly, of tourism models – can be (Fournier, 2018).

Trajectory and governance as key concepts

Whether they are built according to the corporate model, the community-based model or any of the many hybrid possibilities between the two, resorts cannot be considered isolated entities. They are always (albeit differently) connected both to the territories in which they are located in their physical and human dimensions and to general flows (climate change, political and economic events, cultural developments etc.). For this reason, it seems essential to identify the trajectory of which they are part and, especially, to integrate the temporal dimension, in terms of legacy, knowledge and social and cultural systems. In other words, the resorts have to be regarded as always partially singular systems embedded not only in geographical and political contexts but also in a (local and global) history that opens up and/or closes certain possibilities for development (Tissot, 2000; Granet, 2013).

Although this work focuses primarily on recent history, some of the authors did adopt a historical perspective (Williams, Gill and Shepard). In future studies and especially to make a comparison on an international scale, it would be worth it to incorporate the reality of the territories and the societies that inhabit and structure them over the long term. Many resorts, including those located in seemingly empty spaces (at high altitudes), have benefitted from local communities' acceptance and, in part, participation; in so doing, they have found new localised forms of applying classical multi-activity. By contrast, some have rejected such developments and sought to keep control of their territory by influencing the evolution or re-orienting the nature of the developments (villages resort). Depending on these individuals' positioning, acceptance or refusal to participate in the development projects from the state or private actors, these territories have undergone different situations. The reality is actually more complicated because inside the same space, attitudes often diverge and people do not take each other's experiences into consideration. This historical perspective, therefore, in no way contradicts the future of the resorts but, on the contrary, sheds light on it.

The notion of trajectory refers to governance models (Gerbaux et al., 2006; Marcelpoil, 2008). As we have seen, innovation often implies simulating and amplifying the relationships between stakeholders. Nonetheless, in order to successfully adapt to global change, it seems indispensable for a leader to emerge. Indeed, the history of resorts frequently highlights the role of "exceptional" individuals (architect, political leader, manager, patron etc.). Whether this role is amplified or takes on mythical dimensions, there is often a local or external personality behind these scenes who plays a training or cohesive role (for instance in French Alps: P. Lamour in Queyras, G. André at Bonneval-sur-Arc, C. Perrière at Méribel etc.). However, the connection between the personalities and the populations is essential to grasp the reality and the effectiveness of these actions and trajectories beyond imaginaries and representations. Therefore, it seems necessary to explore this interweaving between the individual and the collective in order to best take account of the different shades of the territories' trajectories, especially because, with the diversification of activities, interventions become more complicated but remain crucial. It also remains to be seen which type of leader or elite can be involved in initiating these bifurcations or adaptations while also guiding them over time: Is it an elite citizen? Or someone from the political or economic elite?

What kind of sustainability?

The involvement of local actors, which several authors present as a *conditio sine qua non* for resorts to adapt, refers to issues of sustainability, which are often stripped down to their economic dimensions (Nordin and Svensson). However, in order to be sustainable over time, the resorts have to not only generate profits but also ensure that the local actors are included and that their natural resources are preserved. The few studies that focus on resorts' social sustainability tackle the issue from two angles: professional skills (in particular, the status of seasonal workers) and amenity migrations, that is, the types of people who live on the resort's territory and the conflicts that may arise among the latter. The local populations' involvement in the direction a resort takes has not received much attention, even though demands for participatory democracy are on the rise, at least in the West. However, the resorts would benefit from mobilising a form of collective intelligence to define their project, to avoid potential social conflicts linked, for example, to how resources are used and to take into consideration that they form part of the territory. Sometimes, ecological transition projects take place alongside the resorts in a rationale that is more antithetical than complementary. To conclude these essential inquiries, it is important to get out of an often-binary reading that considers local populations as reserved, if not reactionary, in the face of external developers who are aware of the stakes. Most often, there is a diversity of positions that refer to the long history of villages and families and their integration into the overall economy. It must be noted that considering the habitus of governance in mountain territories is very useful to measure the choices made when accepting, innovating or rejecting, which very often transcends local/external cleavages, as well as political cleavages.

From an environmental point of view, it seems equally complicated to get an overview, especially since there are not very many studies on the environmental dimensions of tourism (Granet-Abisset, 2016). Either the studies are highly programmatic (they announce environmental programmes whose effectiveness in practice and actual implementation are unknown) or they focus on relatively isolated conservation issues, such as wildlife protection of a particular species. Studies assessing the impact of a resort on different levels, the carrying capacity of a territory and the effectiveness of possible protective measures are extremely rare, probably because this type of work requires significant resources. In any case, it is currently difficult to comment on the way in which resorts impact and/or protect their environment. Resorts' sustainability is an issue that should be further explored, especially since it is very likely to vary according to the political and economic model, as well as the time frame, in which the resort operates. Thus, the development model of so-called integrated resorts in France, which have in recent years received harsh criticism for their architectural characteristics or for evicting local populations, nevertheless has to be questioned today. Shouldn't their more concentrated management style be considered given the devastating effects of urban sprawl that are linked to urbanisation? Recent resorts in new destinations such as Sochi or South Korea seem to be more along the lines of a concentrated resort model. In the same vein, there should be an analysis of size scales and the combination of large and small resorts, the latter being less studied but taking up a new place in the environmental perspective.

How resorts are adapting to global change

The imperative to innovate and adapt

Despite the diversity of their themes and approaches, all the articles agree on the fact that numerous changes are raising questions about the viability of winter sports tourism and the model on which it is built. These changes are both environmental (climate change, wildlife endangerment) and cultural (tourists visiting for shorter amounts of time and in more peripheral periods like the autumn or the spring). Tourists are looking to have a meaningful overall experience that includes quality and originality and is not limited to skiing (Macchiavelli), and they also attach more importance to respecting the environment (Macchiavelli, Pröbstl, Won). In addition to the decline in skiing, what has to be discussed are the challenges of adapting and innovating.

The innovations that the resorts are proposing are, above all, technical in kind: making artificial snow at higher temperatures and with fewer resources, improving climate forecast models and grooming techniques etc. (Spandre, 2015). Then there are business innovations such as expanding wellness offers, having festive and cultural activities (in recent offers, there has been a surge in concerts, festivals and other events) and providing alternative practical experiences like night hiking. Lastly, the innovations can also be organisational in kind. To respond to the new expectations that tourists have, the stakeholders have to coordinate their tourism offer to enhance the attractiveness of the area where the resort is located by limiting expenditure on equipment (Macchiavelli). In this regard, the higher-altitude resorts seem to be undergoing the same evolution as resorts on the coast or as part of a broader movement that concerns urban tourism territories or remarkable sites.

The innovations require resorts to be proactive in order to create new needs and new practices. In parallel, resorts also have to adapt, which means they have to comply with new trends that are already established among the consumers. For many years, big and small resorts were encouraged to offer the same activities because they had been designed according to the same development model: one place, one time period, one activity. Today, by contrast, they have to respond to a more fragmented demand – different audiences, expectations, time periods etc. – in search of originality and uniqueness. At the smaller resorts, in particular, there is an expectation that they be "authentic", firmly rooted in their territory and not just the land; for their part, the large resorts have to satisfy the need for authenticity while simultaneously offering large-scale events whose festive character sometimes borders on the excessive.

Forms of adaptation and innovation between globalisation and territorialisation

Paradoxically, while the resorts are expected to stand out, the innovation and adaptation models are becoming more widespread. It is not uncommon for an initially marginal practice to become prevalent. Resorts have responded enthusiastically and have been quick to adopt them. The truly unique nature of the solutions proposed by the resorts should, therefore, be questioned on a large scale. It is recognised that in this field, as in others, cognitive capitalism picks up peripheral or marginal dynamics that first emerged in "recreational laboratories" (Falaix and Corneloup, 2017) and out of recreational "dissidences" (Bourdeau and Lebreton, 2013) and turn them into new norms. These creative processes illustrate alternative theories of innovation: culture-based innovation (KEA, 2009), lead-user innovation (Parsons and Rose, 2009), niche innovation (Schot, 1998) and innovation under constraint (Caniëls and Rietzschel, 2015).

While it is necessary to question the globalisation of innovation and adaptation models, their territorialisation also deserves our attention. The requirements of authenticity, quality and innovation at resorts can become widespread without having the same meaning where they are applied in different regions of the world. There is a lack of data on this issue, but among others, we can look into the forms of responses to the changes affecting skiing around the world. Indeed, although resorts are built everywhere according to the Western model, forms of hybridisation can be observed in countries with a history and/or a physical, political and cultural context that are very different from those in the West.

With regard to requests and responses, there is tension between globalisation and territorialisation in the field of winter sports and more broadly in tourism, as Higham and Hinch illustrate very well in their book entitled *Sport & Tourism. Globalization, Mobility and Identity*

(2009). Far from being resolved, this issue requires further research to better understand the relationship between global challenges and local characteristics.

The failures

In addition to successful adaptations, we have also seen that many resorts, or even regions, have remained stuck in their original model and have had difficulty rethinking their ski-based tourism system. This was essentially the case for the Pyrenees, but also for resorts in British Columbia, as exemplified by Whistler's development model. In general, it seems that the resorts are oriented more towards adaptation than towards real bifurcations, even if a particular time scale may add nuance to this assessment. In this respect, we see that the concept of path dependence, which is linked to a historical approach rooted in a resort's trajectory, is particularly relevant to consider the resort's capacity for resilience and further development. This calls to mind the very complementary notion of socio-technical lock-in (Liebowitz and Margolis, 1995), which refers to the difficulty of breaking out of the irreversibility of certain choices rooted in socio-technical systems. It would also be wise to focus on territories that are located close to resorts, have not been part of the development and, with the new expectations of certain categories of customers looking for less urbanised territories, are turning their periphery into a strength by implementing completely new forms of tourism.

What kind of research agenda can help resorts to adapt?

Thematic blind spots

This book teaches several thematic, theoretical and methodological lessons. Let's start by tackling the first of these. The graph below depicts the themes of the 120 articles that steered the writing of our introduction and the identification of contributors. The bibliographic research was guided by our two main themes: global changes and ski resorts. We chose to limit the research on climate change because of the multitude of articles on this topic, and many of them are very technical. By contrast, having identified that some themes needed more elaboration, we sought to document them. This is the case, in particular, for the themes of "precariousness" and "land tenure". At the same time, we were particularly interested in certain approaches and sub-themes like innovation and adaptation, as well as the issue of summer tourism. Ultimately, the graph does not depict the themes related to the impact of global changes on ski resorts but those that guided the writing of this book.

Graph of the themes addressed in this review of English-language literature (120 articles):

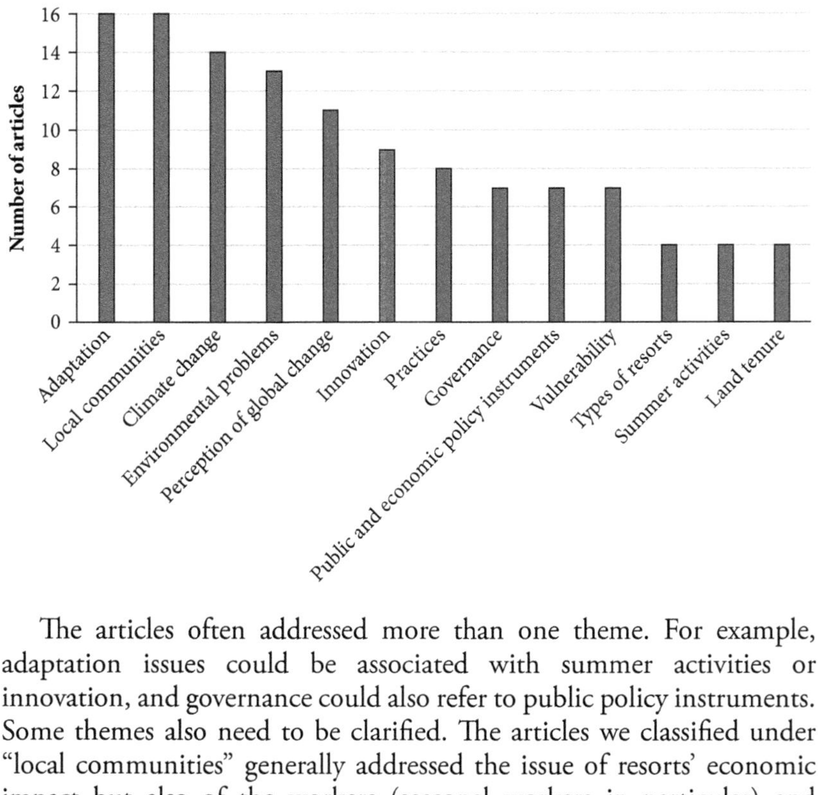

The articles often addressed more than one theme. For example, adaptation issues could be associated with summer activities or innovation, and governance could also refer to public policy instruments. Some themes also need to be clarified. The articles we classified under "local communities" generally addressed the issue of resorts' economic impact but also of the workers (seasonal workers in particular) and their precariousness, as well as land tenure (conflict with second-home owners); environmental problems could be tied to the question of land use and, therefore, to local populations; and tourism profiles could go hand in hand with different types of resorts.

It seemed to us that some of these themes had not previously been sufficiently explored, as confirmed by several of the chapters in the book. These include introducing younger generations to skiing

(Archambault; Yan, Scott and Steiger; Won); the role of real estate and land in the profitability of resorts and its lack of sustainability (Vaccaro and Oriol); and the compatibility between the ski market and environmental protection (Zemla; Nordin and Svensson; Archambault; Pröbstl; Won). Other results have emerged that deserve our attention, such as the advantage that resorts have when they are located close to an urban centre and the efficiency of the transport networks that are connected to it, especially as the lengths of stays tend to decrease, and the climate, which is fluctuating more and more, reduces the possibility of planning a long-term stay (Yan, Scott and Steiger; Archambault; Vaccaro).

The notion of territory as a key concept

From a theoretical point of view, this book is structured around the tension between globalisation (changes) and territorialisation (adaptations). In conclusion, we think it is important to return to the concept of territory. This unit of analysis is a typically French "invention" and is not found as such in the English-language literature, which prefers terms like area and region. Moreover, territory is also an extremely polysemic concept. Within the framework of the Labex ITTEM which started the work on this book and for which the notion is highly significant, territory is defined as a complex system in which global (competition, economy, global changes) and local (cultural, political etc.) phenomena interact with each other, but also as a social construct and a global fact that should be placed in its historical context by incorporating practices, social uses and mutual representation. In France, this notion is linked to the issue of habitability, and it establishes a strong link between decentralised public policies and the inhabitants who are supposed to participate or, at the very least, sign onto a local development project. This territory-oriented approach also applies particularly to densely and formerly populated regions, like the Alpine Arc, where different kinds of land use can be on a collision course and where they have to collaborate in order to limit social conflict.

It has already been mentioned that there is a lack of data on new destinations, but we can question whether the concepts of territory and territorialisation are relevant in these places. In China, for example, resorts can be built in the same way as in industrial areas, without taking the local populations or their economic activities into consideration.

However, although the corporate model currently dominates, according to Yan Fang, the community-based model could grow in the coming years, as the government intends to use skiing as a means to grow certain isolated and rural populations (personal communication). In other words, there is nothing static about the relationship that resorts have (or do not have) with the territory hosting them.

The territory is also the scale at which the French "laboratories of excellence" (Labex), of which ITTEM forms part, are structured. The mission of LabEx ITTEM is to structure research related to mountains, including tourism, so as to make it more visible and to facilitate complementarity with laboratories focused on mountain resorts or tourism, particularly mountain resorts. Territory is also a key concept in research carried out by Labex ITTEM. Among the goals that it had set itself from 2012 to 2019, two related directly to the study of mountain territories from different perspectives. The aim of the first one was to analyse the innovations in mountain territories and to conceptualise what constitutes a "territorialised innovation ecosystem"; in the second, the study of mountain territories involved an analysis of their trajectories over time, an analysis that highlighted the existence of reticular territorialities. It is now necessary to deepen this study of resorts' trajectories and governance models at the international level. A continuation of Labex research between 2020 and 2024 will make it possible to intensify this work, with a view to deepening the notion of transition with respect to the associated mountain resorts and territories.

Methodological approaches

There are several lessons to be drawn from the chapters in this book with regard to methodology. First of all, it is important to note the diversity of approaches, disciplinary roots and concepts used to study winter sports resorts and the challenges they face. While this diversity adds richness to the field of study, it also complicates the drafting of research agendas and sharing of results because so much time is needed. The variety of thematic issues tackled by the disciplines regarding resorts and their trajectories also requires an integrated approach. Of course, the challenge is to deepen knowledge on this or that issue but also – and above all – to articulate these contributions on a territorial scale. This is the perspective of Labex and its third research goal: the development of interdisciplinary concepts and methodologies in the humanities and social sciences that make it possible to understand the transformations of the relationship between man and nature within the framework of global changes. The objective is, by continuing the disciplinary work, on the one hand, to work on interdisciplinarity and this territorialised reading of current dynamics but also, on the other hand, to highlight what in these trajectories of resorts/territories is part of the transition in terms of processes, stakeholders and territorial impact.

Studies on global changes affecting winter sports resorts adopt either a very micrological and qualitative or a macrological and quantitative approach. The micrological/qualitative approach is found primarily in French-language literature. Focusing on one or a few case studies, this approach has the merit of offering precise analyses that nevertheless raise the question of generalisability. By contrast, the macrological/quantitative approach, which is often found in the international English-language literature, lacks contextualisation and is often based on bibliographic sources (Pröbstl; Yan, Scott and Steiger; Macchiavelli). This sometimes makes it difficult to understand the reality of the situation in depth. This is particularly true in the matter of skiers' practices and expectations. As we have seen, this category of actors is understudied, even though the tourist/skier is at the centre of resorts' viability and their adaptation strategies. These two complementary approaches in terms of results should be compared and opened up to dialogue. It is precisely for this reason related to games of scale that thinking should be further developed and various works be carried out. And, by proposing a scale of mesological analysis, a territorial approach might fill this gap and help to establish links between the case studies, which are rich in detail, and the more general statements, whose value lies in their wide applicability. This scale is also part of this aim of increased generalisability to question the processes that are already underway and to do so in the different territorial contexts.

The need for international collaboration

Several texts raised the issue of knowledge production regarding the adaptation of winter sports resorts (Macchiavelli; Wyss and Luthe; Scott and Steiger; Archambault). It raises the question of the position that researchers can or should occupy in this process. To assess the extent of the changes and respond to them, resorts must be supported by professionals in the sector. However, outside of research, the latter are sometimes lacking. If researchers are to take a more prominent role in this process, what form should their participation take? Should they move towards involved research and jointly develop questions, concepts and methods with those working on the ground? Should they spin off their services or continue to provide a public service? These questions were also at the heart of Labex's objectives since experimentation with innovative modes of research-action was its fourth objective. This objective resulted in the construction of a theoretical and methodological system (as well as a hybrid group) that combined managers and researchers. This approach can now be extended to larger, more international spaces.

Labex ITTEM has also set itself the task of identifying resource researchers who can quickly be mobilised when there are social issues relating to the mountains. This book was written in this vein to bring together international specialists in winter sports resorts and the dynamics of adaptation they must deploy in the face of global changes. The next major objective of Labex ITTEM is to contribute to build an international research community that will study the environmental and cultural changes affecting mountain resorts and will do so in a sustainable and harmonised way. The research community exploring the changes and adaptations of mountain resorts is relatively limited, and the case studies, concepts and approaches developed in this field are too disparate to allow generalisable results. In this context, the implementation of studies that have a shared methodology and are deployed at an international scale on several cases considered to represent research questions seems not only realistic, but necessary in the long term. To this end, the Labex ITTEM community will be able to rely on a number of structures with which it is already in contact, in order to solidify research and contribute to a network of expertise on the subject.

References

Bourdeau, Philippe et Lebreton, Florian. 2013. "Les dissidences récréatives en nature: entre jeu et transgression. Exploration liminologique". *EspacesTemps.net* [En ligne] http://www.espacestemps.net/articles/les-dissidences-recreatives-en-nature-entre-jeu-et-transgression.

Boyer, Marc. 2000. *Histoire de l'invention du tourisme XVIe-XIXe siècles: origine et développement du tourisme dans le Sud-Est de la France*. La Tour d'Aigue, Éditions de l'Aube.

Caniëls, Marjolein et Rietzschel, Eric F. 2015. "Organizing creativity: creativity and innovation under constraints". *Creativity and Innovation Management* 24(2): 184–196. DOI: 24. 10.1111/caim.12123.

Falaix, Ludovic et Corneloup, Jean. 2017. "Habitabilité et renouveau paradigmatique de l'action territoriale: l'exemple des laboratoires récréatifs". *L'Information géographique* 4: 78–102.

Flagestad, Arvid, Svensson, Bo et Nordin, Sara. 2006. "Destination Governance and Contemporary Development Models". In *Tourism Local Systems and Networking*. Lazzeretti, Luciana et Petrillo, Clara S. (eds.). Advances in Tourism Research Serie, Elsevier, Oxford.

Fournier, Dorothée. 2018. La glisse réinventée. La construction d'un territoire de sports et de loisirs : l'Oisans des années 1960 à nos jours, Thèse de doctorat, Ingénierie de la Cognition, de l'interaction, de l'Apprentissage et de la création, Université Grenoble-Alpes.

Gerbaux, Françoise et Marcelpoil, Emmanuelle. 2006. La gouvernance des stations de montagne en France : les spécificités du partenariat public-privé, Revue de Géographie Alpine, 94-1, 9–19.

Gigase, Marc, Humair, Cédric et Tissot, Laurent. 2014. *Le tourisme comme facteur de transformations économiques, techniques et sociales (XIX*[e] *– XX*[e] *siècles)*. Neuchâtel: Alphil.

Granet-Abisset, Anne-Marie. 2013. " Tourisme et territoires. Les Alpes, terre d'innovation ou terrain de jeu(x) ?". *Bulletin de l'Académie Delphinale,* 40–53.

Granet-Abisset, Anne-Marie. 2016. "Les transformations apportées aux hautes terres alpines par les aménagements touristiques". In. *Les impacts environnementaux et sociaux de l'aménagement territorial.* Massard-Guilbaud, Geneviève et Fournier, Patrick. PUR.

Higham, James et Hinch, Tom. 2009. *Sport and tourism: globalization, mobility and identity*. Amsterdam and Boston: Butterworth-Heinemann/ Elsevier.

KEA. 2009. *The impact of culture on creativity*. Study prepared for the European Commission. Available at: www.keanet.eu/docs/ impactculturecreativityfull.pdf (accessed 31 January 2019).

Liebowitz, Stan J. and Margolis, Stephen E. 1995. "Path dependence, lock-in and history". *Journal of Law, Economics, and Organization* 11(1): 205–226. DOI:10.2139/ssrn.1706450.

Marcelpoil, Emmanuelle. 2008; Les trajectoires d'évolution des destinations touristiques de montagne, Mémoire d'HDR, Université de Pau et des Pays de l'Adour.

Parsons, Mike and Rose, Mary. 2009. "Lead User Innovation and the UK outdoor Trade Since 1850". Lancaster University Management School, Working paper 2009/028.

Spandre, Pierre. 2015. Observation et modélisation des interactions entre conditions d'enneigements et activité des stations de sports d'hiver dans les Alpes françaises, Thèse de doctorat en Océan, Atmosphère et Hydrologie, Université Grenoble-Alpes.

Schot, Johan W. 1998. "The usefulness of evolutionary models for explaining innovation". *History of Technology* 14(3): 173–200. DOI:10.1080/07341519808581928.

Tissot, Laurent. 2000. *Naissance d'une industrie touristique, Les Anglais et la Suisse au 19e siècle*. Lausanne, Editions Payot.

Tissot, Laurent. 2003. *Construction d'une industrie touristique aux 19ème et 20ème siècles. Perspectives internationales*. Editions Alphil, Neuchatel.

List of Figures

List of Tables

www.ingramcontent.com/pod-product-compliance
Ingram Content Group UK Ltd.
Pitfield, Milton Keynes, MK11 3LW, UK
UKHW020714071026
15590UKWH00011B/600

9 782807 613119